Lecture Notes in Artificial Intell

Edited by J. G. Carbonell and J. Siekmann

Subseries of Lecture Notes in Computer Science

Fumiya Iida   Rolf Pfeifer
Luc Steels   Yasuo Kuniyoshi (Eds.)

Embodied
Artificial Intelligence

International Seminar
Dagstuhl Castle, Germany, July 7-11, 2003
Revised Selected Papers

 Springer

Series Editors

Jaime G. Carbonell, Carnegie Mellon University, Pittsburgh, PA, USA
Jörg Siekmann, University of Saarland, Saarbrücken, Germany

Volume Editors

Fumiya Iida
Rolf Pfeifer
University of Zurich, Artificial Intelligence Laboratory, Department of Informatics
Andreasstr. 15, 8050 Zurich, Switzerland
E-mail: {iida, pfeifer}@ifi.unizh.ch

Luc Steels
Vrije Universiteit Brussel, Artificial Intelligence Laboratory
and Sony Computer Science Laboratory, Paris
Pleinlaan 2, 1050 Brussels, Belgium
E-mail: steels@arti.vub.ac.be

Yasuo Kuniyoshi
University of Tokyo, School of Information Science and Technology
Dept. of Mechano-Informatics, Laboratory for Intelligent Systems and Informatics
Engineering Bldg. 8, 7-3-1, Hongo, Bunkyo-ku, Tokyo 113-8656, Japan
E-mail: kuniyosh@isi.imi.i.u-tokyo.ac.jp

Library of Congress Control Number: 2004094628

CR Subject Classification (1998): I.2

ISSN 0302-9743
ISBN 3-540-22484-X Springer-Verlag Berlin Heidelberg New York

Springer-Verlag is a part of Springer Science+Business Media

springeronline.com

© Springer-Verlag Berlin Heidelberg 2004
Printed in Germany

Typesetting: Camera-ready by author, data conversion by PTP-Berlin, Protago-TeX-Production GmbH
Printed on acid-free paper   SPIN: 11301905   06/3142   5 4 3 2 1 0

Preface

The term "Embodied Artificial Intelligence" designates a rapidly growing, highly interdisciplinary field, uniting researchers from areas as diverse as engineering, philosophy, psychology, computer science, biology, neuroscience, biomechanics, material science, and linguistics. What motivates these researchers to cooperate is the common interest in intelligence, in particular the development of intelligent machines. Another unifying characteristic of the field is the conviction that intelligence must be embodied, must be conceived of in terms of physical agents – biological or artificial – behaving in a real physical and social world. Given this perspective, most of the work involves the design and construction of robots or other kinds of artifacts.

The reason for the very strong transdisciplinary nature of "Embodied Artificial Intelligence" is that intelligence, especially embodied intelligence, is to do with behavior, with real-world interaction, and because we are dealing with physical agents there are many aspects and components involved: materials, morphology, sensors, actuators, energy supply, control, planning, cognition, and perhaps even consciousness. This makes the study of embodied intelligence truly challenging but it is precisely what makes the subject area so unique and fascinating.

In this book we provide a representative collection of papers written by the leading researchers in the field who attended a seminar on "Embodied Artificial Intelligence", held at Schloss Dagstuhl, Germany, July 7–11, 2003. The contributions are all interdisciplinary in nature and are targeted at an interdisciplinary audience. As far as possible, they avoid scientific jargon and do not contain unnecessary technical detail. The authors were all asked to critically review the state-of-the-art in their particular domain, to elaborate the basic principles, and to describe what they consider to be research challenges for the coming years. This gives the book also a certain tutorial flavor so that it can be used for classes as additional reading material.

The first part of the book, "Philosophical and Conceptual Issues", tries to uncover the basic characteristics of "Embodied Artificial Intelligence", and discusses a number of deep issues related to high-level cognition, abstract thinking, and consciousness in an embodied system. How the contributions to this volume are situated within the field is discussed in the overview article on "Embodied Artificial Intelligence – Trends and Challenges". The papers in the second part, "Information, Dynamics, Morphology", deal with one of the basic principles of embodiment, namely the trade-offs and task distributions between morphology, materials, control (computation), and system-environment interaction, or, in other words, with the information theoretic aspects of embodiment. This contrasts with the more standard way of conceptualizing, embodiment, i.e., in physical terms (inertia, forces, torques, control, energy dissipation), thereby largely ignoring the information theoretic implications. The section on "Principles of Embodiment for Real-World Applications" explores how neural systems can be embodied to enable interactions with the real world, and describes a number of cutting-edge applications to the design of robotic arms, hands, and robots moving in the real world. The collection of papers under the heading "Developmental Approaches" all share the vision of mimicking, one way or another, developmental processes of biological systems, and/or they attempt to achieve technological solutions by imitating aspects of development typically using humanoid robots. Finally, "Artificial Evolution and Self-reconfiguration" deals with "population thinking" and discusses on the one hand automated design methods by drawing inspiration from nature, where, in contrast to the usual evolutionary approaches, developmental processes are taken into account; on the other hand, principles of self-reconfiguration are discussed not only in simulation but in the real world.

We would like to thank all of the participants of the seminar, the authors, and the reviewers for their excellent contributions to this volume. We would also like to express our thanks to Prof. Reinhard Wilhelm, the scientific director of the International Conference and Research Center for Computer Science in Schloss Dagstuhl, who suggested that we organize a workshop on "Embodied Artificial Intelligence", and to the organizers of this center for their continuous and professional support of the seminar. Credit also goes to the Executive Editor of the Springer series LNCS/LNAI, Alfred Hofmann, for his helpful comments and support of this publication project.

May 2004   Fumiya Iida
                                                                    Rolf Pfeifer
                                                                     Luc Steels
                                                                Yasuo Kuniyoshi

Table of Contents

III Principles of Embodiment for Real-World Applications

IV Developmental Approaches

V Artificial Evolution and Self-Reconfiguration

Embodied Artificial Intelligence:
Trends and Challenges

Rolf Pfeifer and Fumiya Iida

Artificial Intelligence Laboratory, Department of Informatics, University of Zurich
Andreasstrasse 15, CH-8050 Zurich, Switzerland
{pfeifer,iida}@ifi.unizh.ch

Abstract. The field of Artificial Intelligence, which started roughly half a century ago, has a turbulent history. In the 1980s there has been a major paradigm shift towards embodiment. While embodied artificial intelligence is still highly diverse, changing, and far from "theoretically stable", a certain consensus about the important issues and methods has been achieved or is rapidly emerging. In this non-technical paper we briefly characterize the field, summarize its achievements, and identify important issues for future research. One of the fundamental unresolved problems has been and still is how thinking emerges from an embodied system. Provocatively speaking, the central issue could be captured by the question "How does walking relate to thinking?"

1 Introduction

This conference and this paper are about embodied artificial intelligence. If you search for "embodied artificial intelligence" or "embodied cognition" on the Internet using your favorite search engine, you will find a radically smaller number of entries than if you search for "artificial intelligence" or "cognition". Trying to answer this question of why this might be the case, reveals a lot about the structure of this research field and uncovering its organization is one of the goals of this paper.

Over the last 50 years Artificial Intelligence (AI) has changed dramatically from a computational discipline into a highly transdisciplinary one that incorporates many different areas. Embodied AI, because of its very nature of being about embodied systems in the real physical and social world, must deal with many issues that are entirely alien to a computational perspective: as we will discuss later, physical organisms in the real world, whether biological or artificial, are highly complex and their investigation requires the cooperation of many different areas. The implications of this change in perspective are far-reaching and can hardly be overestimated. In this paper, we will try to outline some of them.

With the fundamental paradigm shift from a computational to an embodied perspective, the kinds of research topics, the theoretical and engineering issues, and the disciplines involved have undergone dramatic changes, or stated differently, the "landscape" has been completely transformed. In the first part of the paper we try to

F. Iida et al. (Eds.): Embodied Artificial Intelligence, LNAI 3139, pp. 1–26, 2004.
© Springer-Verlag Berlin Heidelberg 2004

characterize these changes. In the second part, we will identify the grand challenges in the field and discuss how far researchers have come towards achieving them. Given the enormous diversity, as discussed in the first part, this will necessarily be abstract, somewhat selective and reflect the authors' personal opinion, but we do hope that many people will agree with the our description of how the field is now structured. We conclude with some general comments on the future of the field and applications.

2 The "Landscape"

The landscape of artificial intelligence has always been rugged but it has become even more so over the last two decades. When the field started initially, roughly half a century ago, intelligence was essentially viewed as a computational process. Research topics included abstract problem solving and reasoning, knowledge representation, theorem proving, formal games like chess, search techniques, and – written – natural language, topics normal associated with higher level intelligence. It should be mentioned however, that in the 60s there was a considerable amount of research on robotics in artificial intelligence at MIT, SRI, and CMU. But later on the artificial intelligence research community has not paid much attention to this line of work.

Successes of the Classical Approach

By the mid 1980s, the classical, computational or cognitivistic approach, had grown into a large discipline with many facets and has brought forward many successes in terms of computer and engineering applications. If you start your favorite search engine on the Internet, you are, among many others, employing clever machine learning algorithms. Text processing system utilizes matching algorithms, or algorithms that try to infer user's intentions from the context of what have been done earlier. Controls for appliances using fuzzy logic, embedded systems (as they are employed in fuel injection systems, breaking systems, air conditioners, etc.), control systems for elevators, and trains, natural language interfaces to directory information systems, translation support software, etc., are also among the successes of the classical approach. More recently, data mining systems have been developed that heavily rely on machine learning techniques, and chess programs have been realized that beat 99.99 percent of all humans on earth, a considerable achievement indeed! The development of these kinds of systems, although they have their origin in artificial intelligence, have now become indistinguishable from applied informatics in general: they have become a firm constituent of any computer science department.

Problems of the Classical Approach

However, the original intention of artificial intelligence was not only to develop clever algorithms, but also to understand natural forms of intelligence that have – as argued here – more to do with the interaction with the real world. Alas, as is now generally agreed, the classical approach has not contributed significantly to our understanding of, for example, perception, locomotion, manipulation, everyday speech and conversation, social interaction in general, common sense, emotion, and so on.

Classical approaches to computer vision, for example, have been successful in factory environments, where there are constant lighting conditions, the geometry of the situation is precisely known (i.e. the camera is always in the same place, the objects appear on the conveyer belt always in the same position), and the types of potential objects are known and can therefore be modeled. However, when these conditions do not hold – and in the real world, they are never given, i.e. the distance of objects from the eyes always changes, which is one of the many consequences of moving around, and lighting conditions and orientation also vary continuously – these algorithms can no longer be used. Moreover, objects are often entirely or partially occluded, they move (e.g. cars, people), and they appear against very different and changing backgrounds. Artificial vision systems with capacities similar to human or animal vision, are far from being realized artificially.

A further example where the classical approach could not provide adequate answers is object manipulation. Indeed, animals and humans are enormously skilled at manipulating objects; even very simple animals like insects are masters at manipulation. Or watch a dog chew on a bone, how he controls it with his paws, mouth and tongue: unbelievable. Although there are specialized machines for virtually any kind of manipulation (driving a screw, picking up objects for packaging in production lines, lifting heavy objects in construction sites), the general purpose manipulation abilities of natural systems are to date unparalleled.

Locomotion is another case in point. Animals and humans move with an uncanny flexibility and elegance. We can walk with a bag in one hand, an arm around a friend, up and down the stairs, while looking around, something none of the existing robots can do. And building a running robot is still considered one of the great challenges.

In the classical approach, common sense has been treated at the level of "semantic content" and has been taken to include knowledge such as "cars cannot become pregnant", "objects (normally) don't fly", "people have biological needs" (they get hungry and thirsty), etc. Building systems with this type of common-sense knowledge has been the goal of many classical natural language and problem solving systems like CYC (e.g. Lenat et al., 1986). But there is an important additional aspect of common-sense knowledge, which is to do with bodily sensations and feelings, and this aspect has its origin largely in our embodiment. Take, for example, the word "drinking" and freely associate what comes to mind. Perhaps being thirsty, liquid, cool drink, beer, hot sunshine, the feeling of wetness in your mouth, on the lips, and on your tongue when you are drinking, and the feeling of thirst disappearing as you drink, etc. It is this kind of common sense knowledge and common experience that everyone shares and that forms the basis of robust natural language communication, and it is firmly grounded in our own specific embodiment. And to our knowledge, there are currently no artificial systems, capable of dealing with this kind of knowledge in a flexible and adaptive way.

The last point that we would like to mention here concerns speech systems. While in restricted areas, speech systems can be used, e.g. as an interface to directory information systems, or systems where single word commands can be used (e.g. for robot control, or name databases for mobile phones), in most areas they have only been used with limited success. Speech to text systems have to be tuned to the speaker's voice, and because of the high error rate, typically a lot of post-editing needs to be done on the text produced by the software. This may be one of the reasons why speech systems have not really taken off so far, even though the idea of not having to type any more, of producing text rapidly, is highly appealing. Although some of the systems may have a relatively impressive performance, the fact of the matter remains that there are to date no general purpose natural language systems whose performance even remotely resembles the one of humans in a free format everyday conversation.

Finally, it is interesting to note, It is interesting to note that these more natural kinds of activities (perception, manipulation, speech) are all activities that have, in some very essential ways, to do with complex, "high bandwidth" interaction with the real world. We will come back to this point later on.

Embodied Artificial Intelligence

These failures, largely due to the lack of rich system-environment interaction, have lead some researchers to pursue a different avenue, the one of embodiment. With this change of orientation, the nature of the research questions also began to change. Rodney Brooks, one of the first promoters of embodied intelligence (e.g. Brooks, 1991), started studying insect-like locomotion, building, for example, the six-legged walking robot "Ghengis". So, walking and locomotion in general became important research areas, topics typically associated with low-level sensory-motor intelligence. This is, of course, a fundamental change from studying chess, theorem proving, and abstract problem solving, and it is far from obvious how the two relate to one another, an issue that we will elaborate in detail later. Other subjects that people started investigating have been orientation behavior (i.e. finding one's way in only partially known and changing environments), path-finding, and elementary behaviors such as wall following, and obstacle avoidance.

The perspective of embodiment requires working with real world physical systems, i.e. robots. A crucial aspect of embodiment is that it requires working with real world physical systems, i.e. robots. Computers and robots are an entirely different ball game: computers are neat and clean, they have clearly defined inputs and outputs, and anybody can use them, can program them, and can perform simulations. Computers also have for the better part only very limited types of interaction with the outside world: input is via keyboard or mouse click, and output is via display panel. In other words, the "bandwidth" of communication with the environment is extremely low. Also computers follow clearly defined "input processing" output scheme that has, by the way, shaped the we think about intelligent systems and has become the guiding metaphor of the classical cognitivistic approach. Robots, by contrast, have a much wider sensory-motor repertoire that enables a tight coupling with the outside world and the computer metaphor of input-processing-output can no longer be directly applied.

Building robots requires engineering expertise, which is typically not present in computer science laboratories, let alone psychology departments. So, with the advent of embodiment the nature of the field, artificial intelligence, changed dramatically. While in the traditional approach, because of the interest in high-level intelligence, the relation to psychology, in particular cognitive psychology was very prominent, the attention, at least in the early days of the approach of embodied intelligence, shifted more towards – non-human – biological systems, such as insects, but other kinds of animals as well.

Also, at this point, the meaning of the term "artificial intelligence" started to change, or rather started to adopt two meanings. One meaning stands for GOFAI (Good Old-Fashioned Artificial Intelligence), the traditional algorithmic approach. The other one designates the embodied approach, a paradigm that employs the synthetic methodology which has three goals: (1) understanding biological systems, (2) abstracting general principles of intelligent behavior, and (3) the application of this knowledge to build artificial systems such as robots or intelligent devices in general. As a result, the modern, embodied approach started to move out of computer science laboratories more into robotics and engineering or biology labs.

It is also of interest to look at the role of neuroscience in this context. In the 1970s and early 1980s, as researchers in artificial intelligence started to realize the problems with the traditional symbol processing approach, the field of artificial neural networks, an area that had been around since the 1950s, started to take off – new hope for AI researchers who had been struggling with the fundamental problems of the symbol processing paradigm. Inspiration was drawn from the brain, but only at a very abstract level. In the embodied approach, there was a renewed and much stronger interest in neuroscience because researchers realized that natural neural systems are extremely robust and efficient at controlling the interaction with the real world. As mentioned above, animals can move and manipulate objects with great ease, and they are controlled by – natural – neural networks. In addition, they can move very elegantly, with great speed and with little energy consumption. These impressive kinds of behaviors can only be achieved if the dynamical properties of the neural networks are exploited. This is quite in contrast to the traditional AI approach where mostly static feedforward networks were employed.

Diversification

So, in terms of research disciplines participating in the AI adventure, we see that in the classical approach it was mainly computer science, psychology, philosophy, and linguistics, whereas in the embodied approach, it is computer science and philosophy as before, but also engineering, robotics, biology, and neuroscience (with a focus on dynamics), whereas psychology and linguistics have lost their role as core disciplines. So we see somewhat of a shift from high-level (psychology, linguistics) to more low-level sensory-motor processes, with the neurosciences covering both aspects, sensory-motor and cognitive levels. With this shift, the terms used for describing the research area shifted: researchers working in the embodied approach no longer referred to themselves as working in artificial intelligence but more in robotics, engineering of adaptive systems, artificial life, adaptive locomotion, bio-inspired systems, and neuroinformatics. But more than that, not only have researchers in artificial intelligence moved into neighboring fields, but researchers that have their origins in these other fields started in natural ways to contribute to artificial intelligence. This way, the field on the one hand significantly expanded, but on the other, its boundaries became even more fuzzy and ill-defined than before.

These considerations also provide a partial answer to the question of why we don't get many entries when we type "embodied intelligence" or "embodied artificial intelligence" into one of the search engines: Because the communities started to split and researchers in embodied intelligence started attending other kinds of conferences, e.g. "Intelligent Autonomous Systems, IAS", "Simulation of Adaptive Behavior – From Animals to Animats, SAB", "International Conference on Intelligent Robotics and Systems, IROS", "Adaptive Motion in Animals and Machines, AMAM", "European Conference on Artificial Life, ECAL", "Artificial Life Conference, ALIFE", "Artificial Life and Robotics, AROB", "Evolutionary Robotics, ER", or the various IEEE conferences (International Society of Electrical & Electronics Engineering), etc. Anecdotally speaking, I (Rolf Pfeifer) remember that initially, in the early 90s, when I tried to convince people at AI conferences such as International Joint Conference on Artificial Intelligence (IJCAI), the European Conference on Artificial Intelligence, ECAI, or the German annual AI Conference, that embodiment is not only interesting but essential to understanding intelligence, I mostly got very negative reactions and no real discussion was possible. So, together with many colleagues we turned to other conferences where people were more receptive to these new ideas. More recently, perhaps because of the stagnation in the field of classical AI in terms of tackling the big problems about the nature of intelligence, there has been a growing interest in embodiment and now AI conferences, at least some of them, have started workshops on issues in embodiment. But by and large, the two communities, the classical and the embodied one, are pretty much separate, and will probably remain so for a while.

Biorobotics

There are a number of additional interesting developments worth mentioning here. One is, in the field of embodiment, a renewed interest in high-level cognition. Rodney Brooks, at the time, had forcefully argued that getting insects to walk from scratch took evolution much longer than getting from insects to humans. This implies that creating insects was the really hard problem and after that, moving towards human level intelligence was relatively easy. Thus, so his conclusion, one should first work on insects rather than humans, one should do "biorobotics".

Many people started doing biorobotics and began cooperations with biology laboratories. An excellent example is the work by Dimitrios Lambrinos at the Artificial Intelligence Laboratory in Zurich, who started to cooperate with the world champion in ant navigation, Ruediger Wehner of the University of Zurich. Jointly, they built a series of robots, the Sahabot-Series that mimic long- and short-term navigation behavior of the desert ant Cataglyphis (e.g. Lambrinos et al., 2000). Rodney Brooks cooperated with the famous biologist Holk Cruse of the University of Bielefeld in Germany, who had been studying insect walking for many years and who had found that there is no central control for leg coordination in walking in ants. Brooks implemented Cruse's ideas on an MIT ant-like robot and termed the controller "cru(i)se" control, in honor of the designer, Holk Cruse. There are many examples of such cooperation which have all been very productive (for an excellent collection of papers on biorobotics, see (Webb and Consi, 2000)).

Developmental Robotics

However, after a few years of working on insect like behavior, Brooks started changing research topics. He argued that we have to "think big" and should work towards human level intelligence, and the project "Cog" for the development of a humanoid robot, was born (Brooks and Stein, 1993). He neatly mapped out the necessary steps and stages for achieving human-level intelligence, but due to many problems, after less than 10 years, changed topics again. But the Cog project generated a lot of excitement and many researchers were attracted by the idea of moving towards human-level intelligence, which had also been the target of classical artificial intelligence, and the field of developmental robotics emerged. The term developmental robotics designates the attempt to model aspects of human or primate development using real robots. Its pertinent conferences come under many labels, "Emergence and Development of Embodied Cognition, EDEC", "Epigenetic robotics", "Development of Embodied Cognition, DECO", "International Conference on Development and Learning", etc. This was, of course, a happy turn for those who might have been slightly sad or disappointed by the direction the field took – insects simply are not as sexy as humans! And human intelligence happens to be the most fascinating type of intelligence that we know. But once again, this strand of conferences is separate from the traditional ones in artificial intelligence, and they do not contain the term "embodied intelligence".

Ubiquitous Computing

Another line of development that should be introduced here is the one of ubiquitous computing (Weiser 1993). Computer science has undergone dramatic changes as well. Computing as such, software engineering, the development of algorithms, operating systems, the virtual machine, etc. are topics that we now understand relatively well and it is not clear whether there will be big innovations in these areas in the near future. Rather, it seems that the new challenges are seen in the interaction with the real world. Initially, the field was characterized by the idea of putting sensors every-where, into rooms (mostly cameras, motion detectors), floors (e.g. pressure sensors to detect the position of individuals) objects such as cars, chairs, beds, but also cups, or any kind of devices such as mobile telephones, clothes (e.g. t-shirts, shoes) to meas-ure physiological data of the individual wearing them for sports or medical reasons (the list is in fact endless). More recently, ubiquitous computing has also been inves-tigating actuation, i.e. ways in which systems can influence their environments: con-trol systems for buildings for temperature, humidity, windows, and blinds; cars that automatically apply their breaks when the distance to the car in front gets too small, or – in the medical domain – systems that monitor physiological variables (pulse rate, skin resistance, level of dehydration) and send a message to a physician if necessary. The field of ubiquitous computing is closely related to user interfaces or generally to human-machine interaction.

Even though user interfaces have always been an important topic in computers, the problem, in contrast to robotics, has been the low "bandwidth of communication", as pointed out earlier. In order to increase this "bandwidth", there has been a lot of work on speech, spoken language, to interact with computers, but these efforts, for various reasons, have only been met with very limited success (see our discussion above). Just recently have there been projects for developing more interesting and richer interfaces using, for example, touch, and to some extent vision. There is also work on smell but that has – although very exciting – not yet advanced significantly. The re-search on wearables should be pointed out here as well. What is interesting about these "movements", human-machine interface, wearables, ubiquitous computing, is that now virtually all computer science departments start moving into the real world. They are not doing robotics per se, but many have started hiring engineers and estab-lishing mechanical and electronics workshops where they can build hardware, be-cause now real-world devices with certain sensory-motor abilities need to be con-structed, devices that could be called "robotic devices". So far as we can tell, there has been little theory development, but there is a lot of creative experimentation going on. We feel that the set of design principles that we have developed for embodied systems will be extremely useful in designing such systems (e.g. Pfeifer and Scheier, 1999). For example, the principle of sensory-motor coordination which states that through the – active – interaction with the environment, patterns of sensory stimula-tion are induced that are correlated across sensory modalities, is an important guiding principle, but has, to date, not been applied. We might also say that computer science has now come full circle, from disembodied algorithm to embodied real-world com-puting, or rather real-world interaction, with embodied artificial intelligence as the fore-runner.

Artificial Life and Multi-agent Systems

Another interesting line of development has its origins in the field of Artificial Life, also called Alife for short. The classical perspective of artificial intelligence had a strong focus on the individual, just as psychology, and psychology was the major discipline with which artificial intelligence researchers cooperated at the time. ALife research which has strong roots in biology – rather than psychology – has been focusing on emergence of behavior in large populations of agents, in other words it is interested in what some call multi-agent systems. We deliberately say "that some call multi-agent systems" because normally, in Alife research, the term complex dynamical system is preferred, as it encompasses also physical systems where the individual components only have limited "agent character", e.g. the molecules in the famous Bénard experiment. An agent typically has certain sensory-motor abilities, i.e. it can perceive aspects of the environment, and depending on this information and its own state, performs a particular behavior. Molecules, rocks, or other "dead" physical objects do not have this ability. One point of interest has been the emergence of complex global behavior from simple rules and local interactions. (Langton, 1995)

Modular robotics, a research area that has drawn inspiration from artificial life research, also relates to multi-agent systems, where the individual agents are robotic modules capable of configuring into different morphologies (see the volume by Hara and Pfeifer (2003) for examples of modular robotic systems). One of the goals of this research is to design systems capable of self-repair, a property that all living systems have to some extent. Self-assembly and self-reconfiguration are fascinating topics that will become increasingly important as systems have to operate over extended periods of time in remote, hostile environments. The seminal work by Murata and his coworkers (Murata et al., 2004) demonstrates, how self-reconfiguration can be achieved not only in simulation but with real robotic systems. It should be mentioned, however, that to date, much of the research on self-repair and self-reconfiguration is tightly controlled, rather than being emergent from local interactions.

Evolutionary systems are another example of "population thinking", where the adaptivity of entire populations is studied rather than that of individuals. Because of its close relation to biology, economics has also taken inspiration from multi-agent systems and created the discipline of agent-based economics (e.g. Epstein and Axtell, 1996). Work on self-organization in insect societies, for example, by Jean-Louis Deneubourg of the Université libre de Bruxelles, has attracted many researchers from different fields: "ant intelligence" was one of their slogans (e.g. Bonabeau et al., 1999).

Interestingly, the term multi-agent systems has quickly been adopted by researchers in classical artificial intelligence. However, rather than looking for emergence, they endowed their individual agents with the same types of centralized control that they used for individuals (e.g. Ferber, 1999). As a consequence they could not study emergent phenomena, and a look into the journal "Autonomous Agents and Multi-Agent Systems" shows that the research under the heading "multi-agent systems" typically has different goals and does not focus on emergence. For the better part, the research is geared towards internet applications using software agents.

In robotics there has also been an interest in multi-agent systems. There the problem has been that often only relatively few robots have been available so that it has proved difficult to investigate emergence phenomena in populations. This is illustrated by the rapidly growing "Robocup" or robot soccer community. Initially the robots, for the better part, were programmed directly by the designers in order to win the game. More recently there has been growing interest and significant results in producing scientifically compelling and elegant solutions by incorporating ideas of emergence, but this still remains a big challenge.

One of the important research problems and limitations so far has been the achievement of higher levels of intelligence by the multi-agent community: typically, as in the work of ethologist and Alife researcher Charlotte Hemelrijk, the interest is in emergent hierarchies, group size formation, or migration patterns. Thinking, reasoning, or language, have typically not been topics of interest here. An exception is the work of the group of researchers interested in evolution of communication and evolution of language. An excellent example of this type of research that tries to combine population thinking or multi-agent systems with higher-level processes such as language is the "Talking Heads" experiment by Luc Steels (e.g. Steels, 2001, 2003). In an ingenious experiment he could demonstrate how, for example, a common vocabulary emerges through interaction of agents with their environment and with other agents via a language game. He has also been working on emergence of syntax, but in these experiments many assumptions have to be made to bootstrap the process. In this research strand, many insights have been gained into how communication systems establish themselves and how something like grammar could emerge. Although fascinating and highly promising, the jury is still out on whether this approach will indeed lead to something resembling human natural language.

Because of the fundamental differences in goals, the distributed agents community artificial life style, and the artificial intelligence and robotics community, individual style, have to date remained largely separate.

Summary

In summary, we can see that the landscape has changed significantly: while originally artificial intelligence was clearly a computational discipline, dominated by computer science, cognitive psychology, linguistics, and philosophy, it has turned into a multidisciplinary field requiring the cooperation and talents of many other fields such as biology, neuroscience, engineering (electronic and mechanical), robotics, biomechanics, material sciences, and dynamical systems. And this exciting new transdisciplinary community is now called "embodied artificial intelligence." While for some time, psychology and linguistics have not been at center stage, with the rise of developmental robotics, there has been renewed interest in these disciplines. The ultimate quest to understand and build systems capable of high-level thinking and natural language, and ultimately consciousness, has remained unchanged. Only the path on how to get there is fundamentally different. Although the emergence of ideas of embodiment can be traced back to pre-Socratic thinking and can be found throughout the history of philosophy, the recent developments in artificial intelligence that enable not only the analysis but also the construction of embodied systems, are supplying ample novel intellectual fodder for philosophers. As we will show later, these developments significantly change the image we have of ourselves and our society.

In spite of the multifaceted nature, there is a unifying principle and that is the actual agent to be designed in the context of the synthetic methodology, be it physical in the real world, or simulated in a realistic physics-based simulation. Such agents have a highly integrating function by bringing together results from all these different areas, and allowing concrete testing in an objective way. Moreover, they serve as excellent platforms for transdisciplinary research and communication.

3 State-of-the-Art and Challenges

Given the diversity of embodied artificial intelligence and the ruggedness of the landscape it will be next to impossible to come up with a set of challenges and a characterization of the state-of-the-art that everybody will agree on.

In characterizing the state-of-the-art we will start from the overall challenges that we will organize according to the three time scales ("here and now", ontogenetic, phylogenetic) (see Table 1). These time scales, although clearly identifiable, have important interactions, a point that we will also take into account. Moreover, we will divide our discussion into two parts, theoretical/ conceptual, and engineering. In identifying the challenges and research issues we tried to do a comprehensive survey of the literature and we, in particular, consulted the papers in this volume in order to assess the important trends. By the very nature of this endeavor of identifying challenges, this will be rather subjective and mirrors the personal research interests of the authors.

Table 1. Time scales for understanding and designing agents

time scale	designer commitments
state-oriented "here and now"	"hand design"
learning and development "ontogenetic"	initial conditions; learning and developmental processes
evolutionary "phylogenetic"	evolutionary algorithms; morphogenetic processes

However, we do believe that they reflect, one way or other, the important directions in the field. Nevertheless, we do not expect everyone to agree.

We propose the following "grand challenges" for future research, theoretical understanding of behavior; achieving higher level intelligence; automated design methods (artificial evolution and morphogenesis), and "moving into the real world".

Theoretical Understanding of Behavior

By theoretical understanding of behavior we mean an understanding of how particular behaviors in the real world can be achieved in artificial agents. This may also shed light on how particular behaviors that we observe in nature come about, which is also one of the goals of artificial intelligence research. This goal is mainly to do with the "here and now" time scale, i.e. with the question of the mechanisms underlying behavior. Although a vast body of knowledge has been accumulated this still remains one of the big conundrums.

As outlined in the previous section, many research areas and a host of studies have contributed to this understanding. However, we still don't have, for example, general purpose perceptual systems – human or primate vision is still unparalleled, and we still have an insufficient understanding of motor control, e.g. how we can achieve rapid legged locomotion. For example, there has been a lot of progress in research on humanoid walking robots, especially in Japan (e.g. Sony's QRIO, Honda's Asimo, Kawada's HRP, the University of Tokyo's H-7, to mention but a few). However, although most of these robots show impressive performance, they still walk slower than humans, their walking style looks somewhat unnatural, and research on running is still in its infancy.

One of the issues, and this is one of the challenges, is the fact that most of the research has been focused on control, which has been, and still is, the standard perspective in robotics. Recent work in the area of biomechanics seems to suggest that material and morphological properties, i.e. the intrinsic dynamical properties of the muscle-tendon systems and the specific shapes and material properties of the limbs and the body play an essential role in locomotion (e.g. Blickhan et al., 2003; Kubow and Full, 1999), but also in behavior in general, e.g. object manipulation, posture control, gesturing, etc. These ideas are captured in the theoretical principle of "ecological balance", as outlined by Pfeifer et al., (in press), Hara and Pfeifer (2000), Ishiguro et al., (2003) and earlier in Pfeifer and Scheier (1999), which states that there is a balance or task distribution between morphology, materials, control, and interaction with the environment: Some tasks, e.g. the elastic movement of the knee joint when the foot hits the ground in running, can be taken over by the – elastic – materials, and their trajectories do not need to be explicitly controlled. By morphology we mean the form and structure of an organism and its parts, including the physical nature of the sensors and their distribution. We discuss materials separately, as they play an extraordinary role in agent design.

There is another aspect of ecological balance, namely that there should be a match in the complexity of the sensory, motor and (neural) control systems. Many robotic systems are "unbalanced" in the sense that they are built of hard materials and electrical motors, and thus the control requires an enormous amount of computation. Robot vision systems are also often unbalanced as they are largely algorithmic and do not exploit morphological properties. For example, natural systems don't have cameras but retinas that perform some kind of morphological computation by their non-homogeneous arrangement of the light-sensitive cells. Moreover, generally speaking retinas perform an enormous amount of computation right at the periphery so that the signals that are passed on, are already highly processed. Artificial retinas have been around since the mid-80s (e.g. Mead, 1989), but they are still not widely used in the field. Moreover, vision or perception in general is not a matter of mapping inputs to internal representation, but of sensory-motor coordination, requiring a complex motor system as well. While initially it might seem that taking the motor system into account as well in perception would make the problem harder, when viewed in an ecological context, many problems might in fact be simplified, as demonstrated by the field of active vision or animate vision (e.g. Ballard, 1991). In animate vision, the ability of the agent (the vision system) to move is exploited to make the vision task easier. The development of vision systems, which includes the development of retinas, remains a big challenge. And these vision systems must not be developed in isolation, but in the context of multi-modal systems (see also below, achieving higher level intelligence).

Recently, it has been demonstrated that by exploiting the intrinsic dynamics of an agent, the complexity of the control system can be substantially reduced (e.g. Collins et al., 2001; Iida and Pfeifer, 2004a, b; Wisse and Frankenhuyzen, 2003; Yamamoto and Kuniyoshi, 2001), as articulated in the principle of ecological balance. Thus, in order to achieve rapid locomotion, but also motion in general, material properties must be exploited. In order to achieve real progress, artificial muscles, tendons, and flexible joints must be developed which represents a big engineering challenge. Big strides in this direction have been made by Rudolf Bannasch and his colleagues (Boblan et al., 2004).

Behavior in general requires sensory-motor coordination that again, in natural systems, is achieved by a subtle interplay of morphology (of the sensory and motor systems), materials, control, and interaction with the environment. While the design principles of Pfeifer et al. (in press) do provide intuitions, they are only qualitative in nature. What is needed now, and this is a big challenge, is a more quantitative approach. While it is relatively straightforward to quantify sensory data and to estimate the amount of computation in a controller, little research has been done on quantifying morphology and materials in computational terms. Finding a common currency which is required for a theoretical and quantitative understanding, is an important research issue as it will connect the computational effort (or control) with the contributions of physical, i.e. non-computational aspects of the system (for quantitative research in the field of sensory-motor coordination that will be relevant for these issues using methods from information theory and statistics, see, e.g. Sporns and Pegors, 2004; te Boekhorst et al., 2003) (Lungarella and Pfeifer, 2001). Lichtensteiger (2004), for example, demonstrated how the pre-processing function performed by the morphological arrangement of facets in an insect (or robot) eye, can be measured quantitatively and how a particular arrangement influences learning speed.

In general, there is a definite need for more quantitative methods in order to turn the field into a true scientific discipline. Gaussier et al., for example (Gaussier, et al., 2004) developed a formalism in the form of an algebra for cognitive processes based on the idea of perception-action coupling in autonomous agents. They apply the formalism to demonstrate how facial expressions can be learned and that there is no need to postulate innate mechanisms. Other examples of quantification will be discussed in the section on development.

While we must move towards more quantative methods, there is a certain danger involved: Because of the limitations of formal description, there tends to be a focus on isolated, well-formalizable areas, as we know it from the field of classical robotics and control theory. For example, there is a lot of formal work on path planning and inverse kinematics which lends itself more readily to a formal treatment than, for example, locomotion of complex systems involving materials with different kinds of properties and many degrees of freedom. Formalizing the latter represents a big challenge.

From an engineering perspective, in addition to the materials of the motor system, there are challenges concerning the various sensory modalities: haptics for example, is a very fundamental and rich modality in natural organisms. But the technology is, compared to natural systems, very underdeveloped: low resolution, hard, non-bendable materials, pressure only. However, there are exciting developments towards overcoming these limitations, as illustrated by the soft robotic fingertip with randomly distributed sensors for measuring slip and texture by Hosoda (2004). The development of skin-sensors by which the entire body can be covered represents a big challenge, not so much for artificial intelligence, but for the material sciences, similar to the issue of artificial muscles. At the moment, this is a significant bottleneck: better materials would almost certainly entail a quantum leap in artificial intelligence.

Achieving Higher Level Intelligence

The term "higher level" intelligence is used to designate behavior that is not purely sensory-motor, such as problem solving and reasoning, or generally thinking, natural language, emotion, and consciousness. Note that there is a frame-of-reference issue here: when we say "not purely sensory-motor" it is not really clear whether we are referring to behavior or mechanism. Inspection of the mechanisms underlying so-called non-sensory motor or cognitive behavior yields that almost universally the sensory and motor systems will be involved since in natural systems brains are intrinsically intertwined with embodiment and cannot clearly separated (e.g. Thelen and Smith, 1994). While it is possible in principle to "hand design" agents (see Table 1) endowed with higher level intelligence, all efforts to date have been met with only very limited success. One of the big unresolved issues to date is the one of symbol processing: How is it possible that humans have the capability for symbol processing? More precisely we would have to ask how it is possible that humans can behave in ways that it makes sense to describe their behavior as "symbolic", irrespective of the underlying mechanisms, which might involve explicit symbol processing or not. The question is very broad and of general importance: it is about how organisms can acquire meaning, how they can learn about the real world, and how they can combine what they have learned to generate symbolic behavior, a problem known as the "symbol grounding problem.". There is general agreement that learning will make substantial contributions towards a solution. However, learning alone will not suffice – embodiment must be taken into account as well.

Drawing inspiration from nature, a consensus has emerged that a productive approach might be to mimic at some level of abstraction a developmental process. Development, in contrast to learning, also incorporates growth and maturation of the organism. There is a vast literature on machine learning that might be potentially relevant here for solving the symbol grounding problem, but also for development in general. The book "Re-thinking innateness" has been viewed as a kind of landmark publication, employing a connectionist modeling approach (Elman et al., 1996). While a lot of ideas can be taken from this book, the approach does not deal with embodiment. This is the case with most of the machine learning literature.

As indicated earlier, the impact of taking embodiment into account can hardly be over-estimated. For example, there is the big challenge of general perception in the real world: How come we can recognize objects or faces under large variations of distance, orientation, partial occlusion, and lighting conditions? Again, many people seem to agree that a developmental approach might be useful. One of the basic issues is the fact that agents in the real world do not receive neatly structured input vectors – as is assumed in most simulation studies – but there is a continuously changing stream of sensory stimulation which strongly depends on the agent's current behavior. One way to deal with this issue is by exploiting the embodied interaction with the real world: Through the – physical – interaction with the environment, the agent induces or generates sensory stimulation (e.g. Pfeifer and Scheier, 1999). The thus generated stimulation will typically be more structured, and will contain correlations within and between sensory channels that greatly facilitate the problem of focusing on the relevant stimulation and is in fact the enabler for learning (Lungarella and Pfeifer, 2001; Sporns and Pegors, 2004). A very simple example is grasping and centering which stabilizes and normalizes the visual stimulation of an object on the retina, and at the same time produces correlated haptic and proprioceptive stimulation. This issue is covered in the principle of sensory-motor coordination which may be an important constituent in bootstrapping perception. Achieving general purpose, flexible and adaptive perception in the real world is certainly one of the very grand challenges. This is one of the big research topics in the field of "developmental robotics" or "cognitive robotics" that has recently picked up a lot of momentum. It has been suggested that the principle of sensory-motor coordination should be called more generally the principle of information self-structuring because the agent himself (or itself) interacts in particular ways with the environment to generate proper sensory stimulation.

Now the goal of this new field is not only perception, but development in general. An important direction is and has been imitation learning that seems to play a key role. This research has been inspired by the discovery of mirror neurons in the 1990s (e.g. Dipellegrino et al., 1992; Fadiga et al., 2000; Gallese et al., 1996) which demonstrated that motor and sensory systems are very closely intertwined in the brain. Designing and building a system capable of a wide range of imitation behaviors is certainly another one of the big challenges. Important first steps have demonstrated the in-principle feasibility of this approach (e.g. Kuniyoshi et al., 2004; Jansen et al.,

2004; Yoshikawa et al., 2004). Robots will no longer have to be programmed, but the skills they should acquire can simply be demonstrated. While this ability will certainly improve the sensory-motor behavior of agents, the hope is that it will also contribute to the development of social behavior, and language and communication abilities. For a review of the research in developmental robotics, see Lungarella et al. (2004). One of the challenges for the research on imitation is that direct copying is not possible, because the caregiver has a morphology that considerably differs from the one of the baby, i.e. certain perceptual generalizations will have to be made by the baby in order to interpret the caregiver's action. Over the last few years, there has been increasing consensus that joint attention plays a key role in learning and social development, a topic now being studied in developmental robotics (e.g. Nagai et al., 2003).

Let us briefly discuss a few additional grand challenges in development, acquisition of natural language, consciousness, emotion, and motivation. First steps toward acquisition of natural language, acquisition of a joint vocabulary, has been demonstrated in Luc Steels's ingenious "Talking Heads" experiment. Steels also did some preliminary work on acquisition of syntax, but there is a long way to the final goal of complete natural language development.

Consciousness has always been considered as something like the ultimate criterion for true intelligence. An elusive and fascinating topic that has attracted quite a bit of attention in the field of embodied artificial intelligence. Owen Holland is also having a stab at the future of embodied artificial intelligence and asks the question of whether we will be able to achieve machine consciousness (Holland, 2004). A topic often discussed in investigating consciousness – and in building machine consciousness, are the so-called qualia. Qualia are the subjective sensory qualities like "the redness of red" that accompany our perception. Qualia symbolize the explanatory gap that exists between the subjective qualities of our perception and the physical brain-body system whose states can, in principle, be measured objectively. In our terminology, qualia are closely related to embodiment, to the physical, material, and morphological structure of the sensory systems.

Emotions, another highly controversial topic, also relate to the issue of consciousness and the development of emotional machines is also a topic of interest (for a partial review of an embodied perspective, see e.g. Pfeifer, 2000) . Last but not least, a topic that anyone interested in intelligence and especially development will have to deal with is why an agent does anything in the first place? Why should it learn new things? This question is especially relevant if there are rich task environments with many behavioral possibilities. A chess computer only has one task, i.e. to make the next move, whereas in the real world there are always a host of possibilities – at least for those agents that we are potentially interested in (not for Braitenberg Type 1 vehicles). It is the entire issue of motivation, a topic with an enormous history. Luc Steels and Frederic Kaplan in this volume present two simple but powerful and highly plausible general solutions (Steels, 2004; Kaplan and Oudeyer, 2004). These are all fundamental questions of cognitive science.

In order to make development work, a number of engineering challenges must be resolved. From developmental studies it is known that sensory-motor coordination underlies much of concept development. This requires on the one hand the development of proper actuators: upper torso with head/neck, and arms with hands. Many researchers work with torsos only, but given the importance of locomotion for cognitive development, it would be desirable to have complete agents capable of walking freely in their environments. To date most robots are specialized, either for walking or other kinds of locomotion purposes, or for sensory-motor manipulation, but rarely are they skilled at performing a wide spectrum of tasks. This is due to conceptual and engineering limitations. Actuator technology is a major problem as today mostly electrical motors are employed, whereas – as argued earlier – artificial muscles would be more desirable. Skin sensors for the fingertips, but also for covering the entire body, would be essential for building up something like a body image, and ultimately to bootstrap cognition. Huge transdisciplinary efforts between engineering, biomechanics, and material science will be required to make progress here.

Note that although most people in developmental or cognitive robotics are interested in humanoids, this is by no means the only path. A developmental perspective can be beneficial for all kinds of animal studies.

High-level intelligence cannot only be achieved using a developmental approach, but also, at least theoretically, by means of evolutionary methods. We will discuss them in the subsequent paragraph, but given the state-of-the-art in artificial evolution, we will have to resort to more direct methods such as hand design or developmental approaches for the time being.

Automated Design Methods (Artificial Evolution and Morphogenesis)

Using artificial evolution for design has a tradition in the field of evolutionary robotics (e.g. Nolfi and Floreano, 2001). The standard approach is to take a particular robot and use an evolutionary algorithm to evolve a controller for a particular task. However, if we want to explore morphological issues, and if we want to design entire agents rather than controllers only, we have to devise powerful methods capable of handling these issues. Floreano et al. (2004) provide an excellent overview of the field with many illustrations and experiments.

Because of the many parameters and design considerations involved, automated methods must be employed because humans will no longer be able to "hand design" all aspects of such systems. There is the morphology of the body, the materials, the neural control, the interaction with the environment, and there is the possibility of having several agents, perhaps simpler ones, perform the task collectively. For individual organisms, there have been some initial successful attempts at designing systems by evolutionary means, the main approaches being the parameterization with recursive encoding (e.g. Sims, 1994; Lipson and Pollack, 2000), and those where ontogenetic development is based on abstract models of genetic regulatory networks using cell-to-cell signaling mechanisms (Eggenberger, 1997, 1999; Bongard, 2002, 2003; Bongard and Pfeifer, 2001; Banzhaf, 2004). The advantage of genetic regulatory networks is that they incorporate less of a designer bias and that they allow for incorporation of interaction with the environment during ontogenetic development, developmental plasticity (Bongard, 2003). Moreover, because they encode growth processes, they also, in some sense, contain the mechanisms for self-repair, an essential property of natural systems.

There are a number of challenges, here. First, it is the further development of models genetic regulatory networks to grow creatures of arbitrary complexity and to make the evolution open-ended in the sense that not only the parameters of the genetic regulatory networks can be manipulated, but that the mechanisms themselves are under evolutionary control. Moreover, understanding and controlling the highly involved complex dynamics of genetic regulatory networks will require a lot of research (see Bongard, 2003; Eggenberger, 1999; and Banzhaf, 2004, for some preliminary pertinent research). An important aspect will be the understanding of the emergence of hierarchical structures and modularity of the phenotypes (see also Floreano et al., 2004). Second, the physics-based simulation models need to be augmented to allow for more sophisticated agent-environment interactions. Also, deformable, flexible materials, additional sensors such as "skins" for covering the entire body, or olfaction, as well artificial muscles should be accounted for. Third, along these lines, the task environments must be made much more complex in order to put these design methods to a real test. In this way, we might be able to observe and better understand phenomena of centralization of neural substrate, i.e. the formation of brains. Eventually we might be able to see not only exploitation of physical interaction constraints, but also social ones. Whether the mechanisms of simulated genetic regulatory networks will in fact scale to very complex organisms capable of sophisticated social interaction, is an open question. The grand challenge remains to evolve truly complex creatures capable of communication, language, high-level cognition, and – perhaps – consciousness. Several orders of magnitude of scale will have to be bridged in the process, from molecules to macroscopic organisms. To what extent physically realistic simulations will be sufficient for this purpose, or whether evolution actually must happen in the real world with its indefinite richness, is a deep and currently unanswered issue.

This evolutionary level, designing the evolutionary mechanisms as well as the developmental processes based on genetic regulatory networks, might in fact provide a proper level of formalization of ecological balance. While it is indeed hard to find a common currency for trading computation for materials and morphology, it might turn out to be much easier to formally specify the developmental processes as encoded in the genome. This is because, at this stage, it is still undecided how the tasks will be distributed to control, materials, and morphology for a particular task-environment.

Moving into the Real World

The last grand challenge that we would like to discuss here concerns very generally speaking the "move into the real world." The first significant step in this direction has been the introduction of the notion of embodiment and the insight that true intelligence always requires the interaction with the real world. Embodied artificial intelligence is based on this idea. Building intelligent robots, i.e. robots capable of performing a wide range of tasks, is, as we have argued throughout this paper, hard enough, and the robots we currently are capable of building are not to our satisfaction, and so building robots per se remains a grand challenge in the field.

In designing higher-level intelligence we identified developmental approaches as a potentially suitable method. Development requires growth processes that we can currently only simulate. But there are some tricks that can be applied to make development somewhat more realistic vis-à-vis the real world. One possibility is to start with high-resolution, high-precision systems with many degrees of freedom. Growth, at least in some respects, can then be "simulated" by constraining the systems initially, freezing degrees of freedom, and simulating low resolution, for example, of the vision sensor in software by applying certain kinds of filters. These constraints can successively be released which in some sense reflects an organism's maturational processes (Gómez et al., 2004).

However, biological organisms actually do grow in the real world by means of cell division and cell differentiation, a process that may in fact be essential for the emergence of cognition. Developing growing structures in the real world is one of the great engineering challenges that will require the cooperation of material scientists, engineers, molecular and developmental biologists, and nanotechnology experts. These are, by the way, all disciplines that are not normally associated with artificial intelligence.

If artificial evolutionary processes are not only to be simulated in a computer but performed in the real world, we will need growth processes as well. As mentioned earlier, it is not clear to what extent physics-based simulations will be sufficient for scalable artificial evolution, and to what extent evolution has to rely on processes in the real world. First steps in performing artificial evolution in the real world have been taken already in the 1960s by Ingo Rechenberg who evolved optimal shapes of fuel pipes by actually configuring the physical system "designed" by the evolutionary algorithm (an evolution strategy) and measuring the performance on the real fuel pipe system (Rechenberg, 1973). Another example is the work by Adrian Thompson at the University of Sussex who used FPGAs to test the circuits evolved using a genetic algorithm (Thompson, 1996). FPGAs, in contrast to microprocessors, rather than making a digital simulation of a circuit, actually configure a physical circuit. The results achieved are truly amazing and provides a glimpse at the power of evolution in the real world.

A major step is taken by researchers in the EU-funded PACE (Programmable Artificial Cell Evolution) project by John McCaskill of the Ruhr University Bochum, in Germany, where the goal is to evolve an artificial cell in a chemical laboratory. Using micro-fluidic arrays, carefully controlled chemical reactions can be induced so that cells can be formed and their metabolisms influenced in precise ways. Part of the evolution will be performed in simulation and part in the real world. The goal is to evolve self-replicating cells in the laboratory, an enormous challenge. If successful, this would enable us to perform artificial evolution in the real world and thus we could generate any kind of structure required for performing a particular task. Because the cells can divide we would have actual growth processes in the real world.

Some people like Ray Kurzweil believe that nanotechnology will be the key to engineer growth in the real world. Whether this will materialize we will only know in the future.

Cyborgs could also be viewed as a way to "move into the real world": rather than constraining the neural substrate to function in a dish in isolation, it is connected to either a simulation or to a robot that behaves in the real world and sends its sensory signals back to the neural tissue in the dish (Bakkum et al., 2004). Coupling biological neural tissue to a real world artifact opens up entirely new avenues in man-machine interaction. This research in itself bears many great challenges, the general issue of coupling biological and technical substrate. On the one hand, we can expect to learn something about neural functioning, and on the other we might, in the future, be able to better understand how to control robots by observing the natural neurons. Medical applications in prosthetics (e.g. Yokoi et al., 2004), are of course obvious candidates for practical applications.

Finally, coming back to the research on self-repair, self-assembly, and self-reconfiguration discussed in the "Landscape" section, a big challenge, conceptually and from an engineering perspective, is the development of such systems in the real world. Again, while simulation of processes of self-repair, for example, represents a challenge and is far from being straight-forward, the ultimate challenge will be the transfer to the real world. Murata and his collaborators (2004) have demonstrated first ideas using modular robotic systems.

4   Conclusions, the Future, and Applications

The challenges outlined are big challenges and we must not expect to reach them in the near future. However, it is important to keep the long-term visions in mind when thinking about the next steps. The difficulty of research in any field, but in particular in artificial intelligence, is to map the big visions and challenges onto concrete, doable steps. We have also tried to outline what researchers in the field are currently attempting to do and what they are planning for the near future. And the papers presented in this volume provide an excellent starting point.

Let us now return to the initial question of what thinking has to do with walking – the symbol grounding problem – and reflect on how the challenges outlined in the paper will contribute to this question which metaphorically summarizes the goals of embodied artificial intelligence. In the early phases of embodied artificial intelligence, many people were working on navigation and orientation out of a conviction that locomotion and orientation are somehow the underlying driving forces in the development of cognition, in the evolution of the brain. This is corroborated by the question asked by the famous Oxford neuroscientist Daniel Wolpert "Why don't plants have brains?". And he suggested that the answer might actually be quite simple: "Plants don't have to move!" Because of the "embodied turn", researchers started working with robots, and because they were readily available and easy to use, wheeled robots were the tools of choice. Navigation in the real world is a challenging problem and there has been much exciting research in robotics in general (e.g. Bellot et al., 2004, who introduce the new method of Bayesian Programming) and in bio-logically inspired approaches in particular (e.g. Hafner, 2004). While there was a lot of progress – researchers were forced to deal with the intricacies of the interaction with the real world, such as noise, imprecisions, change, unpredictability – there were also some intrinsic problems with the approach. Remember that one of the aspects of the principle of ecological balance is the match in complexity of sensory, motor, and neural systems. Because it is easy to put a high-resolution camera on a robot, and because wheeled robots only have few degrees of freedom of actuation, many experimental designs were "unbalanced": complex sensory systems, very simple motor systems. As a result of these unbalanced designs, these systems had a relatively uninteresting physical dynamics. One implication is that the algorithms used for control were largely arbitrary: Even though they were mostly biologically inspired, they were arbitrary with respect to the robot's own dynamics; one algorithm can be exchanged by another, achieving essentially the same behavior. Something was missing and many suspected that this is a complex sensory-motor level with an interesting and rich dynamics.

As a consequence a number of researchers started working on complex body dynamics (e.g. Kuniyoshi et al, 2004; Iida and Pfeifer, 2004a; Proc. of the Int. Workshop on Adaptive Motion in Animals and Machines, AMAM-2003). This shift was interpreted by critics but also by people sympathetic to these developments, as a move away from the goal of understanding and building cognitive systems. However, and this is one of the big insights from embodied artificial intelligence, the exact opposite was the case: It turned out that a rich complex body dynamics is the foundation, the prerequisite for something like symbol processing to develop (see, e.g. Okada et al., 2003; Iida and Pfeifer, 2004b; Kuniyoshi et al., 2004). So what happened is that what seemed like a deviation from the road to cognition, turned out to be necessary. This view is also compatible with Núñez (2004) who argues that even very abstract mathematical concepts have their origins, are grounded, in our embodiment which provides the basis for metaphors. Because these metaphors must be sufficiently rich for bootstrapping interesting concepts, the embodiment must reflect this richness. Of course, at the moment, this is all speculation that must be corroborated by many experiments. But at the risk of being entirely wrong, let us speculate a little further.

There is another, unexpected idea that emerges from this research. The question of symbol grounding always entails the question of how it is possible that something like discrete symbol processing can emerge from a completely continuous dynamical system, such as a human. Rich, complex dynamics also implies many attractor states and transitions between them. Attractor states are, within the continuous dynamics, objectively identifiable, discrete states, that can, of course, also be identified by the agent itself (or himself), given the proper neural system. Once identified, the agent can start using them, for example, for planning purposes (e.g. Okada et al., 2003; Kuniyoshi et al., 2004). It is interesting to note that a complex intrinsic sensory-motor dynamics implies that the neural control is no longer arbitrary, but has to be "in tune" with the physical substrate, quite in contrast to wheeled robots. Ishiguro and his colleagues (2004) have provided a beautiful demonstration, theoretically and in a robot case study, of how control and body dynamics in a complex agent have to be coupled.

If coupled properly, control is not only simpler, but the entire system tends to be more energy-efficient. Lungarella and Berthouze (2004) in a robotics case study convincingly demonstrate that a judicious – non-arbitrary – choice of parameters coupling the neural and body dynamics facilitates the acquisition of motor skills in a developing organism. Whether these ideas on dynamics will ultimately lead to high-level cognition or to conscious agents, whether in this way we can achieve the goals set out by Holland (2004), is an entirely open question.

Tom Ziemke in his contribution (2004) quotes from Gerald Edelman "It is not enough to say that the mind is embodied: one has to say how." (Edelman, 1992). Bootstrapping it from complex body dynamics might be part of the answer.

In their current state, evolutionary studies are, for the time being, restricted to providing ideas on the distribution of morphology, materials, control, and interaction with the environment. More varied and taxing task environments will be necessary to investigate agents with more complex sensory-motor dynamics on top of which cognition can bootstrap. But some of recent approaches demonstrate definite progress in this direction (e.g. (Bongard, 2003)). However, as alluded to in the previous section, in order to achieve truly complex organisms, it may be necessary to couple the artificial evolutionary process to the real world.

To conclude, just few words about applications. While the classical approach has created many applications in terms of clever algorithms that are now widely used, the embodied approach seems to be more limited. The major applications have been in the entertainment and educational areas. As this paper demonstrates, the field is just beginning to develop a basic understanding and there are many big challenges lying ahead. We could also add a challenge, namely to exploit these technologies for practical applications in industry, the environment, and services for the benefit of society.

Research on humanoid robots has an interesting side-effect, so to speak. Humanoids require the development of sophisticated body parts, legs, arms, hands, etc., that can potentially be used, at least to some extent, as prosthetic devices. The fascinating research by Yokoi et al. (2004) and by Boblan et al. (2004) points in this direction. The ground breaking research by Potter and his co-workers (Bakkum et al., 2004) might eventually be employed for interfacing these devices smoothly with humans – an additional intriguing perspective.

As outlined in the section of ubiquitous computing, a better understanding of embodied intelligence will lead to many applications in terms of so-called embedded systems, i.e. systems that autonomously interact with the real world, not only through sensing, but also by influencing the world without human intervention. These systems are not robots in the restricted sense of the word (they are very different from humanoid robots, for example), but they have many of their characteristics in terms of intelligent, autonomous interaction with the environment. These kind of systems, also called "robotic devices" are already present in many technical applications (cars, airplanes, household appliances, elevators, etc.), but by augmenting their "intelligence", so to speak, many more applications will become possible. This way, the ideas that embodied artificial intelligence has spurred will spread to numerous scientific and technological areas for the benefit of society.

Acknowledgments. We would like to thank the scientific director of the International Conference and Research Center for Computer Science, Prof. Reinhard Wilhelm, for suggesting this conference, and the Swiss National Science Foundation for supporting the research presented in this paper, grant # 20-68198.02 ("Embodied Artificial Intelligence"). We would also like to thank the members of the Artificial Intelligence Laboratory of the University of Zurich for numerous stimulating discussions on this topic. Credit also goes to Max Lungarella for his many thoughtful comments on this paper.

References

Bakkum, D.J., Shkolnik, A.C., Ben-Ary, G., Gamblen, P., DeMarse, T.B., and Potter, S.M. (2004). Removing some 'A' from AI: Embodied cultured networks (this volume)

Ballard, D. (1991). Animate vision. Artificial Intelligence, 48, 57-86.

Banzhaf, W. (2004). On evolutionary design, embodiment, and artificial regulatory etworks (this volume).

Boblan, I., Bannasch, R., Schwenk, H., Miertsch, L., and Schulz, A. (2004). A human like robot hand and arm with fluidic muscles: Biologically inspired construction and functionality. (this volume)

Bellot, D., Siegwart, R., Bessière, P., Tapus, A., Coué, C., and Diard, J. (2004). Bayesian modeling and reasoning for real-world robotics: Basics and examples (this volume).

Blickhan, R., Wagner, H., and Seyfarth, A. (2003). Brain or muscles?, Rec. Res. Devel. Biomechanics, 1, 215-245.

Bonabeau, E., Dorigo, M., and Theraulaz, G. (1999). Swarm intelligence: from natural to artificial systems. New York, N.Y.: Oxford University Press.

Bongard, J.C. (2003). Incremental approaches to the combined evolution of a robot's body and brain. Unpublished PhD thesis. Faculty of Mathematics and Science, University of Zurich.

Bongard, J.C. (2002). Evolving modular genetic regulatory networks. In Proc. IEEE 2002 Congress on Evolutionary Computation (CEC2002). MIT Press, 305-311.

Bongard, J.C., and Pfeifer, R. (2001). Repeated structure and dissociation of genotypic and phenotypic complexity in artificial ontogeny. In L. Spector et al. (eds.). Proc. of the Sixth European Conference on Artificial Life, 401-412.

Brooks, R. A. (1991). Intelligence Without Reason. Proceedings of the 12th International Joint Conference on Artificial Intelligence (IJCAI-91), pp. 569–595.

Brooks, R.A., and Stein, L.A. (1993). Building brains for bodies. Memo 1439, Artificial Intelligence Lab, MIT, Cambridge, Mass.

Collins, S.H., Wisse, M., and Ruina, A. (2001). A three-dimensional passive-dynamic walking robot with two legs and knees. The International Journal of Robotics Research, 20, 607-615.

Dipellegrino G, Fadiga L, Fogassi L, Gallese V, Rizzolatti, G (1992). Understanding motor events - a neuro-physiological study. Exp Brain Res 91: 176-180.

Edelman, G.E. (1992). Bright air, brilliant fire. On the matter of the mind. New York: Basic Books.

Eggenberger, P. (1997). Evolving morphologies of simulated 3d organisms based on differential gene expression. In: P. Husbands, and I. Harvey (eds.). Proc. of the 4th European Conference on Artificial Life. Cambridge, Mass.: MIT Press.

Eggenberger, P. (1999). Evolution of three-dimensional, artificial organisms: simulations of developmental processes. Unpublished PhD Dissertation, Medical Faculty, University of Zurich, Switzerland.

Elman, J.L, Bates, E.A., Johnson, H.A., Karmiloff-Smith, A., Parisi, D., and Plunkett, K. (1996). Rithinking innateness: A connectionist perspective on development. Cambridge, Mass.: MIT Press.

Epstein, J.M. and Axtell, R.L. (1996). Growing artificial societies: social science from the bottom up. Cambridge, Mass.: MIT Press.

Fadiga L, Fogassi L, Gallese V, Rizzolatti G (2000) Visuomotor neurons: Ambiguity of the discharge or 'motor' perception? Int J Psychophysiol 35: 165-177.

Ferber, J. (1999). Multi-agent systems. Introduction to distributed artificial intelligence. Addison-Wesley.

Floreano, D., Mondada, F., Perez-Uribe, A., and Roggen, D. (2004). Evolution of embodied intelligence (this volume).

Gallese, V., Fadiga, L., Fogassi, L., and Rizzolatti G. (1996). Action recognition in the premotor cortex. Brain 119: 593-60.

Gaussier, P., Prepin, K., and Nadel, J. (2004). Toward a cognitive system algebra. Application to facial expression learning and imitation (this volume).

Gómez, G., Lungarella, M., Eggenberger Hotz, P., Matsushita, K. and Pfeifer, R. (2004). Simulating development in a real robot: on the concurrent increase of sensory, motor, and neural complexity. The 4th annual workshop of Epigenetic Robotics (EPIROBOT04), (in press).

Hafner, V. (2004). Agent-environment interaction in visual homing (this volume).

Hara, and R. Pfeifer (eds.) (2003). Morpho-functional machines: the new species – designing embodied intelligence. Tokyo: Springer-Verlag.

Hara, F., and Pfeifer, R. (2000). On the relation among morphology, material and control in morpho-functional machines. In Meyer, Berthoz, Floreano, Roitblat, and Wilson (eds.): From Animals to Animats 6. Proceedings of the sixth International Conference on Simulation of Adaptive Behavior 2000, 33-40.

Holland, O. (2004). The future of embodied artificial intelligence: Machine consciousness? (this volume).

Hosoda, K. (2004). Robot finger design for developmental tactile interaction. Anthropomorphic robotic soft fingertip with randomly distirbuted receptors (this volume).

Iida, F. and Pfeifer, R. (2004a) "Cheap" Rapid locomotion of a quadruped robot: Self-stabilization of bounding gait. F. Groesn et al. (eds.). Intelligent Autonomous Systems 8. IOS Press, 642-649.

Iida, F., and Pfeifer, R. (2004b). Self-stabilization and behavioral diversity of embodied adaptive lcomotion (this volume).

Ishiguro, A., and Kawakatsu, T. (2003). How should control and body systems be coupled? A robotic case study (this volume).

Janssen, B., de Boer, B., and Belpaeme, T. (2004). You did it on purpose! Towards intentional embodied agents (this volume).

Kaplan, F., and Oudeyer, P.-Y. (2004). Maximizing learning progress: an internal reward system for development (this volume).

Kubow, T. M., and Full, R. J. (1999). The role of the mechanical system in control: a hypothesis of self-stabilization in hexapedal runners, Phil. Trans. R. Soc. Lond. B, 354, 849-861.

Kuniyoshi, Y., Yorozu, Y., Ohmura, Y., Terada, K., Otani, T., Nagakubo, A., and Yamamoto, T. (2004). From humanoid embodiment to theory of mind (this volume).

Lambrinos, D., Möller, R., Labhart, T., Pfeifer, R., Wehner, R. (2000). A mobile robot employing insect strategies for navigation. Robotics and Autonomous Systems, 30, 39-64.

Lenat, D., Prakash, M., and Shepher, M. (1986). CYC: Using common sense knowledge to overcome brittleness and knowledge acquistion bottlenecks.AI Magazine, vol. 6, issue 4, 65-85.

Langton, C. G. (1995). Artificial life: an overview. Cambridge, Mass.: MIT Press.

Lipson, H., and Pollack J. B. (2000), Automatic design and manufacture of artificial life forms. Nature, 406, 974-978.

Lichtensteiger, L. (2004). The need to adaptv and its implications for embodiment (this volume).

Lungarella, M., and Berthouze, L. (2004). Robot bouncing: On the synergy between neural and body dynamics (this volume).

Lungarella, M. and Pfeifer, R. (2001). Information-theoretic analysis of sensory-motor data. In Proc. of the IEEE-RAS International Conference on Humanoid Robots, 245-252.

Lungarella, M., Metta, G., Pfeifer, R. and Sandini, G. (2003). Developmental robotics: a survey. Connection Science, 15 (4), 151-190.

Mead, C.A. (1989). Analog VLSI and neural systems. Reading, Mass.: Addison-Wesley.

Murata, S., Kamimura, A., Kurokawa, H., Yoshida, E., Tomita, K., and Kokaji, S. (2004). Self-reconfigurable robots: platforms for emerging functionality (this volume).

Nagai, Y., Hosoda, K., and Asada, M. (2003). Joint attention emerges through bootstrap learning, Proc. of the 2003 IEEE/RSJ International Conference on Intelligent Robots and Systems (IROS2003), 168-173.

Nolfi, S. and Floreano, D. (2001). Evolutionary robotics: the biology, intelligence, and technology of self-organizing machines. Cambridge, MA: MIT Press.

Núñez, R. (2004). Do real numbers really move? The embodied cognitive foundations of mathematics (this volume).

Okada, M., Nakamura, D., and Nakamura, Y. (2003). On-line and hierarchical design methods of dynamics based information processing system. Proc. of the 2003 IEEE/RJS Int. Conference on Intelligent Robots and Systems, 954-959.

Pfeifer, R. (2000). On the role of embodiment in the emergence of cognition and emotion. In H. Hatano, N. Okada, and H. Tanabe (eds.). Affective minds. Amsterdam: Elsevier, 43-57.

Pfeifer, R., Iida, F., and Bongard, J. (2004). New robotics: design principles for intelligent systems. Artificial Life (in press).

Pfeifer, R., and Scheier, C. (1999). Understanding intelligence. Cambridge, Mass.: MIT Press.

Rechenberg, I. (1973). Evolution strategies: optimization of technical systems with principles from biological evolution (in German). Stuttgart, Germany: Frommann-Holzboog.

Sims, K. (1994a). Evolving virtual creatures. Computer Graphics, 28, 15-34.

Sporns, O., and Pegors, T.K. (2004). Information-theoretical aspects of embodied artificial intelligence (this volume).

Steels, L. (2001). Language games for autonomous agents. IEEE Intelligent Systems, Sept/Oct issues.

Steels, L. (2003). Evolving grounded communication for robots. Trends in Cognitive Sciences, 7 (7), 308-312,

Steels, L. (2004). The autotelic principle (this volume).

te Boekhorst, R., Lungarella, M., and Pfeifer, R. (2003). Dimensionality reduction through sensory-motor coordination. Proc. of the 10th Int. Conf. on Neural Information Processing (ICONIP'03), p.496-503, LNCS 2174.

Thelen, E., and Smith, L. (1994). A dynamic systems approach to the development of cognition and action. Cambridge, Mass.: MIT Press.

Thompson, A. (1996). Silicon evolution. In J.R. Koza et al. (Eds.). Genetic Programming 1996: Proc. of the First Annual Conference, Cambridge, Mass.: MIT Press, 444-452.

Webb B. and Consi R. C. (2000). Biorobotics -Methods & application-, Cambridge, Mass.: MIT Press.

Weiser, M. (1993). Hot topics: Ubiquitous computing, IEEE Computer.

Wisse, M and Frankenhuyzen, J.van, (2003) Design and Construction of MIKE; a 2D autonomous biped based on passive dynamic walking, Proceedings of the 2nd International Symposium on Adaptive Motion of Animals and Machines, Kyoto, March.4-8, 2003.

Yamamoto, T. and Kuniyoshi, Y. (2001). Harnessing the robot's body dynamics: a global dynamics approach. Proc. of 2001 IEEE/RSJ International Conference on Intelligent Robots and Systems (IROS2001), pp. 518-525, Hawaii, USA.

Yokoi, H. Arieta, A.H., Katoh, R., Yu, W., Watanabe, I., and Mruishi, M. (2004). Mutual adaptation in a prosthetic application (this volume).

Yoshikawa, Y., Asada, M., and Hosoda, K. (2004). Towards imitation learning from a view point of an internal observer (this volume).

Ziemke, T. (2004). Embodied AU as science: Models of embodied cognition, embodied models of cognition, or both? (this volume).

Embodied AI as Science: Models of Embodied Cognition, Embodied Models of Cognition, or Both?

Tom Ziemke

University of Skövde, School of Humanities and Informatics,
PO Box 408, 54128 Skövde, Sweden
`tom@ida.his.se`

Abstract. This paper discusses the identity of embodied AI, i.e. it asks the question exactly what it is that makes AI research *embodied*. From an engineering perspective, it is fairly clear that embodied AI is about robotic, i.e. physically embodied systems. From the scientific perspective of AI as building models of natural cognition or intelligence, however, things are less clear. On the one hand embodied AI seems to be about physically embodied, i.e. robotic models of cognition. On the other hand the term 'embodied' seems to signify the type of intelligence modeled and/or the conception of (embodied) cognition that is underlying the modeling. In either case, it appears that embodied AI, as it currently stands, might be too narrowly conceived since each of these perspectives is addressed only partially.

1 Introduction

> *"It is not enough to say that the mind is embodied;*
> *One has to say how."* [11]

Although more than a decade old now, the above quote summarizes fairly well what this paper is about. It will be argued here that, although, practically by definition, research in embodied AI emphasizes the importance of embodiment for cognitive processes, from a cognitive-scientific perspective it does not take the concept sufficiently seriously. In particular, in our opinion, many researchers, driven by engineering rather than scientific concerns and/or in an attempt to distinguish embodied AI from its traditional predecessor, overemphasize the importance of physical embodiment when it comes to scientific modeling of cognition. Being physical, however, is only one aspect that distinguishes natural embodied cognizers from the computer programs of traditional, cognitivist AI? Hardly surprising therefore, richer conceptions and discussions of embodiment can be found in, other research fields, such as cognitive linguistics and philosophy of mind. Hence, when it comes to embodied AI as cognitive-scientific modeling, it remains unclear, and is hardly ever discussed in the field, what conception of embodied cognition researchers are committed to.

On the one hand, much of embodied AI and its emphasis on physically embodied models is very compatible with the view of robotic functionalism [15], according to which embodiment is about *symbol grounding* or, more generally speaking, *representation grounding*, whereas cognition/thought can still be conceived of as computa-

F. Iida et al. (Eds.): Embodied Artificial Intelligence, LNAI 3139, pp. 27–36, 2004.
© Springer-Verlag Berlin Heidelberg 2004

tion, i.e. syntactically driven internal manipulation of representations. In a nutshell, this is the core and "central research focus" of embodied AI according to a recent review of the field in the *Artificial Intelligence* journal [1], which has subsequently been rejected as too narrow [5]. On the other hand, much of the rhetoric in the field of embodied AI, in particular its rejection of traditional notions of representation, suggests sympathy for more radical notions of embodied cognition that view *all* of cognition as embodied or body-based. This is what in Section 3 will be referred to as the position(s) of "full embodiment" [23] or "radical embodiment" [8]. This paper does not try to argue for one or the other of these views (although it is hardly a secret that we favor the second one), but it simply argues that embodied AI researchers have to realize that there are at least two different views that should not be conflated. Or, to paraphrase and extend the above introductory quote [11]: It is no longer enough for embodied AI researchers to say that (artificial) intelligence has to be embodied; but one has to be more specific concerning what that means.

The rest of this paper is structured as follows. The following section further addresses the problematic identity of embodied AI, i.e. the question what it is that makes it embodied. Section 3 then briefly summarizes different conceptions of embodied cognition and some distinctions that might be useful to import into embodied AI research. Section 4 then discusses the implications for embodied AI as cognitive-scientific modeling.

2   Background: What Is *Embodied* AI Anyway?

2.1   Motivation

This paper has actually been directly motivated by discussions at and about the Dagstuhl workshop on *Embodied AI*. Mentioning the workshop afterwards to other researchers who had not participated frequently triggered reactions such as *"But, I am working on embodied AI, why didn't I know about this workshop?"* (or *"..., why wasn't I invited?"*) or *"I didn't know there was an embodied AI community"* or *"What the heck is embodied AI?"* or *"Is there any difference between embodied AI and X?"*, where X could be, e.g., (intelligent or cognitive) robotics or (traditional) AI. There are at least two possible explanations for these reactions: (1) what embodied AI is, or is about, is simply not particularly well defined, or (2) it is in fact well defined, but the definition is only well known within a very limited community.

That explanation (1) is at least partly true was also indicated by discussions at the workshop itself, i.e. among the participants who, naturally, as experts might be supposed to have some level of agreement concerning what embodied AI is, and more specifically, exactly what it is that makes it *embodied*. For example, right after a talk that argued that mathematical cognition, although it might seem abstract at a first glance, in fact is embodied in the sense that it is based, more or less directly, on bodily experience, another participant in a discussion argued that the activity of an air traffic controller was situated, but *not* embodied, i.e. that the body was not involved to any significant degree (presumably because there is no, or only little, overt movement involved). The fact that there are different notions of embodiment is hardly surprising in itself. After all, many central terms in the cognitive sciences, such as 'intelligence', 'cognition', 'agency', 'autonomy' or 'life', are to some degree controversial and still far from being well-defined. What is surprising, however, is that none of the workshop participants reacted (until long after) to either of the above claims, although they are based on diametrically opposed positions, namely that *all* human cognitive processes are embodied or body-based, or that only some of them are, respectively.

This example clearly shows that even within the embodied AI community there are in fact very different conceptions of embodiment, and perhaps consequently embodied AI.[1] As mentioned above, there is not necessarily anything wrong with this - quite the opposite, different conceptual and theoretical frameworks within a field can in many cases lead to fruitful discussions. In the embodied AI community, however, these differences are rarely addressed more than superficially. Fields such as cognitive linguistics, phenomenology and philosophy of mind, on the other hand, seem to take embodiment much more seriously, which has led to richer and more varied conceptions of embodiment as the basis of, for example, meaning and phenomenal experience (e.g. [17, 34, 47]). However, one does not have to look at 'deep' philosophical questions to realize that the treatment of embodiment in embodied AI is somewhat shallow.

A more pragmatic problem with embodied AI, or in fact embodied cognitive science in general, is that it seems to be much more defined in terms of what it argues *against*, i.e. traditional AI[2] and the computer metaphor for mind, than what it argues *for* - a fact commonly pointed out by opponents of embodied theories. That means, many embodied AI researchers reject the idea that intelligence and cognition can be explained in purely computational terms, but it is left unclear exactly what the alternative is. Characteristic for the field is, for example, the statement that "intelligence cannot merely exist in the form of an abstract algorithm but requires a physical instantiation, a body" [27]. There are two problems with this: Firstly, being physical can at most be a *necessary* condition for intelligence (which, by the way, is contradicted by some proponents of embodied AI [13, 28]). That means, probably nobody believes that chairs and tables are intelligent, or make better models of intelligence than computer programs for that matter, just because they are physical. Secondly, it is unclear exactly which view concerning (dis-) embodiment this is in opposition to (except for dualism, perhaps). As discussed in more detail elsewhere [6], even proponents of hardcore computationalism would hardly dispute that computer programs require *some* physical instantiation or realization. After all, Newell and Simon, for example, did not include the word 'physical' in their *Physical* Symbol Systems Hypothesis [22] for no reason, but they were of course aware of the need for some form of what is now called 'grounding' (e.g. [1, 15, 37]), although it perhaps never played a crucial role in their theories.

---

[1] However, most embodied AI researchers, including the author, probably share the intuitive and somewhat unscientific conviction that, as reviewer 1 formulated it, "embodied AI is AI done right, i.e. exploring intelligence and cognition by paying attention to the biological, sensorimotor, evolutionary and developmental bases".

[2] As reviewer 2 pointed out, what exactly constitutes 'traditional AI' is of course equally ill-defined as what constitutes embodied AI, especially since some traditional AI systems, e.g. the robot Shakey, are/were embodied in at least the physical sense (cf. Section 4).

2.2 Embodied AI: Science Versus Engineering

To some extent the somewhat unclear commitment to embodiment seems to arise from the fact that embodied AI has the ambition to combine science and engineering, and that physical embodiment is not equally important in both of them, or at least not important for the same reasons. As several authors have pointed out, AI generally can be viewed from at least two different, though intertwined perspectives: that of *engineering*, mostly concerned with the design of artifacts (robots in the case of embodied AI), and that of *science*, mostly concerned with the understanding of natural systems. Furthermore, the latter can of course be broken down according to the different scientific fields that use robots and/or other autonomous agents as modeling tools, for example, cognitive science (e.g. [2, 25, 27]), neuroscience (e.g. [29, 35]), or the study of animal behavior (e.g. [39, 40]).

While these distinctions appear fairly obvious, they receive surprisingly little attention in discussions of methodology in the field of embodied AI, where overly general statements such as *"simulations are useless"* or *"Khepera robots are not real robots"* or *"existence proofs are not sufficient"* often can be heard. While from an engineering point of view all of these statements might very well be correct, they do not necessarily apply equally generally to the scientific use of autonomous agents as models of natural organisms. Steels, for example, explained the skepticism towards simulations as follows:

> The goal is to build artifacts that are "really" intelligent, that is, intelligent in the physical world, not just intelligent in a virtual world. This makes unavoidable the construction of robotic agents that must sense the environment and can physically act upon the environment, particularly if sensorimotor competences are studied. This is why researchers insist so strongly on the construction of physical agents ... Performing simulations of agents ... is, of course, an extremely valuable aid in exploring and testing out certain mechanisms, the way simulation is heavily used in the design of airplanes. But a simulation of an airplane should not be confused with the airplane itself. [31]

Obviously, Steels had a point there, and nobody would seriously question the view that simulations, however good they are, cannot fully capture the complexities of the physical world. Hence, simulations certainly have limited value in robot engineering. Furthermore, it can very well be argued that physically embodied, robotic systems make better models of animal behavior in cases where a real robot can made to interact with (roughly) the same physical environment as the modeled animal, as in the case of Webb's robot models of cricket phonotaxis [39], which could successfully be tested with real crickets (sounds), or the Pfeifer Lab's Sahabot [19], which was actually tested in the Tunisian desert environments inhabited by the ant species whose navigation behavior it was supposed to model.

However, it is far from clear to what degree this can be generalized to other cases of more general or abstract modeling. Are, for example, Vogt's robotic models of adaptive language games [37], by virtue of their physical embodiment, better scientific models than Steels' partly physical, partly simulated Talking Heads [32] or their fully simulated counterparts [38]? After all, neither the robot bodies used nor their environments in any of the experiments have much of a similarity to their counterparts in human adaptive language games. Although from an engineering point of view the physical models certainly appear more interesting, from a scientific perspective there seems to be no strong reason why they *necessarily* should make better models. Quite the opposite, as argued in more detail elsewhere [44], in many cases simulations, despite their obvious limitations, might have an important, complementary role to play, due to the fact that they allow for more extensive, more systematic and more replicable experimentation, which simply takes less time in simulation, as well as for experiments, e.g. with evolving robot morphologies (e.g. [4]), that can only be carried out in very limited form on real robots.

Just as an aside, concerning the role of existence proofs, one should also distinguish between engineering and scientific modeling. While from an engineering point of view existence proofs certainly are of limited value (e.g. nobody would want to fly in an airplane that has been tested successfully once or twice), from a cognitive science point of view they can be very valuable in the development of theories. Much connectionist cognitive modeling research, for example, has been concerned with providing concrete examples of neural networks exhibiting properties such as systematicity (e.g. [3,14]), which on purely theoretical grounds they had been argued not to be able to exhibit [12]. This is just one example, where existence proofs constrain and thus aid the development of cognitive-scientific theories. For this type of research, both physical and simulated robots, with their respective benefits and drawbacks, are useful tools in agent-based modeling [30], paying more attention to the interaction of agents and environments than traditional computational cognitive modeling of mostly internal processes.

3   Notions of Embodiment

The aim of this section is to briefly overview some distinctions in conceptions of embodiment that might be useful to import into discussion of embodied AI, in particular for the purpose of clarifying differences in theoretical frameworks and commitments in the field that usually remain hidden under a superficial agreement on (physical) 'embodiment'.

Nunez made a useful distinction between trivial, material, and full embodiment [23]. *Trivial embodiment* simply is the view that "cognition and the mind are directly related to the biological structures and processes that sustain them". Obviously, this is not a particularly radical claim, and consequently few cognitive scientists would reject it (dualist philosophers of consciousness, on the other hand, might). According to Nunez, this view further "holds not only that in order to think, speak, perceive, and feel, we need a brain – a properly functioning brain in a body – but also that in order to genuinely understand cognition and the mind, one can't ignore how the nervous system works" [23].

Material embodiment makes a stronger claim, but it is only about the interaction of internal cognitive processes with the environment, i.e. the issue of grounding, and thus considers reference to the body to be only required for accounts of low-level sensorimotor processes In Nunez's terms: "First, it sees cognition as a decentralized phenomenon, and second it takes into account the constraints imposed by the complexity of real-time bodily interactions performed by an agent in a real environment" [23].

Full embodiment, finally, is the view that the body is involved in *all* forms of human cognition, including seemingly abstract activities, such as language or mathematical cognition [18]. In Nunez's own words:

Full embodiment explicitly develops a paradigm to explain the objects created by the human mind themselves (i.e., concepts, ideas, explanations, forms of logic, theories) in terms of the non-arbitrary bodily-experiences sustained by the peculiarities of brains and bodies. An important feature of this view is that the very objects created by human conceptual structures and understanding (including scientific understanding) are not seen as existing in an transcendental realm, but as being brought forth through specific human bodily grounded processes. [23]

In a similar vein, Clark distinguished between the positions of *simple embodiment* and *radical embodiment* [8]. According to the former, traditional cognitive science can roughly remain the same; i.e. theories are merely constrained, but not essentially changed by embodiment. This is similar to Nunez's view of material embodiment. The position of radical embodiment, on the other hand, very much compatible with Nunez's full embodiment, is, as Clark formulated it, "radically altering the subject matter and theoretical framework of cognitive science" [8].

More recently, Wilson distinguished between six views of embodied cognition [41], of which only the last one requires full or radical embodiment whereas the first five might be considered variations or aspects of material embodiment: (1) cognition is situated i.e. it occurs "in the context of task-relevant inputs and outputs", (2) cognition is time-pressured, (3) cognition is for the control of action, (4) we off-load cognitive work onto the environment, e.g. through epistemic actions [16], i.e. manipulation of the environment 'in the world', rather than 'in the head', (5) the environment is actually part of the cognitive system, e.g. according to Clark and Chalmers' notion of the 'extended mind' [9], and (6) 'off-line' cognition is body-based, which according to Wilson is the "most powerful claim" [41].

Finally, we have elsewhere [6, 43, 45] distinguished between the following views of embodiment and what kind of body it actually requires:

- the view of embodiment as *structural coupling* between agent and environment, which does not necessarily require a physical body (e.g. [10, 13, 24);
- the view of *historical embodiment* as the result of a history of structural coupling and the resulting (mutual) adaptation of an agent to its ecological niche, which again does not necessarily require a physical body (e.g. [28]);
- *physical embodiment*, in the sense discussed above, commonly found in the embodied AI literature;
- *'organismoid' embodiment,* i.e. the view that cognition not only depends on a physical body, but that (organism-like) morphology plays a crucial role, a view also commonly found in embodied AI (e.g. [4, 27]); here we can further distinguish between the claim that the body mediates between internal processes and the environment (e.g. computational properties of materials that substitute of internal processing [26]), which is more in line with material embodiment, and the claim that the key to the embodiment of cognition is the sharing of neural circuitry between sensorimotor and more 'abstract', cognitive processes, which is more in line with full/radical embodiment and Wilson's sixth claim;
- *organismic (or organismal) embodiment*, i.e. the view that at least some aspects of mind (e.g. self and phenomenal experience) crucially depend on the autopoietic, i.e. self-creating and –maintaining, organization of living bodies (e.g. 20, 21, 33, 36, 43, 46, 47).

4 Discussion: Implications for Embodied AI as Science

Raising the question of different conceptions of embodiment in discussions of embodied AI is sometimes dismissed as a philosophical issue of limited value to the practice of embodied AI research. It should be noted, however, that the questions raised in this paper, although they overlap with philosophical issues, are not themselves questions of philosophy, but questions of scientific methodology and practice, i.e. the kind of questions that any scientific community has to ask itself, e.g. what defines and sustains a field as such, and the need for shared conceptions and agreed-upon terminology.

It has been pointed out in this paper that the identity of embodied AI, i.e. what it is that makes a particular type of AI research (or several, in this case) 'embodied', is far from clear. As mentioned before, from an engineering perspective, it seems fairly obvious that embodied AI is about robotic, i.e. physically embodied systems. From the scientific perspective of AI as building models of natural cognition or intelligence, however, things are less clear.

On the one hand embodied AI seems to be about physically embodied, i.e. robotic models of cognition. This matches the engineering perspective very well, and it allows us to distinguish the approach of embodied AI from its traditional, cognitivist counterpart which predominantly used computer programs as models of cognition. However, if physical, robotic models of cognition is what embodied AI is about then one might ask why there is very little interaction between embodied AI research and the work of the type carried out, for example, in Reiter's *Cognitive Robotics Group* at the University of Toronto[3] which uses traditional, symbolic AI techniques, such as situation calculus, in real robotic systems, i.e. carrying on the type of AI that started with Stanford's Shakey project in the 1960s. It seems quite obvious that, despite the use of physically embodied robots, not many embodied AI researchers would consider this an example of embodied AI. In fact the type of symbolic knowledge representation used in this type of AI is rejected outright by many proponents of embodied AI. The use of physically embodied robots then, after all, does not, at least not by itself, seem to be a distinguishing feature of embodied AI.

On the other hand, for many embodied AI researchers, the term *embodied* seems to signify the conception of (embodied) cognition that is underlying their work. This then supposedly is the reason why the work of Reiter's group, for example, would not count as embodied AI, because it supposedly is not based on a theoretical framework that conceives of cognition as embodied. However, this is not unproblematic either, since, as discussed in the previous section, in some sense(s) the work of Reiter's group could very well be characterized as guided by the notions of simple, trivial or material embodiment. Is then perhaps the conception of radical or full embodiment, i.e. the view that all of cognition is embodied or body-based, what distinguishes embodied AI from non-embodied AI? Well, this does not seem to match the practice of embodied AI very well, as discussed above, since clearly not everybody in the field, perhaps not even a majority, would subscribe to a fully or radically embodied view of cognition, as previously illustrated by the case of air traffic control. Furthermore, if embodied AI was actually dedicated to building models of fully/radically embodied cognition, the community would have to ask itself why it has so little interaction with

---

[3] For details see http://www.cs.toronto.edu/cogrobo/.

work of the type carried out by, for example, Lakoff, Feldman and Shastri's *Neural Theory of Language* group at Berkeley[4] that builds neuro-computational models of embodied cognition, in the full/radical sense, but sees no need for physically embodied, robotic models. Is this not embodied AI, because it deals with non-embodied models?

Since neither the use of physically embodied models nor the modeling of embodied theories of cognition seems to properly characterize the identity of embodied AI as cognitive-scientific modeling, one might ask if perhaps it is the combination of the two? That means, one might want to characterize embodied AI as the use of (physically) embodied systems in the modeling of embodied theories of cognition. But again, both of these would require a substantial re-definition of what we consider embodied AI today, because, as discussed above, either you use a simple conception of embodiment and thereby include robotically-grounded-symbol-systems-type AI, which clearly is incompatible with current mainstream embodied AI, or you use the conception of radical/full embodiment as a theoretical framework, which would exclude much of what is currently considered embodied AI.

Finally, embodied AI does of course not necessarily have to adopt any coherent definition or theoretical framework, but can continue as the pluralistic research field that it currently is, addressing to some degree both computational and physically embodied models of both non-embodied and embodied theories of cognition. However, this runs risk of confirming the old criticism that embodied AI is defined only in terms of what it is *against* (traditional AI, which itself is not well-defined either), rather than what it is *about*, and it might in fact be worth considering to further open up the field for research that is currently not considered embodied AI. Whatever the future of the field, embodied AI as a scientific endeavor would certainly benefit from further clarification of its own theoretical foundations and commitments.

Acknowledgements. The author would like to thank Rafael Nunez since many of the points raised in this paper stem from discussions with him in Dagstuhl. Collaboration with Ron Chrisley and Jessica Lindblom also helped to clarify some of the issues involved. Thanks also to the anonymous reviewers who provided useful feedback on the first version of this paper.

References

1. Anderson, M. (2003). Embodied Cognition: A Field Guide. *Artificial Intelligence*, 149(1), 91-130.
2. Berthouze, L. & Ziemke, T. (2003). Epigenetic Robotics: Modelling Cognitive Development in Robotic Systems. *Connection Science*, 15(4), 147-150.
3. Bodén, M. & Niklasson L. (2000). Semantic systematicity and context in connectionist networks, *Connection Science*, 12(2), 1–31.

---

[4] For details see http://www.icsi.berkeley.edu/NTL/.

4. Buason, G.; Bergfeldt, N. & Ziemke, T. (in press). Brains, Bodies, and Beyond: Competitive Co-Evolution of Robot Controllers, Morphologies and Environments. *Genetic Programming and Evolvable Machines*, to appear.
5. Chrisley, R. (2003). Embodied Artificial Intelligence. *Artificial Intelligence*, 149(1), 131-150.
6. Chrisley, R. & Ziemke, T. (2002). Embodiment. In: *Encyclopedia of Cognitive Science* (pp. 1102-1108). London: Macmillan Publishers.
7. Clark, A. (1997). *Being There*. Cambridge, MA: MIT Press.
8. Clark, A. (1999). An embodied cognitive science? *Trends in Cognitive Science*, 9, 345-351.
9. Clark, A. & Chalmers, D. (1998).The Extended Mind. *Analysis* 58 (1), 7-19.
10. Dautenhahn, K.; Ogden, B. & Quick, T. (2002). From embodied to socially embedded agents. *Cognitive Systems Research*, 3(3), 397-428.
11. Edelman, G. (1992). *Bright Air, Brilliant Fire: On the Matter of the Mind*. Penguin.
12. Fodor, J. & Pylyshyn, Z. (1988). Connectionism and cognitive architecture: A critical analysis. *Cognition*, 28, 3-71.
13. Franklin, S. A. (1997). Autonomous agents as embodied AI. *Cybernetics and Systems*, 28(6), 499-520.
14. Hadley, R.F; Rotaru-Varga, A.; Arnold, D.V. & Cardei, V.C. (2001). Syntactic systematicity arising from semantic predictions in a Hebbian-competitive network. *Connection Science*, 13(1), 73-94.
15. Harnad, S. (1989). Minds, Machines and Searle. *Journal of Theoretical and Experimental Artificial Intelligence*, 1(1), 5-25.
16. Kirsh, D., & Maglio, P. (1994). On distinguishing epistemic from pragmatic action. *Cognitive Science*. 18, 513-549.
17. Lakoff, G. & Johnson, M. (1999). *Philosophy in the Flesh*. New York: Basic Books.
18. Lakoff, G. & Nunez, R. (1999). *Where Mathematics Comes From*. New York: Basic Books.
19. Lambrinos, D.; Marinus, M.; Kobayashi, H.; Labhart, T.; Pfeifer, R. & Wehner, R. (1997). An autonomous agent navigating with a polarized light compass. *Adaptive Behavior*, 6(1):131-161.
20. Maturana, H. R. & Varela, F. J. (1980). *Autopoiesis and Cognition*. Dordrecht: Reidel.
21. Maturana, H. R. & Varela, F. J. (1987). *The Tree of Knowledge - The Biological Roots of Human Understanding*. Boston, MA: Shambhala.
22. Newell, A. & Simon, H. (1976). Computer Science as Empirical Inquiry: Symbols and Search. *Communications of the ACM*, 19, 113-126.
23. Nunez, R. (1999). Could the Future Taste Purple? Reclaiming Mind, Body and Cognition. *Journal of Consciousness Studies*, 6(11-12), 41-60.
24. Quick, T.; Dautenhahn, K.; Nehaniv, C. & Roberts, G. (1999). On Bots and Bacteria: Ontology Independent Embodiment. In: Floreano, D. et al. (eds.) *Proceedings of the Fifth European Conference on Artificial Life*. Heidelberg: Springer Verlag.
25. Pfeifer, R. (1995). Cognition – Perspectives from autonomous agents. *Robotics and Autonomous Systems*, 15, 47-70.
26. Pfeifer, R. (2000). On the role of morphology and materials in adaptive behavior. In: Meyer, J.A.; Berthoz, A.; Floreano, D.; Roitblat, H. & Wilson, S.W. (eds.) *From animals to animats6* (pp. 23-32). Cambridge, MA: MIT Press.
27. Pfeifer, R. & Scheier, C. (1999). *Understanding Intelligence*. Cambridge, MA: MIT Press.
28. Riegler, A. (2002). When is a Cognitive System Embodied? *Cognitive Systems Research*, 3(3), 339-348.
29. Ruppin, E. (2002). Evolutionary autonomous agents: A neuroscience perspective. *Nature Reviews Neuroscience*, 3(2), 132-142.
30. Schlesinger, M. & Parisi, D. (2001). The agent-based approach: A new direction for computational models of development. *Developmental Review*, 21, 121-146.
31. Steels, L. (1994). The artificial life roots of artificial intelligence. *Artificial Life*, 1, 75-110.

32. Steels, L. (1999). *The Talking Heads Experiment.* Antwerpen: Laboratorium.
33. Stewart, J. (1996). Cognition = Life: Implications for higher-level cognition. *Behavioral Processes,* 35, 311-326.
34. Varela, F. J.; Thompson, E. & Rosch, E. (1991). *The Embodied Mind.* Cambridge, MA: MIT Press.
35. Voegtlin, T. & Verschure, P. (1999). What can robots tell us about brains? *Reviews in Neuroscience,* 10(3-4), 291-310.
36. von Uexküll, J. (1982). The Theory of Meaning. *Semiotica,* 42(1), 25-82.
37. Vogt, P. (2002). The physical symbol grounding problem. *Cognitive Systems Research,* 3(3) 429-457.
38. Vogt, P. (2003). THSim v3.2: The Talking Heads simulation tool In: Banzhaf, W.; Christaller, T.; Dittrich, P. & Kim, J.T & Ziegler, J. (eds.) *Advances in Artificial Life - Proceedings of the 7th European Conference on Artificial* Life. Heidelberg: Springer.
39. Webb, B. (2000). What does robotics offer animal behaviour? *Animal Behavior,* 60, 545-558.
40. Webb, B. (2001). Can robots make good models of biological behaviour? *Behavioral and Brain Sciences,* 24(6).
41. Wilson, M. (2002). Six views of embodied cognition. *Psychonomic Bulletin and Review,* 9(4), 625-636.
42. Ziemke, T. (2001). The Construction of 'Reality' in the Robot. *Foundations of Science,* 6(1), 163-233.
43. Ziemke, T. (2001). Are Robots Embodied?. In: Balkenius, C.; Zlatev, J.; Brezeal, C.; Dautenhahn, K. & Kozima, H. (eds.) *Proceedings of the First International Workshop on Epigenetic Robotics: Modelling Cognitive Development in Robotic Systems* (pp. 75-83). Lund University Cognitive Studies, vol. 85, Lund, Sweden.
44. Ziemke, T. (2003). On the Role of Robot Simulations in Embodied Cognitive Science. *AISB Journal,* 1(4), 389-399.
45. Ziemke, T. (2003). What's that thing called embodiment? In: Alterman, R. & Kirsh, D. (eds.) *Proceedings of the 25th Annual Conference of the Cognitive Science Society* (pp. 1305-1310). Mahwah, NJ: Lawrence Erlbaum.
46. Ziemke, T. & Sharkey, N. E. (2001). A stroll through the worlds of robots and animals. *Semiotica,* 134(1-4), 701-746.
47. Zlatev, J. (2002). Meaning = Life (+ Culture) - An outline of a unified biocultural theory of meaning. *Evolution of Communication,* 4 (2), 175-199.

The Future of Embodied Artificial Intelligence: Machine Consciousness?

Owen Holland

Department of Computer Science
University of Essex
Wivenhoe Park
Essex CO4 3SQ
owen@essex.ac.uk

Abstract. The idea that internal models of the world might be useful has generally been rejected by embodied AI for the same reasons that led to its rejection by behaviour based robotics. This paper re-examines the issue from historical, biological, and functional perspectives; the view that emerges indicates that internal models are essential for achieving cognition, that their use is widespread in biological systems, and that there are several good but neglected examples of their use within embodied AI. Consideration of the example of a hypothetical autonomous embodied agent that has to execute a complex mission in a dynamic, partially unknown, and hostile environment leads to the conclusion that the necessary cognitive architecture is likely to contain separate but interacting models of the body and of the world. This arrangement is shown to have intriguing parallels with new findings on the infrastructure of consciousness, leading to the speculation that the reintroduction of internal models into embodied AI may lead not only to improved machine cognition but also, in the long run, to machine consciousness.

1 In the Beginning

"Behaviorism and the acceptance of the norms of the natural sciences in psychology greatly restricted for a generation or more the range of behavioural phenomena with which the psychologist, as scientist, was willing to concern himself....There has been considerable relaxation of this austerity in the past decade, although not without misgiving and apology." So begins the classic paper 'The simulation of human thought' (Newell and Simon 1959), and it begins that way because it was presented to an audience of psychologists. It outlines the approach, assumptions, and achievements of early artificial intelligence, and, in particular, it makes it clear that the overall aim of the enterprise was "...the explanation of complex human behavior." The point of the opening remarks was that psychology itself was only just ready to accept the validity of studying such behaviour. Of course, thinking, consciousness, and other complex human behaviour had been the central subjects of early psychology, but the methods used, and the standards of explanation, were no longer acceptable: "Those

F. Iida et al. (Eds.): Embodied Artificial Intelligence, LNAI 3139, pp. 37–53, 2004.
© Springer-Verlag Berlin Heidelberg 2004

who regard thinking as the core of psychological enquiry, and who urge a return to concern with it, don't want to turn the clock back." (Newell and Simon 1959)

It is interesting to see how careful Newell and Simon were to define the limits of what they were attempting to do. At the end of the paper, they remark: "...(W)e wish to record our conviction that it is no longer necessary to talk about the theory of higher mental processes in the future tense. There now exist tools sharp enough to cut into the tough skin of the problem, and these tools have already produced a rigorous, detailed explanation of a significant area of human symbolic behaviour". (Newell and Simon 1959) With the benefit of hindsight, we can now see the inevitability of the rise of embodied artificial intelligence; even if they had been correct in their explanation of higher mental processes, there was a need to deal with those aspects of behaviour that were not ultimately symbolic, and also with those mental processes which could not be classified as 'higher'. The key error of the inheritors of the early AI tradition was to assume that higher level symbolic processing – representing knowledge as statements within a language, and reasoning over those representations – could account for all behaviour, not just for the behaviour identified by Newell and Simon. As the artificial systems became less abstract and more concrete, and as they became embodied in mobile robots in the real world, it became clear that something was wrong. As system resources became faster and more powerful, and something was still obviously amiss, it became clear that what was wrong was something fundamental – not merely passing problems of implementation.

In the seventies, the Stanford Cart, along with Shakey, represented the state of the art in robotics. Rod Brooks, aware of the work of Grey Walter thirty years earlier, which anticipated much that has been painfully rediscovered in the last two decades (Holland 2003b), records his feelings while helping with research on the Cart: "I could not help feeling disappointed. Grey Walter had been able to get his tortoises to operate autonomously for hours on end, moving about and interacting with a dynamically changing world and with each other. Here, at the center of high technology, a robot relying on millions of dollars of equipment did not appear to operate nearly as well. Internally, it was doing more than Grey Walter's tortoises had ever done – it was building accurate three-dimensional models of the world and formulating detailed plans within these models. But to an external observer all that internal cogitation was hardly worth it." (Brooks 2002)

2   The Development of Embodied AI

Brooks went on to develop a fruitful stream of ideas, the ideas that laid the foundations of the modern view of embodied systems. Internal models and representations of the state of the external world were seen as both unnecessary and harmful. They were rejected both by reasoned argument (Brooks 1991), and by polemic. They were shown to be unnecessary by the construction and demonstration of a series of robots that were fast, responsive, robust, and apparently capable of everything of which the robots of the old school were incapable. The new ideas went into space, and into war. In time, the techniques for milking high performance and apparent intelligence out of simple robots developed into a sophisticated technology, with elegantly expressed principles, and some stunning examples of their deliberate application (Pfeifer and Scheier 1999). Much of this flowed from the exploitation of the nature and consequences of embodiment. Carefully chosen structures – morphologies and materials – could in effect process information, filtering and controlling without filters or controllers. Robots with bodies and simple behaviours closely coupled to the environment could produce sophisticated emergent behaviour. Even better, many of these phenomena had parallels in biology. Indeed, some of the most spectacular exploitations were inspired by observations of biological systems and natural behaviour. The development of artificial evolution in various forms showed how intelligent behaviour could best arise not from the evolution of control systems (brains) alone, but from the simultaneous evolution of control systems, sensor placements, morphology, and materials.

An excellent and elegant example of many of these trends is the work of Lichtensteiger and Salomon (2000). The facet distributions of many insect compound eyes have been observed to vary considerably between species. One of the main tasks performed by the compound eye is the guidance and control of flight, using optic flow information. It is believed that certain arrangements of facets can simplify the control problem, enabling the use of simpler control architectures and offering some additional operational advantages; furthermore, the optimisation of facet distribution for different task environments may account for much of the observed inter-species variation in eye morphology. Lichtensteiger built a physical model of a compound eye in which the placement of the 16 individual facets could be adjusted by means of motors and gears. The outputs of the facets fed into a fixed controller which was used for a variety of tasks – for example, for estimating the distance of the system from a light of a given orientation. Artificial evolution was then applied to the positioning of the facets, with the fitness function being the quality of the solution produced by the controller. The eventual distribution of the facets resembled that found in insects, and varied appropriately with the requirements of the different tasks. The evolved morphology of the model eye was thus shown to play a critical part in filtering, and therefore processing, the visual information in a biologically significant and task dependent manner.

3   Arrested Development?

Despite this torrent of good news, there were doubters, because there were doubts. Although nothing about embodiment demands it, most of the robots were predominantly reactive. One of the catchphrases of the period was 'the world is its own best model'. Why build and maintain an internal model of a world that was directly available to the sensors? If the information was in the environment, the robots could exploit it quickly, economically, reliably, and robustly. But what if it wasn't? What if it had been there, but had disappeared, or had been hidden? What if it was now too far away to affect the immediate action? Dealing with these contingencies is the province of cognition (Clark and Grush 1999) and though the new embodied systems were freely and uncontroversially described as intelligent, they were not always as readily described as cognitive. True, some tasks previously thought to require cognition were shown to be achievable without it, just as collective tasks thought to be difficult or impossible to coordinate without explicit communication or memory turned out to be easy when the communication and memory were placed in the environment in a way that enabled the operation of stigmergy. The robot Toto

(Mataric 1992) could build and exploit a kind of map of an environment, but did so in a way that was so closely coupled to immediate sensing and movement that it seemed to belong entirely in the behaviour based camp. No limits were placed on the intelligence that would eventually be achieved without representation – but it was to be understood that the process of developing such intelligence would take some time, and so the approach should not be judged solely on the basis of its limited performance to date.

Early challenges to the new approach came not from outside, but from inside. The behavior based, closely coupled, grounded, embodied stuff dealt well with immediate contingencies, but could not deal with situations where mapping and simple planning were more suitable. What could be more natural than to combine the two techniques, superimposing a deliberative/planning layer on top of a reactive robot, yet representing the result as predominantly behaviour based? These hybrid architectures were built and demonstrated, but they did not seem to have the impact of the purer designs. At about the same time (or earlier, according to Brooks) some proponents of classical GOFAI style robots had added what looked very like a reactive layer to support their designs. In 1997, Stein wrote of these endeavours: "Unfortunately, much work along these lines has has taken the approach to be expected of computer scientists: layering separate 'thinking' systems on top of ...robotic bodies. Since the two typically need a separate component to mediate between them, the resulting (prevailing) approach to cognitive robotics has come to be known as a Three Level Architecture, or TLA....Although it is built out of two reasonably successful fields of research – robotics and traditional artificial intelligence – this approach to cognitive robotics has been severely limited...." The details of some of these schemes, along with an analysis of their limitations, can be found in Horswill and Stein's earlier review (1994).

Animals do not seem to suffer from the problems associated with integrating fundamentally different subsystems. Nothing is more quick and agile than a primate in a tree, yet this same primate will have clearly demonstrable cognitive abilities – and even some ability to manipulate symbols, though this may never occur in the wild. The exploitation of embodiment is clearly necessary, and the embodied AI community has shown how to do it. But something is missing from the current vision – something to do with cognition, with a system taking account of things unseen. Adding the two modes of operation together does not seem to work. Is there some way of starting from embodiment, using all the principles and tricks we know, and ending up in truly cognitive territory?

4   Models and the Rehabilitation of Representation

The most eye-catching features of the early behaviour based robots were their speed and responsiveness. They were not fast and responsive just because they did little calculation, but because what little calculation they did was directed towards the simple connection of sensing (or perception) and movement (or action). The work was done by the interaction between the two, mediated by the body and the environment. Speed is good in many walks of life. In biology, it is often critical. And accuracy is also important. However, there was no pressure on behaviour based robots to increase their speed, and the accuracy of their movement was never held at a premium – indeed, their motion control was often made deliberately poor to show that their success in a noisy and dynamic world did not depend on precisely calibrated components. If speed and accuracy had been more important, then perhaps the road to cognition would have been mapped out two decades ago.

As noted above, many animals are quick and agile; we take it for granted when we see squirrels or monkeys racing through treetops. But this speed and accuracy of movement is not intrinsic to the components. Human sensors and neurons are slow, and they and muscles have very variable characteristics. If a fast reaching movement was carried out by simply stimulating the neurons controlling the muscles in some predetermined pattern, the result would vary significantly on different occasions. One way of coping with such variability in nature or in engineering is by using a feedback controller – measuring the difference between the actual movement and that specified by the pattern of stimulation, and adding some appropriate stimulation to correct for the error. Unfortunately it takes a relatively long time for the visual system to respond, and so a fast movement could become very inaccurate indeed by the time a correction is issued. One way of dealing with this in an engineering system is to use a forward model – a system component that, when given information about the sequence of stimulation, produces an output corresponding to the arm configuration that will result. In what sense will it 'correspond'? One option is to produce an output duplicating the sensory input that the final configuration would produce. If the forward model component produces its output before the arm actually reaches its end point, then the forward model will have *predicted* the sensory input that will come from the final configuration. Such a prediction can be put to a number of uses, but in the present context the most important is that, instead of generating a late feedback correction using the difference between the final sensed configuration and that specified by the pattern of motor stimulation, the system can generate an early correction by using the predicted instead of the final configuration. This will result in a smoother correction, and a more accurate final result.

This is an engineering solution. Is there any evidence that such systems are present in animals? There is a wealth of such evidence, and it is a very active field of research (see review by Wolpert and Ghahrahmani [2000]). Moreover, forward models are not just found in primates, but occur also in insects (see review by Webb [2004]). The inputs to the neural structures forming the forward models are the well known efference copies or corollary discharges – copies of the volleys of impulses sent to the muscles or to the neurons controlling them. But the key thing to notice is the significance of the outputs: they are, in effect, copies of the sensory inputs that will occur as a consequence of those muscle-controlling volleys. In the sense identified by Clark and Grush (1999), they are potential cognitive elements, able to 'stand for' something not currently present.

Improving the accuracy of fast movements is only one possible use for forward modelling – the reviews mentioned above note several possibilities, and examples of their use. If the sensory inputs do not match those predicted by the forward model, then something in the environment must have changed. One of the major functions of the cerebellum seems to be to predict the sensory consequences of movement, probably in order to detect changes in the environment. Such a system automatically corrects for self-movement. Only one of these benefits may have channeled the evolution of forward models, but once the forward model had evolved, the other benefits would be potentially available.

5   One Step Further

A recent paper by Grush (2002) develops the ideas set out above in some detail. More importantly, it extends them in a direction that suddenly enables a plethora of potentially cognitive applications. A forward model (or emulator, in his terminology) driven by efference copies predicts the sensory inputs that will be produced by those efference copies, but the system as a whole receives both real sensory input and the predicted sensory input. What would happen if the real sensory inputs were gated off somehow? The sensory input to the system would then consist only of the predicted input. Suppose now that the system is driven by efference copies, but that the motor commands from which those copies originate are gated off so that they have no effect on the body. If the forward model was good and complete enough, the evolution of the system would predict the sensory inputs expected from those motor commands, were they to have been carried out. If the forward model is modelling only proprioceptive inputs – signals from tendons, joints, and muscles – then the system will produce what Grush calls motor imagery: the full proprioceptive repertoire that would be produced by the movement that would have occurred if the motor command had not been gated off.

But the real meat of Grush's thought is in his extension of his emulator theory to exteroceptive sensing. He uses the example of vision because two robots implementing something very like his theory were built fifteen years ago by Mel (Mel, 1986, 1988). The terms in which Mel described his work are very different from those used by Grush, but the reinterpretation is convincing. In a nutshell, the system learns a forward model which includes visual sensory input, and so by disconnecting both the real sensory input and the real motor output, the system is able to operate in a closed system in which visual imagery of the external world is generated. The system is therefore able to produce the sensory input corresponding to what would occur in the external world if a 'hypothetical' movement were carried out. The whole point of Murphy, of course, was to learn off-line in this closed environment, and then demonstrate this learning in the real situation.

Grush is therefore able to make the case for an incremental and almost inevitable progression from the simple requirements for enabling fast and accurate movement to a system able to engage in the most demanding cognitive activity, imagination. And during the whole exposition, representation is dealt with at the most primitive sensory level; there is no requirement for anything beyond the familiar sensory-motor level; and the emphasis throughout is on the body, sensors, and effector control. If our simple robots, in their exploitation of embodiment, were required to make faster and more accurate movements than their circuitry and effectors could manage with only feedback control, then we would have to equip them with the capacity for forward modelling. Once we had done that, a few small and ideologically unobjectionable steps would carry them to the threshold of cognition. Our fear of the world model would vanish. I believe this is the path that could and should lead embodied artificial intelligence into the area of cognition.

6   Are We There Yet?

As in the case of Grey Walter, there are always precursors to significant movements, and we can see one in Stein's MetaToto (1991, 1994). It was based on Mataric's Toto (Mataric 1992), the behaviour based robot mentioned above that was capable of building a "world model" – a dynamic graph with nodes corresponding to environmental landmarks defined in sensory terms (gross sonar readings and compass readings, augmented by odometry records from previously encountered landmarks). Toto could find its way to a goal location by using spreading activation in the landmark graph to find the direction, and following the indicated path until the specified landmark was reached.

The language used in Stein's 1991 paper is quite non-partisan:

"Toto's landmark representation and goal-driven navigation are cognitive tasks, involving internal representation of the external environment. This represents a qualitative advance in the capabilities of subsumption-based robots. Nonetheless, this internal representation is available only through interaction with the world....This paper is concerned with a concrete example of the integration of higher-level cognitive AI and lower-level robotics. Robotic systems are embodied: their central tasks concern interaction with the immediately present world. In contrast, cognition is concerned with objects that are remote – in distance, in time, or in some other dimension. We exploit the architecture of a particular robotic system to perform a cognitive task, by *imagining* the subjects of our cognition.

We suggest that much of the abstract information that forms the meat of cognition is used not as a central model of the world, but as virtual reality. The self-same processes that robots use to explore and interact with the world form the interface to this information. The only difference between interaction with the actual world and with the imagined one is the set of sensors providing the lowest-level interface." (Stein 1991)

MetaToto is simply Toto modified through "...the creation and integration of an imagination system." This is a photographed floor plan of an environment, a means of translating position on the floor plan to simulated sonar inputs, and a means of changing position according to Toto's motor outputs. Toto explores the floor plan as it would a real environment, creating landmarks from the simulated sonar data as it goes. When transferred to the real environment for the first time, MetaToto is able to go directly to a commanded goal landmark. The crudeness of the sonar modelling (ray tracing) does not matter; the overall model is good enough. "The inaccuracy of MetaToto's imagination (is) little worse than the variance between two runs of the actual robot..."

The paper ends with the hope that: "...experiences with MetaToto will lead to more sophisticated use of imagination and virtual sensing, and to the development of truly embodied forms of cognition." It is perhaps significant that Grush is able to accommodate MetaToto within his framework. In this collection, Ziemke's recent work deals with similar issues. However, where Toto's imagination was enabled by the external provision of the floor plan of the environment, Ziemke's robot must rely on its previous experience of the real environment. But the main thrust is similar: outputs for controlling movement produce simulated sensor inputs which correspond to the inputs that would have been produced by a real movement in the real environment; these inputs elicit further movement commands, and so on, with the chaining of successive actions leading to an extended sequence.

7   Where Do We Go from Here?

Reaching cognition through simple internal modelling seems a large enough step to many within the field of embodied artificial intelligence, and there may be a danger that progress will stop too soon. MetaToto and Murphy may have attained some form of cognition, and in the right way, through exploiting the principles of the discipline, but they have not derived any spectacular benefits from so doing. Perhaps we should look at the progress that modelling seems to have enabled within the natural world.

In biology, the keenest advocate of the utility of modelling for prediction is Richard Dawkins. In 1976 he set out his view that animals can be regarded as machines designed to ensure the survival and propagation of their genes, and remarked: "Survival machines that can simulate the future are one jump ahead of survival machines who can only learn on the basis of overt trial and error." (Dawkins 1976). More recently, he noted that an animal's body "...represents a kind of prediction that the future will resemble the past, in broad outline. The animal is likely to survive to the extent that this turns out to be true. And simulation models of the world allow the animal to act as if in anticipation of what that world is likely to throw its way on the next few seconds, hours, or days." (Dawkins 1998)

The force and utility of this kind of active modelling has been best expressed by the philosopher, Dan Dennett, in his description of a hypothetical creature, the last of three in an evolutionary sequence (Dennett 1995). The first, the Darwinian creature, is the basic model. Its responses to its environment are specified by its genes; those examples with genes producing bad responses die, and those with genes for good responses survive to breed, eventually producing a population with better responses. The second, the Skinnerian creature, is capable of learning, and as a result becomes capable of producing better responses if it is not killed by an early bad response. The one of greatest interest is the Popperian creature, which is able to preselect its responses so that those likely to kill it are inhibited:

"But how is this preselection in Popperian agents to be done? Where is the feedback (about the quality of the proposed action) to come from? It must come from a sort of *inner environment* – an inner something-or-other that is structured in such a way that the surrogate actions it favours are more often than not the very actions the real world would also bless, if they were performed. In short, the inner environment, whatever it is, must contain lots of *information* about the outer environment and its regularities....we must be very careful not to think of this inner environment as simply a replica of the outer world, with all its physical contingencies reproduced....The information about the world has to be there, but it also has to be structured in such a way that there is a nonmiraculous explanation of how it got there, how it is maintained, and how it actually achieves the preselective effects that are its *raison d'etre*." (Dennett 1995, pp375-6).

The forward modelling idea provides the predictive element, but Dennett emphasises that prediction alone is not enough – that feedback is needed 'about the quality of the proposed action'. This aspect of the idea is hugely underexplored.

8   Let's Start Again

We can combine the biological and robotic perspectives by considering the problems of an autonomous embodied agent in a complex, occasionally novel, dynamic, and hostile world. It is useful to think of the agent as having a mission – some long term task or set of tasks that it is required to achieve. The details and complexity of the mission do not matter. Animals vary enormously in their behavioural repertoires and lifestyles, but they all have exactly one mission, and it is the same one for everything from bacteria to elephants: to propagate their genetic material, either directly or indirectly, as effectively as possible. Survival, often thought to be at the root of animal behaviour, is of course merely the means to the end of the universal mission of successful reproduction.

In order to maximise its chances of achieving the mission, the agent should at all times produce the action most likely to lead to the achievement of the mission. How can this be arranged, given that the agent is in a 'complex, occasionally novel, dynamic, and hostile' world? There are only a small number of possibilities:

- *By being preprogrammed for every possible contingency.* This is of course impossible, given the presumed finite means available to the agent, and the unqualified complexity and novelty of the world. However, it is possible for the agent to be preprogrammed for some sensed subset of contingencies – its Merkwelt – but this will necessarily be suboptimal, in that the actions selected will not always be those most likely to lead to success. Invertebrates probably operate like this.
- *By having learned the consequence for the achievement of the mission of every possible action or action sequence in every contingency, and by selecting the best available?* This is not possible for the obvious variants of the reason given above, and for another reason: given that the world is hostile, some of the possible actions tried during learning will lead to the destruction or disabling of the agent, as with Dennett's Skinnerian creature. But it might perform better than an agent with invariant responses.
- *By being able, through learning or otherwise, to predict the consequences of actions, by being able to evaluate those consequences for their likely contribution to the mission, and by selecting a relatively good course of action?* This, the strategy of Dennett's Popperian creatures, seems to be the only strong contender for arriving at the 'best' action. What we need to investigate within embodied AI is the range of architectures capable of delivering adequate combinations of prediction, evaluation, and selection.

It is outside the scope of this brief paper to examine these three features in any detail, but what we can do is to comment on the central method, that of simulation using internal models. Assuming that prediction is to be achieved through simulation, what exactly has to be simulated? The answer is both obvious and interesting: whatever affects the mission, and nothing else. An embodied agent can only affect the world through the actions of its body in and on the world, and the world can only affect the mission by affecting the agent's body. The agent therefore needs to simulate only those aspects of its body that affect the world in ways that affect the mission, and only those aspects of the world that affect the body in ways that affect the mission. How does the body affect the world? To some extent through its passive properties, but mainly by being moved through and exerting force on the world, with appropriate speed and accuracy. How does the world affect the body? Through the spatially distributed environment (through which the body must move) and through the properties of the objects in it (food, predators, poisons, prey, competitors, falling coconuts, etc.) What is interesting about these observations is that what needs to be simulated is determined more by the agent's embodiment than by anything else.

What is needed for simulation? A minimal statement might be: Some process or structure interpretable as a state of the world that, when operated on by some process or structure interpretable as an action, yields an outcome interpretable as and corresponding to the consequence of that action. (In fact, 'simulation' seems to imply more than this predictive function - perhaps that not only the interpreted outcome, but also the underlying processes, must correspond more directly in some way to whatever is being simulated.)   These structures or processes are best referred to as 'internal models', because they are like working models rather than static representations, and because the term was used in this sense by Craik (1943) and later by Johnson-Laird (1983) and others.

So we require a model that includes the body, and how it is controlled, and the spatial aspects of the world, and the (kinds of) objects in the world, and their spatial arrangement. But consider: The body, and its controller, is always present and available, and changes slowly, if at all. When it moves, it is usually because it has been commanded to move. The world, however, is different. It is 'complex, occasionally novel, dynamic, and hostile'. It is only locally available, and may contain objects of known and unknown kinds in known and unknown places. How should all this be modelled? As a single model containing body, environment, and objects? Or as a model of the body coupled to and interacting with the other modelled components? From the point of view of both biology and computer science, it seems overwhelmingly probable that the best solution will involve some separation of these components, since they seem to be of fundamentally different types. A sensibly designed simulation engine would define a space, place types of objects within it, and control a body moving through the space and interacting with the objects. Interestingly enough, if we adopt this scheme, then we are saying that, in order to behave intelligently by predicting the consequences of events, an embodied agent should run a simulation of itself coupled to a simulation of the world. However, by most current criteria of embodiment, the internally simulated agent - the internal agent model - can also be regarded as being embodied! Given that much of embodied AI is carried out in simulation, it would seem difficult to object to an internal model that was itself an embodied agent.

We seem to have arrived at a view of the type of system likely to be able to deliver the cognition missing from current embodied AI. It is more complex than most existing systems, but the possibility of achieving a true cognitive capability, and the accompanying liberation from the limitations of purely reactive systems, surely justify the extra cost. However, recent developments elsewhere in cognitive science seem to imply that architectures containing separate but interacting internal models of the body and of the environment may have the potential to deliver much more than this.

9   Another Psychological Revolution

We began with the remarks of Newell and Simon in 1959 that the climate within psychology had only recently permitted the investigation of topics such as thinking. Another such change has taken place within psychology in the last ten years, and there are some signs that this liberalisation is enabling and stimulating research outside psychology. The change is this: it is now acceptable to study, not just thinking, but consciousness itself. What is more, the early indications are that 'the sciences of the artificial', in Simon's phrase, may have a contribution to make.

In 1994, Francis Crick recounted the remark of a psychologist to a younger colleague: "It's all right to be interested in consciousness, but get tenure first." Tenured or not, a lot of people have got interested since then. There has been the foundation of the Association for the Scientific Study of Consciousness, with an annual conference, and determinedly high scientific standards. The huge biennial meeting in Tucson, held at the Center for the Study of Consciousness, is called 'Towards a Science of Consciousness'. There are journals, and workshops, and there has been an exponential rise in the rate of publication of all sorts of books on consciousness. But what significance does this hold for the field of embodied artificial intelligence?

10   The New View of Consciousness

What has happened in the last decade has not simply been a resumption of philosophical and psychological discussions about consciousness, but an explosion of new findings that have transformed our view of the phenomenon. So what is consciousness? Again, Crick offers some help:

"Everyone has a rough idea of what is meant by consciousness. It is better to avoid a precise definition of consciousness because of the dangers of premature definition. Until the problem is understood much better, any attempt at a formal definition is likely to be either misleading or overly restrictive, or both." (Crick 1994)

The diversity of content in a typical consciousness conference shows that the community behaves as if his advice had been accepted. Is progress being made towards understanding the problem? That is certainly true in one sense, because, on the new view, consciousness is certainly a problem. At the time at which Newell and Simon were writing, it was still generally accepted that consciousness 'worked', in some sense. Conscious thought, especially intellectually challenging areas such as problem solving, showed the mind working as an integrated, harmonious whole, and an artificially intelligent system would be expected to simulate the operations apparently controlled by consciousness. None of that is true any more. As Nørretranders puts it:

"Consciousness is a peculiar phenomenon. It is riddled with deceit and self-deception; there can be consciousness of something we were sure had been erased by an anaesthetic; the conscious I is happy to lie up hill and down dale to achieve a rational explanation for what the body is up to; sensory perception is the result of a devious relocation of sensory input in time; when the consciousness thinks it determines to act, the brain is already working on it; there appears to be more than one version of consciousness present in the brain; our conscious awareness contains almost no information but is perceived as if it were vastly rich in information. Consciousness is peculiar.' (Nørretranders 1998 p286).

Nørretranders is not a scientist, but a journalist. In this context, that is an advantage, because he is able to react freely to the new view of consciousness. What has happened? There is no single cause. Better medical care and diagnosis means that unusual injuries and illnesses are survived and reported. The effects of many of these are fascinating – some of the cases described in the books of Oliver Sacks spring to mind – and much of that fascination is with instances of bizarre cognition rather than changes in consciousness. But what is surely odd is that perfectly normal consciousness can coexist with bizarre views of the world. An individual is paralysed on one side after a stroke, but denies it. However, if ice-cold water is squirted into the ear on the same side, she talks freely of her paralysis, and asks why people are asking her about something so obvious. Twelve hours later she is denying the paralysis again – and is also denying having admitted to it following the ice water treatment! (Ramachandran and Blakeslee 1998 pp144-5).

Psychologists experimenting on normal individuals have exposed huge differences between what we believe is happening, and objective reality. Voluntary action, the bedrock of civil society, turns out not to be willed and then executed, but first executed and then attributed to ourselves (Wegner 2002). Our subjective 'now' is running about half a second late, but is somehow backdated so that everything seems coherent (Libet 1989). Under some circumstances we fail to notice huge changes in our surroundings – changes as great as the person we are talking to being replaced by a different person (Simons and Levin 1998). One classic demonstration of what is known as inattentional blindness involves a film of an impromptu basketball game between a team dressed in white and one dressed in black; the viewers are asked to count the number of times the white team catch the ball. At the end of the film, members of the audience are asked if they saw anything odd during the film, and most answer that they did not. The film is run again, and they see that, halfway through, a woman in a gorilla suit runs onto the set and does her best to attract attention, jumping up and down in the centre of the screen and waving her arms (Simons and Chabris 1999). The exotic syndrome of multiple personality seems to be real, to the extent that the brain scans of an individual differ when different personalities are 'in charge' (Reinders 2003). Some people blinded through brain injury will report that they are completely blind – yet they can point quite reliably to light stimuli (Weiskrantz 1986).

The upshot of findings such as these is this: it is no longer possible to believe that consciousness is some ideal and perfect process, but rather that, under most conditions and for most people, it presents a consistent illusion of being such a process. What is the utility of consciousness, and why, when examined closely, does it seem so ramshackle and bizarre? Surely no self-respecting engineer would ever design such a system. How and why did it evolve?

11   Models, Bodies, and Consciousness

In 1976, Dawkins, in his section on the utility of modelling, allowed himself to speculate:

"The evolution of the capacity to simulate seems to have culminated in subjective consciousness...Perhaps consciousness arises when the brain's simulation of the world becomes so complete that it must include a model of itself." (Dawkins 1976).

The brain may or may not model itself, but it is very likely to model more than just the characteristics of the limbs. The neuroscientist Ramachandran takes a commonsense view:

"...(I)t is always obvious to you that there are some things you can do and others you cannot given the constraints of your body and of the external world. (You know you can't lift a truck...) Somewhere in your brain there are representations of all these possibilities, and the systems that plan commands...need to be aware of this distinction between things they can and cannot command you to do....To achieve all this, I need to have in my brain not only a representation of the world and various objects in it but also a representation of myself, including my own body within that representation....In addition, the representation of the external object has to interact with my self-representation...." (Ramachandran and Blakeslee 1998 p249).

This is reminiscent of part of Grush's view. Grush distinguishes between two types of emulation: modal (which is tied to a particular modality); and amodal, in which an emulation of an environment is run, and sensory consequences are established by carrying out some kind of measurement on that emulation.

Ramachandran's notion of some kind of separateness between the model of the body and the model of the environment is supported by much of the modern work in consciousness. Perhaps more importantly for those who believe in the importance of embodiment, the model of the body appears to be very closely linked to consciousness itself. One of the major figures in consciousness studies is Antonio Damasio; he has played a key role is showing that consciousness, far from being some abstract and ethereal state, is very tightly linked to bodily events. In Damasio (1999), he proposes a neurologically based theory of consciousness in which the development of a primitive body-centred self structure plays a crucial role. The theory itself is complex, but his hypothesis is well summarised by Churchland (2002) in a paper examining self-representation in nervous systems:

"...the self/nonself distinction, though originally designed to support coherencing, is ultimately responsible for consciousness. According to this view, a brain whose wiring enables it to distinguish between inner-world representations and outer-world representations and to build a metarepresentational model of the relation between outer and inner entities is a brain enjoying some degree of consciousness .... Conceivably, as wiring modifications enable increasingly sophisticated simulation and deliberation, the self-representational apparatus becomes correspondingly more elaborate, and therewith the self/not-self apparatus. On this hypothesis, the degrees or levels of conscious awareness are upgraded in tandem with the self-representational upgrades." (Churchland 2002 p 310).

The recent development of this bodycentric point of view is emphasised in Watt's review of Damasio's ideas: "...consciousness requires that the brain must represent not just the object, not just a basic self structure, *but the interaction of the two*....This is still an atypical foundation for a theory of consciousness, given that until recently, it was implicitly assumed that the self could be left out of the equation. There has been a recent sea change on this crucial point..." (Watt 2000).

For once, there is also some philosophical support: Thomas Metzinger (2000, 2003) has proposed a theory of consciousness, rooted in phenomenological analysis, that is explicitly based around the concept of the self-model:

"The phenomenal self is a virtual agent perceiving virtual objects in a virtual world...I think that 'virtual reality' is the best technological metaphor which is currently available as a source for generating new theoretical intuitions...heuristically the most interesting concept may be that of 'full immersion'....(the phenomenal self-model)...is a plastic multimodal structure that is plausibly based on an innate and 'hardwired' model of the spatial properties of the system (e.g. a 'long-term body image'...) while being functionally rooted in elementary bioregulatory processes...." (Metzinger 2000).

The consensus that seems to be building is strikingly consonant with the development of ideas within embodied artificial intelligence: the structure at the centre of consciousness is the physical body. It is also consonant with the direction in which I believe embodied artificial intelligence should progress: at the heart of the mechanism is not just the body in the environment, it is a model of the body in a model of the environment. A bit of such modelling can give our machines cognition; will more of the same give them consciousness?

12 An Objection Dismissed

To many within the field of embodied AI, this easy talk of self models and world models will evoke the ghost of GOFAI. Did not the experience of the robotics pioneers show that the problems of building, maintaining, and using internal models of external reality were insuperable? Did not Brooks' classic paper 'Intelligence without Representation' demonstrate the fundamental error at the heart of the model-based approach? The answer to both questions is yes – and no. Yes, the pioneers built models, and yes, they and their robots ran into trouble, but the models, expressed as sentences in formal languages, were simply inappropriate and inadequate for the purpose of controlling robots in real time in dynamic and partially known environments. Control engineers routinely build and use models for controlling large and complex plants – aircraft, power stations, oil refineries, etc. – but they use techniques very different from those of the computer scientists who developed classical artificial intelligence. We have seen that the brain also uses models in a variety of contexts and for a variety of purposes, but no one nowadays seriously believes that the brain's models consist solely of symbols processed through symbolic manipulation. It will be impossible to progress to intelligence and consciousness without the use of models, but the models must be of types appropriate to their applications, and it is certain that there will have to be several different types of model used within a single system. To rule out the use of models altogether is probably a more serious error than to be mistaken in the choice of model; the pioneers were at least half right, and we will not improve on their efforts unless and until we acknowledge this.

One of the strengths of the embodied AI community is their acceptance of biological inspiration. As it happens, some of the most recent data coming from consciousness studies give us some intriguing information about the representation of the visual world. It is clear to every normally sighted adult that the world we see is immensely rich in information, and that we are able to apprehend this information as an enormously detailed unity, a task which would challenge any artificial system. Unfortunately, this is another of the illusions perpetrated by consciousness, and is usually referred to as 'The Grand Illusion' (Noë 2002); in fact, as Nørretranders puts it in the passage quoted earlier, "...our conscious awareness contains almost no information but is perceived as if it were vastly rich in information". Our brain does not contain a detailed and fully featured representation of the external world, but it does contain the information needed to gather that information very rapidly using the senses. The world is still its own best model, but consciousness studies appear to tell us that the best way to use that model is with the aid of a much less detailed model that guides the sensors in interrogating the world. Should embodied AI investigate this strategy? And do the internal models involved in consciousness really operate at this low level of detail? This is an open question; it is at least possible that they do.

13 Machine Consciousness

Whatever the mechanism of consciousness turns out to be, it is clear that there is a movement towards investigating consciousness by building systems and seeing how closely they match the human data. This parallels Newell and Simon's programme in relation to intelligence almost fifty years ago. There have been several international workshops on the subject (see www.machineconsciousness.org); there are regular sessions at consciousness conferences; the first books have started to appear (Haikonen 2003, Holland 2003a); and there are funded projects.

Already, certain themes are emerging. The study of imagination in robots, pioneered by Stein (1987), is beginning to take off – at a recent workshop, delegates interested in this were the largest single group. Aleksander's work, using analogues of the primate visual system on a mobile robot, is perhaps the best known. (Aleksander et al: 1999, 2001). Other groups are investigating unembodied software agents (Franklin 2003) and electronic hardware implementations (Haikonen 2003).

Given the close association between the modern view of consciousness, embodiment, and modelling, and the early involvement of roboticists in the emerging field of machine consciousness, it seems likely that embodied artificial intelligence will at least contribute to this field, and may eventually dominate or subsume it. Its role would be exactly as Newell and Simon considered the role of their enterprise to be in 1959: the explanation of complex human behaviour. But it will be important to enter the field with a good enough background in consciousness studies, because the old view of consciousness is dead in the water. We have no understanding of the functional benefits of consciousness – but in this field we are in possession of some ideas that may be fundamental to its emergence: embodiment, and modelling. We should promote these ideas vigorously, because there is at least some danger that Good Old Fashioned Artificial Intelligence may see machine consciousness as an attractive piece of unexplored territory. And, like the psychologists referred to by Newell and Simon, we don't want to turn the clock back.

References

Aleksander, I., Dunmall, B., and Del Frate, V. (1999) Neurocomputational Models of Visualisation: A Preliminary Report. IWANN (1) 1999 : 798-805

Aleksander, I., Morton, H., and Dunmall, B. (2001) Seeing is Believing: Depictive Neuromodeling of Visual Awareness. IWANN (1) 2001: 765-771

Brooks, R.A. (1991) Intelligence without Representation. Artificial Intelligence 47 139-159

Brooks, R.A. (2002) Robot: the Future of Flesh and Machines. Penguin Books, London

Churchland, P.S. (2002) Self-Representation in Nervous Systems. Science, vol 296, pp 308-310.

Clark, A. and Grush, R. (1999) Towards a cognitive robotics. Adaptive Behavior, 7(1):5-16 Cambridge University Press.

Craik, K.J.W. (1943) The Nature of Explanation. Cambridge University Press.

Crick, F. (1994). The Astonishing Hypothesis: The Scientific Search for the Soul. New York: Charles Scribner's Sons.

Damasio, A.R. (1999) The Feeling of What Happens: Body and Emotion in the Making of Consciousness. Harcourt Brace & Co.

Dawkins, R. (1976) The Selfish Gene. Oxford University Press, Oxford

Dawkins, R. (1998) Unweaving the Rainbow. Allen Lane, The Penguin Press

Dennett, D.C. (1995) Darwin's Dangerous idea: Evolution and the Meanings of Life. Allen Lane, The Penguin Press

Franklin, S. (2003) IDA: A Conscious Artifact? Journal of Consciousness Studies, Volume 10, No. 4-5, April-May 2003

Grush, R. (2002) An introduction to the main principles of emulation: motor control, imagery, and perception. Technical Report, Philosophy, UC San Diego: to appear in Behavioral and Brain Sciences.

Haikonen, P.O. (2003) The Cognitive Approach to Conscious Machines, Imprint Academic.

Holland, O. [Ed.] (2003a) Machine Consciousness, Imprint Academic.

Holland, O. (2003b) Exploration and high adventure: the legacy of Grey Walter. Philosophical Transactions of the Royal Society Volume 361 (2003) Number 1811 pp. 2082-2022

Horswill, I.D., and Stein, L.A. (1994) Life after Planning and Reaction. AAAI Fall Symposium on the Control of Intelligent Systems. New Orleans.

Johnson-Laird, P.N. (1983) Mental Models: Towards a Cognitive Science of Language, Inference, and Consciousness. Harvard University Press, Cambridge MA.

Libet, B. (1989) The timing of a subjective experience. Behavioral and Brain Sciences 12, p.183-85.

Lichtensteiger, L. and Salomon, R. (2000) The Evolution of an Artificial Compound Eye by Using Adaptive Hardware. In: Proceedings of the 2000 Congress on Evolutionary Computation, San Diego, CA, USA, pp. 1144-1151

Mataric, M. (1992) Integration of representation into goal-driven behavior-based robots. IEEE Transactions on Robotics and Automation 8(3)

Mel, B.W. (1986) A connectionist learning model for 3-D mental rotation, zoom, and pan. In Proc. of the 8th Annual Conf. of the Cognitive Science Soc., pp 562-71

Mel, B.W. (1988) Murphy: A robot that learns by doing. In Neural Information Processing Systems, pp544-553, American Institute of Physics, New York

Metzinger, T. (2000) The subjectivity of subjective experience: a representationalist analysis of the first person perspective. In The Neural Correlates of Consciousness Ed. Metzinger T. MIT Press, Cambridge MA

Metzinger, T. (2003) Being No One: The Self-Model Theory of Subjectivity. MIT-Bradford.

Newell, A.. and Simon, H.A. (1959) The simulation of human thought. The RAND Corporation, Mathematics Division, Paper P-1734. Later published in: Current Trends in Psychological Theory, Pittsburgh: University of Pittsburgh Press (1961), 152-179.

Noë, A. [Ed.] (2002) Is the Visual World a Grand Illusion? Imprint Academic.

Nørretranders, T. (1998) The User Illusion: Cutting Consciousness down to Size. Trans. J. Sydenham. Allen Lane, Penguin Press.

Pfeifer, R. and Scheier, C. (1999) Understanding Intelligence. MIT Press, Cambridge, MA.

Ramachandran, V.S. and Blakeslee, S. (1998) Phantoms in the Brain: Human Nature and the Architecture of the Mind. Fourth Estate, London.

Reinders, A.A.T.S. et al. (2003) One brain, two selves. NeuroImage, 20, 2119 – 2125.
Simons, D.J. and Levin, D.T (1998) Failure to detect changes to people during real-world interaction. Psychonomic Bulletin and Review, vol 4, p 644 (1998)
Simons, D. J., & Chabris, C. F. (1999). Gorillas in our midst: Sustained inattentional blindness for dynamic events. Perception, 28, 1059-1074
Stein, L.A. (1991) Imagination and situated cognition, AI Memo 1277, MIT AI lab.
Stein, L.A. (1994) Imagination and situated cognition, Journal of Experimental and Theoretical Artificial Intelligence 6:393-407
Stein, L.A. (1997) PostModular Systems: Architectural Principles for Cognitive Robotics. Cybernetics and Systems, 28(6):471-487, September 1997
Watt, D.F. (2000) At the Intersection of Emotion and Consciousness II: A review of "The Feeling of What Happens". Journal of Consciousness Studies Vol 7 (1)
Webb, B. (2004) Neural mechanisms for prediction: do insects have forward models? Trends in Neurosciences, April 2004
Wegner, D.M. (2002) The Illusion of Conscious Will. MIT Press
Weiskrantz, L. (1986) Blindsight. OUP, Oxford.
Wolpert, D.M. and Ghahramani, Z. (2000) Computational principles of movement neuroscience. Nature Neuroscience Supp. vol 3 1212-1217

Do *Real* Numbers Really Move?
Language, Thought, and Gesture: The Embodied
Cognitive Foundations of Mathematics

Rafael Núñez

Department of Cognitive Science
University of California, San Diego
nunez@cogsci.ucsd.edu

Abstract. Robotics, artificial intelligence and, in general, any activity involving computer simulation and engineering relies, in a fundamental way, on mathematics. These fields constitute excellent examples of how mathematics can be applied to some area of investigation with enormous success. This, of course, includes embodied oriented approaches in these fields, such as Embodied Artificial Intelligence and Cognitive Robotics. In this chapter, while fully endorsing an embodied oriented approach to cognition, I will address the question of the nature of mathematics itself, that is, mathematics not as an application to some area of investigation, but as a human conceptual system with a precise inferential organization that can be investigated in detail in cognitive science. The main goal of this piece is to show, using techniques in cognitive science such as cognitive semantics and gestures studies, that concepts and human abstraction in general (as it is exemplified in a sublime form by mathematics) is ultimately embodied in nature.

1 A Challenge to Embodiment: The Nature of Mathematics

Mathematics is a highly technical domain, developed over several millennia, and characterized by the fact that the very entities that constitute what Mathematics is are idealized mental abstractions. These entities cannot be perceived directly through the senses. Even, say, a point, which is the simplest entity in Euclidean geometry, can't be actually *perceived*. A point, as defined by Euclid is a dimensionless entity, an entity that has only location but no extension. No super-microscope will ever be able to allow us to actually perceive a point. A point, after all, with its precision and clear identity, is an idealized abstract entity. The imaginary nature of mathematics becomes more evident when the entities in question are related to *infinity* where, because of the finite nature of our bodies and brains, no direct experience can exist with the infinite itself. Yet, infinity in mathematics is essential. It lies at the very core of many fundamental concepts such as limits, least upper bounds, topology, mathematical induction, infinite sets, points at infinity in projective geometry, to mention only a few. When studying the very nature of mathematics, the challenging and intriguing question that

F. Iida et al. (Eds.): Embodied Artificial Intelligence, LNAI 3139, pp. 54–73, 2004.
© Springer-Verlag Berlin Heidelberg 2004

comes to mind is the following: if mathematics is the product of human ideas, how can we explain the nature of mathematics with its unique features such as precision, objectivity, rigor, generalizability, stability, and, of course, applicability to the real world? Such question doesn't represent a real problem for approaches inspired in platonic philosophies, which rely on the existence of transcendental worlds of ideas beyond human existence. But this view doesn't have any support based on scientific findings and doesn't provide any link to current empirical work on human ideas and conceptual systems (it may be supported, however, as a matter of faith, not of science, by many Platonist scientists and mathematicians). The question doesn't pose major problems to purely formalist philosophies either, because in that worldview mathematics is seen as a manipulation of meaningless symbols. The question of the origin of the meaning of mathematical ideas doesn't even emerge in the formalist arena. For those studying the human mind scientifically, however (e.g., cognitive scientists), the question of the nature of mathematics is indeed a real challenge, especially for those who endorse an *embodied* oriented approach to cognition. How can an embodied view of the mind give an account of an abstract, idealized, precise, sophisticated and powerful domain of ideas if direct bodily experience with the subject matter is not possible?

In *Where Mathematics Comes From*, Lakoff and Núñez (2000) give some preliminary answers to the question of the cognitive origin of mathematical ideas. Building on findings in mathematical cognition, and using mainly methods from Cognitive Linguistics, a branch of Cognitive Science, they suggest that most of the idealized abstract technical entities in Mathematics are created via human cognitive mechanisms that extend the structure of bodily experience (thermic, spatial, chromatic, etc.) while preserving the inferential organization of these domains of bodily experience. For example, linguistic expressions such as "send her my *warm* helloes" and "the teacher was very *cold* to me" are statements that refer to the somewhat abstract domain of Affection. From a purely literal point of view, however, the language used belongs to the domain of Thermic experience, not Affection. The meaning of these statements and the inferences one is able to draw from them is structured by precise mappings from the Thermic domain to the domain of Affection: Warmth is mapped onto presence of affection, Cold is mapped onto lack of affection, X is warmer than Y is mapped onto X is more affectionate than Y, and so on. The ensemble of inferences is modeled by one conceptual metaphorical mapping, which in this case is called AFFECTION IS WARMTH[1]. Research in Cognitive Linguistics has shown that these phenomena are not simply about "language," but rather they are about thought. In cognitive science the complexities of such abstract and non/literal phenomena have been studied through mechanisms such as conceptual metaphors (Lakoff & Johnson, 1980; Sweetser, 1990; Lakoff, 1993; Lakoff & Núñez, 1997; Núñez & Lakoff, in press; Núñez, 1999, 2000), conceptual blends (Fauconnier & Turner, 1998, 2002; Núñez, in press), conceptual metonymy (Lakoff & Johnson, 1980), fictive motion and dynamic schemas (Talmy, 1988, 2003), and aspectual schemas (Narayanan, 1997).

---

[1] Following a convention in Cognitive Linguistics, the name of a conceptual metaphorical mapping is capitalized.

Based on these findings Lakoff and Núñez (2000) analyzed many areas in mathematics, from set theory to infinitesimal calculus, to transfinite arithmetic, and showed how, via everyday human embodied mechanisms such as conceptual metaphor and conceptual blending, the inferential patterns drawn from direct bodily experience in the real world get extended in very specific and precise ways to give rise to a new emergent inferential organization in purely imaginary domains[2]. For the remainder of this chapter we will be building on these results as well as on the corresponding empirical evidence provided by the study of human speech-gesture coordination. Let us now consider a few mathematical examples.

2 Limits, Curves, and Continuity

Through the careful analysis of technical books and articles in mathematics, we can learn a good deal about what structural organization of human everyday ideas have been used to create mathematical concepts. For example, let us consider a few statements regarding limits in infinite series, equations of curves in the Cartesian plane, and continuity of functions, taken from mathematics books such as the now classic *What is Mathematics?* by R. Courant & H. Robbins (1978).

a)   Limits of infinite series
In characterizing limits of infinite series, Courant & Robbins write:

   "We describe the behavior of s_n by saying that the sum s_n *approaches* the limit 1 as
   n tends to infinity, and by writing
       $1 = 1/2 + 1/2^2 + 1/2^3 + 1/2^4 + ...$" (p. 64, our emphasis)

Strictly speaking, this statement refers to a sequence of discrete and motionless partial sums of s_n (real numbers), corresponding to increasing discrete and motionless values taken by n in the expression $1/2^n$ where n is a natural number. If we examine this statement closely we can see that it describes some facts about numbers and about the result of discrete operations with numbers, but that there is *no motion* whatsoever involved. No entity is actually *approaching* or *tending* to anything. So, why then did Courant and Robbins (or mathematicians in general, for that matter) use dynamic language to express static properties of static entities? And what does it mean to say that the "sum s_n approaches," when in fact a sum is simply a fixed number, a result of an operation of addition?

b)   Equations of lines and curves in the Cartesian Plane
Regarding the study of conic sections and their treatment in analytic geometry, Courant & Robbins' book says:

   "The hyperbola *approaches* more and more nearly the two straight lines $qx \pm py$
   $= 0$ as we *go out farther and farther* from the origin, but it never actually
   reaches these lines. They are called the asymptotes of the hyperbola." (p. 76,
   our emphasis).

---

[2] The details of how conceptual metaphor and conceptual blending work go beyond the scope of this piece. For a general introduction to these concepts see Lakoff & Núñez (2000, chapters 1-3), and the references given therein.

And then the authors define hyperbola as "the locus of all points P the *difference* of whose distances to the two points $F(\sqrt{(p^2 + q^2)}, 0)$ and $F'(-\sqrt{(p^2 + q^2)}, 0)$ is $2p$." (p. 76, original emphasis).

Strictly speaking, the definition only specifies a *"locus of all points P"* satisfying certain properties based exclusively on arithmetic differences and *distances*. Again, no entities are actually moving or approaching anything. There are only statements about static differences and static distances. Besides, as Figure 1 shows, the authors provide a graph of the hyperbola in the Cartesian Plane (bottom right), which in itself is a static illustration that doesn't have the slightest insinuation of motion (like symbols for arrows, for example). The figure illustrates the idea of *locus* very clearly, but it says nothing about motion. Moreover the hyperbola has *two* distinct and separate *loci*. Exactly which one of the two is then "the" moving agent (3rd person singular) in the authors' statement *"the* hyperbola *approaches* more and more nearly the two straight lines $qx \pm py = 0$ as we *go out farther and farther* from the origin"?

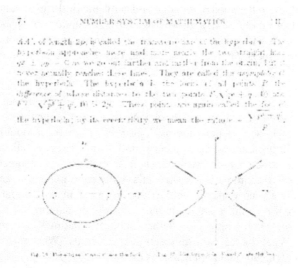

Fig. 1. Original text analyzing the hyperbola as published in the now classic book *What is Mathematics?* by R. Courant & H. Robbins (1978).

c)   Continuity

Later in the book, the authors analyze cases of continuity and discontinuity of trigonometric functions in the real plane. Referring to the function $f(x) = \sin 1/x$ (whose graph is shown in Figure 2) they say: "... since the denominators of these fractions increase without limit, the values of x for which the function $\sin(1/x)$ has the values 1, -1, 0, will cluster nearer and nearer to the point $x = 0$. Between any such point and the origin there will be still an infinite number of *oscillations* of the function" (p. 283, our emphasis).

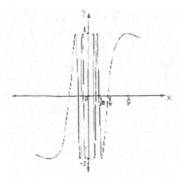

Fig. 2. The graph of the function $f(x) = \sin 1/x$.

Once again, if, strictly speaking, a function is a mapping between elements of a set (coordinate values on the x-axis) with one and only one of the elements of another set (coordinate values on the y-axis), all what we have is a static correspondence between points on the x-axis with points on the y-axis. How then can the authors (or mathematicians in general) speak of "*oscillations* of the function," let alone an infinite number of them?

These three examples show how ideas and concepts are described, defined, illustrated, and analyzed in mathematics books. You can pick your favorite mathematics books and you will see similar patterns. You will see them in topology, fractal geometry, space-filling curves, chaos theory, and so on. Here, in all three examples, static numerical structures are involved, such as partial sums, geometrical loci, and mappings between coordinates on one axis with coordinates on another. Strictly speaking, absolutely no dynamic entities are involved in the formal definitions of these terms. So, if no entities are really moving, why do authors speak of "approaching," "tending to," "going farther and father," and "oscillating"? Where is this motion coming from? What does dynamism mean in these cases? What role is it playing (if any) in these statements about mathematics facts?

We will first look at pure mathematics to see whether we can find answers to these questions. Then, in order to get some deeper insight into them, we will turn into human language and real-time speech-gesture coordination.

3 Looking at Pure Mathematics

Among the most fundamental entities and properties the above examples deal with are the notion of real number and continuity. Let us look at how pure mathematics defines and provides the inferential organization of these concepts.

In pure mathematics, entities are brought to existence via formal definitions, formal proofs (theorems) or by axiomatic methods (i.e., by declaring the existence of some entity without the need of proof. For example, in set theory the axiom of infinity assures the existence of infinite sets. Without that axiom, there are no infinite sets). In the case of real numbers, ten axioms taken together, fully characterize this number system and its inferential organization (i.e., theorems about real numbers). The following are the axioms of the real numbers.

1. Commutative laws for addition and multiplication.
2. Associative laws for addition and multiplication.
3. The distributive law.
4. The existence of identity elements for both addition and multiplication.
5. The existence of additive inverses (i.e., negatives).
6. The existence of multiplicative inverses (i.e., reciprocals).
7. Total ordering.
8. If x and y are positive, so is $x + y$.
9. If x and y are positive, so is $x \bullet y$.
10. The Least Upper Bound axiom.

The first 6 axioms provide the structure of what is called a *field* for a set of numbers and two binary operations. Axioms 7 through 9, assure ordering constraints. The first nine axioms fully characterize ordered fields, such as the rational numbers with the operations of addition and multiplication. Up to here we have already a lot of structure and complexity. For instance we can characterize and prove theorems about all possible numbers that can be expressed as the division of two whole numbers (i.e., rational numbers). With the rational numbers we can describe with any given (finite) degree of precision the proportion given by the perimeter of a circle and its diameter (e.g., 3.14; 3.1415; etc.). We can also locate along a line (according to their magnitude) any two different rational numbers and be sure (via proof) that there will always be infinitely many more rational numbers between them (a property referred to as density). With the rational numbers, however, we can't "complete" the points on this line, and we can't express with infinite exactitude the magnitude of the proportion mentioned above ($\pi = 3.14159 \ldots$). For this we need the full extension of the real numbers. In axiomatic terms, this is accomplished by the tenth axiom: the Least Upper Bound axiom. All ten axioms characterize a complete ordered field.

In what concerns our original question of where is motion coming from in the above mathematical statements about infinite series and continuity, we don't find any answer in the first nine axioms of real numbers. All nine axioms simply specify the existence of static properties regarding binary operations and their results, and properties regarding ordering. There is no explicit or implicit reference to motion in these axioms. Since what makes a real number a real number (with its infinite precision) is the Least Upper Bound axiom, it is perhaps this very axiom that hides the secret of motion we are looking for. Let's see what this axiom says:

10. Least Upper Bound axiom: every nonempty set that has an upper bound has a least upper bound.

And what exactly is an upper bound and a least upper bound? This is what pure mathematics says:

<u>Upper Bound</u>

b is *an upper bound* for S if
$x \leq b$, for every x in S.

Least Upper Bound

b_0 is a *least upper bound* for S if

• b_0 is an upper bound for S, and

• $b_0 \leq b$ for every upper bound b of S.

Once again, all what we find are statements about motionless entities such as universal quantifiers (e.g., for every x; for every upper bound b of S), membership relations (e.g., for every x in S), greater than relationships (e.g., $x \leq b$; $b_0 \leq b$), and so on. In other words, there is absolutely no indication of motion in the Least Upper Bound axiom, or in any of the other nine axioms. In short, the axioms of real numbers, which are supposed to completely characterize the "truths" (i.e., theorems) of real numbers don't tell us anything about a sum "approaching" a number, or a number "tending to" infinity (whatever that may mean!).

Let's try continuity. What does pure mathematics say about it?

Mathematics textbooks define continuity for functions as follows:

• A function f is continuous at a number a if the following three conditions are satisfied:

  1. f is defined on an open interval containing a,
  2. $\lim_{x \to a} f(x)$ exists, and
  3. $\lim_{x \to a} f(x) = f(a)$.

Where by $\lim_{x \to a} f(x)$ what is meant is the following:

Let a function f be defined on an open interval containing a, except possibly at a itself, and let L be a real number. The statement

$$\lim_{x \to a} f(x) = L$$

means that $\forall\, \varepsilon > 0,\ \exists\, \delta > 0,$

such that if $0 < |x - a| < \delta,$

then $|f(x) - L| < \varepsilon.$

As we can see, pure formal mathematics defines continuity in terms of limits, and limits in terms of

• static universal and existential quantifiers predicating on static numbers (e.g., $\forall\, \varepsilon > 0, \exists\, \delta > 0$), and

• on the satisfaction of certain conditions which are described in terms of motionless arithmetic difference (e.g., $|f(x) - L|$) and static smaller than relations (e.g., $0 < |x - a| < \delta$).

That's it. Once again, these formal definitions don't tell us anything about a sum "approaching" a number, or a number "tending to" infinity, or about a function "oscillating" between values (let alone doing it infinitely many times, as in the function $f(x) = \sin 1/x$).

But this shouldn't be a surprise. Lakoff & Núñez (2000), using techniques from cognitive linguistics showed what well-known contemporary mathematicians had already pointed out in more general terms (Hersh, 1997; Henderson, 2001):

• The structure of human mathematical ideas, and its inferential organization, is richer and more detailed than the inferential structure provided by formal definitions and axiomatic methods. Formal definitions and axioms neither fully formalize nor generalize human concepts.

We can see this with a relatively simple example taken from Lakoff & Núñez (2000). Consider the function $f(x) = x \sin 1/x$ whose graph is depicted in Figure 3.

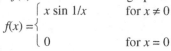

$$f(x) = \begin{cases} x \sin 1/x & \text{for } x \neq 0 \\ 0 & \text{for } x = 0 \end{cases}$$

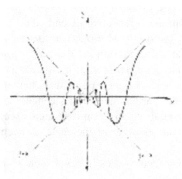

Fig. 3. The graph of the function $f(x) = x \sin 1/x$.

According to the ε - δ definition of continuity given above, this function is continuous at every point. For all x, it will always be possible to find the specified ε's and δ's necessaries to satisfy the conditions for preservation of closeness. However, according to the everyday notion of continuity —*natural continuity* (Núñez & Lakoff, 1998)— as it was used by great mathematicians such as Kepler, Euler, and Newton and Leibniz, the inventors of infinitesimal calculus in the 17th Century, this function is *not* continuous. According to the inferential organization of natural continuity, certain conditions have to be met. For instance, in a naturally continuous line we are supposed to be able to tell how long the line is between to points. We are also supposed to be able to describe essential components of the motion of a point along that line. With this function we can't do that. Since the function "oscillates" infinitely many times as it "approaches" the point (0, 0) we can't really tell how long the line is between two points located on the left and right sides of the plane. Moreover, as the function approaches the origin (0, 0) we can't tell, say, whether it will "cross" from the right plane to the left plane "going down" or "going up." This function violates these basic properties of natural continuity and therefore it is not continuous. The function $f(x) = x \sin 1/x$ is thus ε - δ continuous but it is not naturally continuous. The point is that the formal ε - δ definition of continuity doesn't capture the inferential organization of the human everyday notion of continuity, and it doesn't generalize the notion of continuity either.

The moral here is that what is characterized formally in mathematics leaves out a huge amount of inferential organization of the human ideas that constitute mathematics. As we will see, this is precisely what happens with the dynamic aspects of the expressions we saw before, such as "approaching," "tending to," "going farther and farther," "oscillating," and so on. Motion, in those examples, is a genuine and constitutive manifestation of the nature of mathematical ideas. In pure mathematics, however, motion is not captured by formalisms and axiomatic systems.

4 Embodied Cognition

It is now time to look, from the perspective of embodied cognition, at the questions we asked earlier regarding the origin of motion in the above mathematical ideas. In the case of limits of infinite series, motion in "the sum s_n *approaches* the limit 1 as n *tends* to infinity" emerges *metaphorically* from the successive values taken by n in the sequences as a whole. It is beyond the scope of this chapter to go into the details of the mappings involved in the various underlying conceptual metaphors that provide the required dynamic inferential organization (for details see Lakoff & Núñez, 2000). But we can at least point out to some of the many conceptual metaphors and metonymies[3] involved.

- There are conceptual metonymies in cases such as a partial sum *standing for* the entire infinite sum;
- there are conceptual metaphors in cases where we conceptualize the sequence of these metonymical sums as a *unique trajector*[4] moving in space (as it is indicated by the use of the 3rd person singular in *the* sum s_n approaches);
- there are conceptual metaphors for conceiving infinity as a single location in space such that a metonymical n (standing for the entire sequence of values) can "tend to;"
- there are conceptual metaphors for conceiving 1 (not as a mere natural number but as an infinitely precise *real* number) as the result of the infinite sum; and so on.

Notice that none of these expressions can be *literal*. The facts described in these sentences don't exist in any real perceivable world. They are metaphorical in nature. It is important to understand that these conceptual metaphors and metonymies are not simply "noise" added on top of pre-defined formalisms. They are in fact *constitutive* of the very embodied ideas that make mathematical ideas possible. It is the inferential organization provided by our embodied understanding of "approaching" and "tending to" that is at the core of these mathematical ideas.

In the case of the hyperbola, the moving agent is one holistic object, the hyperbola in the Real plane. This object, which has two distinctive separate parts, is conceptualized as one single trajector metaphorically moving away from the origin. Via conceptual metonymies and metaphors similar to the ones we saw for the case of infinite

---

[3] A conceptual metonymy is a cognitive mechanism that allow us to conceive a part of a whole standing for the whole, as when we say *Washington and Paris have quite different views on these issues*, meaning the governments of two entire nations, namely, United States and France.

[4] In cognitive linguistics, "trajector" is a technical term used to refer to the distinct entity that performs the motion traced by a trajectory. The trajector moves against a background called "landscape."

series, the hyperbola is conceived as a trajector tracing the line, which describes the geometrical locus of the hyperbola itself. In this case, of course, because we are dealing with real numbers, the construction is done on non-countable infinite ($> \aleph_0$) discrete real values for x, which are progressively bigger in absolute terms. The direction of motion is stated as moving away from the origin of the Cartesian coordinates, and it takes place in both directions of the path schemas defined by the two branches of the hyperbola, *simultaneously*. The hyperbola not "reaching" the asymptotes is the cognitive way of characterizing the mathematically formalized fact that there are no values for x and y that satisfy equations

$$qx \pm py = 0 \text{ and } (x^2/p^2) - (y^2/q^2) = 1$$

Notice that characterizing the hyperbola as "not reaching" the asymptotes provides the same *extensionality* (i.e., it gives the same resulting cases) as saying that there is an "absence of values" satisfying the above equations. The inferential organization of these two cases, however, is cognitively very different[5].

Finally, in what concerns our "oscillating" function example, the moving object is again one holistic object, the trigonometric function in the Real plane, constructed metaphorically from non-countable infinite ($> \aleph_0$) discrete real values for x, which are progressively smaller in absolute terms. In this case motion takes place in a specific manner, towards the origin from two opposite sides (i.e., for negative and positive values of x) and always between the values $y = 1$ and $y = -1$. As we saw, a variation of this function, $f(x) = x \sin(x)$, reveals deep cognitive incompatibilities between the dynamic notion of continuity implicit in the example above and the static ε-δ definition of continuity coined by Weierstrass in the second half of the 19th century (based on quantifiers and discrete Real numbers) and which has been adopted ever since as "the" definition of what Continuity really is (Núñez & Lakoff, 1998; Lakoff & Núñez, 2000). These deep cognitive incompatibilities between dynamic-wholistic entities and static-discrete ones may explain important aspects underlying the difficulties encountered by students all over the world when learning the modern technical version of the notions of limits and continuity (Núñez, Edwards, and Matos, 1999).

5 Fictive Motion

Now that we are aware of the metaphorical (and metonymical nature) of the mathematical ideas mentioned above, I would like to analyze more in detail the dynamic component of these ideas. From where do these ideas get motion? What cognitive mechanism is allowing us to conceive static entities in dynamic terms? The answer is *fictive motion*.

---

[5] In order to clarify this point, consider the following two questions: (a) What Alpine European country does not belong to the European Union?, and (b) What is the country whose currency is the Swiss Franc? The extensionality provided by the answers to both questions is the same, namely, the country called "Switzerland." This, however, doesn't mean that we have to engage in the same cognitive activity in order to correctly answer these questions.

Fictive motion is a fundamental embodied cognitive mechanism through which we unconsciously (and effortlessly) conceptualize static entities in dynamic terms, as when we say *the road goes along the coast*. The road itself doesn't actually move anywhere. It is simply standing still. But we may conceive it as moving "along the coast." Fictive motion was first studied by Len Talmy (1996), via the analysis of linguistic expressions taken from everyday language in which static scenes are described in dynamic terms. The following are linguistic examples of fictive motion:

- The Equator *passes through* many countries
- The boarder between Switzerland and Germany *runs along* the Rhine.
- The California coast *goes all the way down* to San Diego
- After Corvisart, line 6 *reaches* Place d'Italie.
- Right after *crossing* the Seine, line 4 *comes to* Chatêlet.
- The fence *stops* right after the tree.
- Unlike Tokyo, in Paris there is no metro line that *goes around* the city.

Motion, in all these cases, is fictive, imaginary, not real in any literal sense. Not only these expressions use verbs of action, but they also provide precise descriptions of the quality, manner, and form of motion. In all cases of fictive motion there is a *trajector* (the moving agent) and a *landscape* (the background space in which the trajector moves). Sometimes the trajector may be a real object (e.g., the *road* goes; the *fence* stops), and sometimes it is an imaginary entity (e.g., the *Equator* passes through; the *boarder* runs). In fictive motion, real world trajectors don't move but they have the potential to move or the potential to enact movement (e.g., a car moving along that road). In Mathematics proper, however, the trajector has always a metaphorical component. That is, the trajector as such can't be literally capable or incapable of enacting movement, because the very nature of the trajectory is imagined via metaphor (Núñez, 2003). For example, a point in the Cartesian Plane is an entity that has location (determined by its coordinates) but has no extension. So when we say "point *P moves* from *A* to *B*" we are ascribing motion to a metaphorical entity that only has location. First, as we saw earlier, entities which have only location (i.e., points) don't exist in the real world, so, as such, they don't have the potential to move or not to move in any literal sense. They simply don't exist in the real world. They are metaphorical entities. Second, literally speaking, point *A* and point *B* are distinct *locations*, and no point can change location while preserving its identity. That is, the trajector (point *P*, uniquely determined by its coordinates) can't preserve its identity throughout the process of motion from *A* to *B*, since that would mean that it is changing the very properties that are defining it, namely, its coordinates.

We now have a basic understanding of how conceptual metaphor and fictive motion work, so we are in a position to see the embodied cognitive mechanisms underlying the mathematical expressions like the ones we saw earlier. Here we have similar expressions:

- sin $1/x$ *oscillates* more and more as *x approaches* zero
- $g(x)$ never *goes beyond* 1
- If there exists a number *L* with the property that $f(x)$ *gets closer and closer to* *L* as *x* gets larger and larger; $\lim_{x \to \infty} f(x) = L$.

In these examples Fictive Motion operates on a network of precise *conceptual metaphors*, such as NUMBERS ARE LOCATIONS IN SPACE (which allows us to conceive numbers in terms of spatial positions), to provide the inferential structure required to conceive mathematical functions as having motion and directionality. Conceptual metaphor generates a purely imaginary entity in a metaphorical space, and fictive motion makes it a moving trajector in this metaphorical space. Thus, the progressively smaller numerical values taken by x which determine numerical values of sin $1/x$, are via the conceptual metaphor NUMBERS ARE LOCATIONS IN SPACE conceptualized as spatial locations. The now metaphorical spatial locus of the function (i.e., the "line" drawn on the plane) now becomes available for fictive motion to act upon. The progressively smaller numerical values taken by x (now metaphorically conceptualized as locations progressively closer to the origin) determine corresponding metaphorical locations in space for sin $1/x$. In this imaginary space, via conceptual metaphor and fictive motion now sin $1/x$ can "oscillate" more and more as x "approaches" zero.

In a similar way the infinite precision of real numbers themselves can be conceived as limits of sequences of rational numbers, or limits of sequences of nested intervals. Because, as we saw, limits have conceptual metaphor and fictive motion built in, we can now see the fundamental role that these embodied mechanisms play in the constitution of the very nature of the real numbers themselves.

6 Dead Metaphors?

Up to now, we have analyzed some mathematical ideas through methods in cognitive linguistics, such as conceptual metaphor, conceptual metonymy, and fictive motion. We have studied the inferential organization modeling *linguistic expressions*. But so far no much has been said of actual people speaking, writing, explaining, learning, or gesturing in real-time when involved in mathematical activities. The analysis so far has been almost exclusively at the level of written and oral linguistic expressions. We must know whether there is any psychological (and presumably neurological) reality underlying these linguistics expressions. The remaining task now is to show that all these cases are not, as some scholars have suggested, mere instances of so-called *dead metaphors*, that is, expressions that once in the past had a metaphorical dimension but that now, after centuries of usage, have lost their metaphorical component becoming "dead." Dead metaphorical expressions are those that have lost their psychological (and cognitive semantic) original reality, becoming simply new "lexical items." Perhaps in the cases we have seen in mathematics, what once was a metaphorical expression has now become a *literal* expression whose meaningful origin speakers of English don't know anymore (very much like so many English words whose Latin or Greek etymology may have been known by speakers at a certain point in history, but whose original meaning is no longer evoked by speakers today). Is this what is happening to cases such as "approaching" limits, "oscillating" functions, or hyperbolae not "reaching" the asymptotes? Maybe, after all, all what we have in the mathematical expressions we have examined, is simply a story of dead metaphors, with no psychological (or neurological) reality whatsoever. As we will see, however, the study of human *gesture* provides embodied convergent evidence showing that this is not the case at all. Gesture studies, via a detailed investigation of real-time cognitive and linguistic production, bodily motion (mainly hands and arms), and voice inflection, show that the conceptual metaphors and fictive motion involved in the mathematical ideas analyzed above, far from being dead, do have a very embodied psychological (and presumably neurological) reality.

7 Gesture as Cognition

Human beings from all cultures around the world gesture when they speak. The philosophical and scientific study of human language and thought has largely ignored this simple but fundamental fact. Human gesture constitutes the forgotten dimension of thought and language. Chomskian linguistics, for instance, overemphasizing syntax, saw language mainly in terms of abstract grammar, formalisms, and combinatorics, you could study by looking at written statements. In such a view there was simply no room for meaningful (semantic) "bodily production" such as gesture. In mainstream experimental psychology gestures were left out, among others, because being produced in a spontaneous manner, it was very difficult to operationalize them, making rigorous experimental observation on them extremely difficult. In mainstream cognitive science, which in its origins was heavily influenced by classic artificial intelligence, there was simply no room for gestures either. Cognitive science and artificial intelligence were heavily influenced by the information-processing paradigm and what was taken to be essential in any cognitive activity was a set of body-less abstract rules and the manipulation of physical symbols governing the processing of information. In all these cases, gestures were completely ignored and left out of the picture that defined what constituted genuine subject matters for the study of the mind. At best, gestures were considered as a kind of epiphenomenon, secondary to other more important and better-defined phenomena.

But in the last decade or so, this scenario has changed in a radical way with the pioneering work of A. Kendon (1980), D, McNeill (1992), S. Goldin-Meadow & C. Mylander (1984), and many others. Research in a large variety of areas, from child development, to neuropsychology, to linguistics, and to anthropology, has shown the intimate link between oral and gestural production. Finding after finding has shown, for instance, that gestures are produced in astonishing synchronicity with speech, that in children they develop in close relation with speech, and that brain injuries affecting speech production also affect gesture production. The following is a (very summarized) list of nine excellent sources of evidence supporting (1) the view that speech and gesture ae in reality two facets of the same cognitive linguistic reality, and (2) an embodied approach for understanding language, conceptual systems, and high-level cognition:

1) <u>Speech accompanying gesture is universal</u>. This phenomenon is manifested in all cultures around the world. Gestures then provide a remarkable "back door" to lin-

guistic cognition (McNeill, 1992; Iverson & Thelen 1999; Núñez & Sweetser, 2001).

2) Gestures are less monitored than speech, and they are, to a great extent, unconscious. Speakers are often unaware that they are gesturing at all (McNeill, 1992)

3) Gestures show an astonishing synchronicity with speech. They are manifested in a millisecond-precise synchronicity, in patterns which are specific to a given language (McNeill, 1992).

4) Gestures can be produced without the presence of interlocutors. Studies of people gesturing while talking on the telephone, or in monologues, and studies of conversations among congenitally blind subjects have shown that there is no need of visible interlocutors for people to gesture (Iverson & Goldin-Meadow, 1998).

5) Gestures are co-processed with speech. Studies show that stutterers stutter in gesture too, and that impeding hand gestures interrupts speech production (Mayberry and Jaques, 2000).

6) Hand signs are affected by the same neurological damage as speech. Studies in neurobiology of sign language show that left hemisphere damaged signers manifest similar phonological and morphological errors as those observed in speech aphasia (Hickok, Bellugi, and Klima, 1998).

7) Gesture and speech develop closely linked. Studies in language acquisition and child development show that speech and gesture develop in parallel (Iverson & Thelen 1999; Bates & Dick, 2002).

8) Gesture provide complementary content to speech content. Studies show that speakers synthesize and subsequently cannot distinguish information taken from the two channels (Kendon, 2000).

9) Gestures are co-produced with abstract metaphorical thinking. Linguistic metaphorical mappings are paralleled systematically in gesture (McNeill, 1992; Cienki, 1998; Sweetser, 1998; Núñez & Sweetser, 2001).

In all these studies, a careful analysis of important parameters of gestures such as handshapes, hand and arm positions, palm orientation, type of movements, trajectories, manner, and speed, as well as a careful examination of timing, indexing, preservation of semantics, and the coupling with environmental features, give deep insight into human thought[6]. An important feature of gestures is that they have three well-defined phases called preparation, stroke, and retraction (McNeill, 1992). The stroke is in general the fastest part of the gesture's motion, and it tends to be highly synchronized with speech accentuation and semantic content. The preparation phase is the motion that precedes the stroke (usually slower), and the retraction phase is the motion observed after the stroke has been produced (usually slower as well), when the hand goes back to a resting position or to whatever activity it was engaged in.

With these tools from gesture studies and cognition, we can now analyze mathematical expressions like the ones we saw before, but this time focusing on the gesture production of the speaker. For the purposes of this chapter, an important distinction

---

[6] An analysis of the various dimensions and methodological issues regarding the scientific study of gestures studies is beyond the scope of this chapter. For details see references mentioned above.

we need to make concerns the gestures that refer to real objects in the real world, and gestures that refer to some abstract idea that in itself doesn't exist in the real world. An example of the first group is shown in Figure 4, which shows renowned physicist Professor Richard Feynman giving a lecture on physics of particles at Cornell University many years ago. In this sequence he is talking about particles moving in all directions at very high speeds (Figure 4, a through e), and a few milliseconds later he completes his utterance by saying "once in a while hit" (Figure 4f). The action shown in the first five pictures correspond to the gesture characterizing the random movements of particles at high speeds. The precise finger pointing shown in figure 4f occurs when he says "once in a while hit" (the stroke of the gesture). The particle being indexed by the gesture is quite abstract and idealized, in the sense that it doesn't preserve some properties of the real referent, such as the extremely high speed at which particles move, for instance. But the point here is that although Prof. Feynman's talk was about a very abstract domain (i.e., particle physics), it is still the case that with his finger he is *indexing* a "particle," an object with location, extension, and mass, which does exist in the real world. The trajector in this dynamic scene is, an extremely small and fast object, but nonetheless a real entity in the real world.

Fig. 4. Professor Richard Feynman giving a lecture on physics. He is talking about particles moving in all directions at very high speeds (a through e), which "once in a while hit" (f).

Now, the gestures we are about to analyze below are similar in many respects, but they are even more abstract. In these cases the entities that are indexed with the various handshapes are purely imaginary entities, like points and numbers in mathematics. Figure 5, for instance shows a professor of mathematics lecturing on convergent sequences in a university level class. In this particular situation, he is talking about a case in which the real values of an infinite sequence do not get closer and closer to a single real value as *n* increases, but "oscillate" between two fixed values. His right hand, with the palm towards his left, has a handshape called *baby O* in American Sign

Language and in gesture studies, where the index finger and the thumb are touching and are slightly bent while the other three fingers are fully bent. In this gesture the touching tip of the index and the thumb are metaphorically indexing a metonymical value standing for the values in the sequence as *n* increases (it is almost as if the subject is carefully holding a very tiny object with those two fingers). Holding that fixed handshape, he moves his right arm horizontally back and forth while he says "oscillating."

Fig. 5. A professor of mathematics lecturing on convergent sequences in a university level class. Here he is referring to a case in which the real values of a sequence "oscillate" (horizontally).

Hands and arms are essential body parts involved in gesturing. But often it is also the entire body that participates in enacting the inferential structure of an idea. In the following example (Figure 6) a professor of mathematics is lecturing on some important notions of calculus at a university level course. In this scene he is talking about a particular theorem regarding monotone sequences.

As he is talking about an unbounded monotone sequence, he is referring to the important property of "going in one direction." As he says this he is producing frontwards iterative unfolding circles with his right hand, and at the same time he is walking frontally, accelerating at each step (Figure 6a through 6e). His right hand, with the palm toward his chest, displays a shape called *tapered O* (Thumb relatively extended and touching the upper part of his extended index finger bent in right angle, like the other fingers), which he keeps in a relatively fixed position while doing the iterative circular movement. A few milliseconds later he completes the sentence by saying "it takes off to infinity" at the very moment when his right arm is fully extended and his hand shape has shifted to an extended shape called *B spread* with a fully (almost over) extension, and the tips of the fingers pointing frontwards at eye-level.

It is important to notice that in both cases the blackboard is full of mathematical expressions containing formalisms like the ones we saw earlier (e.g., existential and universal quantifiers $\exists$ and $\forall$): formalisms, which have no indication of, or reference to, motion. The gestures (and the linguistic expressions used), however, tell us a very different conceptual story. In both cases, these mathematicians are referring to fundamental dynamic aspects of the mathematical ideas they are talking about. In the first example, the oscillating gesture matches, and it is produced synchronically with, the linguistic expressions used. In the second example, the iterative frontally-unfolding circular gesture matches the inferential structure of the description of the iteration involved in the increasing monotone sequence, where even the entire body moves forwards as the sequence unfolds. Since the sequence is unbounded, it "takes off to infinity," idea which is precisely characterized in a synchronous way with the full frontal extension of the arm and the hand.

Fig. 6. A professor of mathematics at a university level class talking about an unbounded monotone sequence "going in one direction" (a through e), which "takes off to infinity" (f).

The moral we can get from these gesture examples is two fold.
- First, gestures provide converging evidence for the psychological and embodied reality of the linguistic expressions analyzed with classic techniques in cognitive linguistics, such as metaphor and blending analysis. In these cases gesture analyses show that the metaphorical expressions we saw earlier are not cases of dead metaphors. The above gestures show, in real time, that the dynamism involved in these ideas have full psychological and cognitive reality.
- Second, these gestures show that the fundamental dynamic contents involving infinite sequences, limits, continuity, and so on, are in fact *constitutive* of the inferential organization of these ideas. Formal language in mathematics, however, is not as rich as everyday language and cannot capture the full complexity of the inferential organization of mathematical ideas. It is the job of embodied cognitive science to characterize the full richness of mathematical ideas.

8 Conclusion

We can now go back to the original question asked in the title of this chapter: Do real numbers really move? Since fictive motion is a real cognitive mechanism, constitutive of the very notion of a real number, the answer is yes. Real numbers are metaphorical entities (with a very sophisticated inferential organization), and they do move, metaphorically. But, of course, this was not the main point of this chapter. The main point was to show that even the most abstract conceptual system we can think of, mathematics(!), is ultimately embodied in the nature of our bodies, language, and cognition. It follows from this that if mathematics is embodied in nature, then *any* abstract conceptual system is embodied.

Conceptual metaphor and fictive motion, being a manifestation of extremely fast, highly efficient, and effortless cognitive mechanisms that preserve inferences, play a fundamental role in bringing many mathematical concepts into being. We analyzed several cases involving dynamic language in mathematics, in domains in which, according to formal definitions and axioms in mathematics, no motion was supposed to exist at all. Via the study of gestures, we were able to see that the metaphors involved in the linguistic metaphorical expressions were not simply cases of "dead" linguistic expressions. Gesture studies provide real-time convergent evidence supporting the psychological and cognitive reality of the embodiment of mathematical ideas, and their inferential organization. Building on gestures studies we were able to tell that the above mathematics professors, not only were using metaphorical linguistic expressions, but that they were in fact, in real time, thinking dynamically!

For many, mathematics is a timeless set of truths about the universe, transcending our human existence. For others, mathematics *is* what is characterized by formal definitions and axiomatic systems. From the perspective of our work in the cognitive science *of* mathematics (itself), however, a very different view emerges: Mathematics doesn't exist outside of human cognition. Formal definitions and axioms in mathematics are themselves created by human ideas (although they constitute a very small and specific fraction of human cognition), and they only capture very limited aspects of the richness of mathematical ideas. Moreover, definitions and axioms often neither formalize nor generalize human everyday concepts. A clear example is provided by the modern definitions of limits and continuity, which were coined after the work by Cauchy, Weierstrass, Dedekind, and others in the 19th century. These definitions are at odds with the inferential organization of natural continuity provided by cognitive mechanisms such as fictive and metaphorical motion. Anyone who has taught calculus to new students can tell how counter-intuitive and hard to understand the epsilon-delta definitions of limits and continuity are (and this is an extremely well-documented fact in the mathematics education literature). The reason is (cognitively) simple. Static epsilon-delta formalisms neither formalize nor generalize the rich human dynamic concepts underlying continuity and the "approaching" of locations.

By finding out that real numbers "really move," we can see that even the most abstract, precise, and useful concepts human beings have ever created are ultimately *embodied*.

References

1. Bates, E. & F. Dick. (2002). Language, Gesture, and the Developing Brain. *Developmental Psychobiology*, 40(3), 293-310.
2. Cienki, A. (1998). Metaphoric gestures and some of their relations to verbal metaphoric expressions. In J-P Koenig (ed.) *Discourse and Cognition*, pp. 189-204. Stanford CA: CSLI Publications.
3. Courant, R. & Robbins, H. (1978). *What is Mathematics?* New York: Oxford.
4. Fauconnier, G. & M. Turner. (1998). Conceptual Integration Networks. *Cognitive Science* 22:2, 133-187.
5. Fauconnier, G. & M. Turner. (2002). *The Way We Think: Conceptual Blending and the Mind's Hidden Complexities*. New York: Basic Books.
6. Goldin-Meadow, S. & C. Mylander. (1984). Gestural communication if deaf children: The effects and non-effects of parental input on early language development. *Monographs of the Society for Research in Child Development*, 49(3), no, 207.
7. Henderson, D. (2001). *Experiencing geometry*. Upper SaddleRiver, NJ: Prentice Hall.
8. Hersh, R. (1997). *What is mathematics*, really? New York: Oxford Univ. Press.
9. Hickok, G., Bellugi, U., and Klima, E. (1998). The neural organization of language: Evidence from sign language aphasia. *Trends in Cognitive Sciences*, 2(4), 129-136.
10. Iverson, J. & S. Goldin-Meadow. (1998). Why people gesture when they speak. *Nature* 396, Nov. 19, 1998. p. 228.
11. Iverson, J. & E. Thelen, E. (1999). In R. Núñez & W. Freeman (Eds.), *Reclaiming cognition: The primacy of action, intention, and emotion*, pp. 19-40. Thorverton, UK: Imprint Academic.
12. Kendon, A. (1980). Gesticulation and Speech: Two aspects of the process of utterance. In M. R. Key (ed.), *The relation between verbal and nonverbal communication*, pp. 207-227. The Hague: Mouton.
13. Kendon, A. (2000). Language and gesture: unity or duality? In D. McNeill (Ed.), *Language and gesture* (pp. 47-63). Cambridge: Cambridge University Press.
14. Lakoff, G. (1993). The contemporary theory of metaphor. In A. Ortony (Ed.), *Metaphor and Thought* (2nd ed.), pp. 202-251. Cambridge: Cambridge University Press.
15. Lakoff, G. & M. Johnson. (1980). *Metaphors we live by*. Chicago: University of Chicago Press.
16. Lakoff, G., & R. Núñez (1997). The metaphorical structure of mathematics: Sketching out cognitive foundations for a mind-based mathematics. In L. English (ed.), *Mathematical Reasoning: Analogies, Metaphors, and Images*. Mahwah, N.J.: Erlbaum.
17. Lakoff, G. and Núñez, R. (2000). *Where Mathematics Comes From: How the Embodied Mind Brings Mathematics into Being*. New York: Basic Books.
18. McNeill, D. (1992). *Hand and Mind: What Gestures Reveal About Thought*. Chicago: Chicago University Press.
19. Mayberry, R. & Jaques, J. (2000) Gesture production during stuttered speech: insights into the nature of gesture-speech integration. In D. McNeill (ed.) *Language and Gesture*. Cambridge, UK: Cambridge University Press.
20. Narayanan, S. (1997). Embodiment in Language Understanding: Sensory-Motor Representations for Metaphoric Reasoning about Event Descriptions. Ph.D. dissertation, Department of Computer Science, University of California at Berkeley.
21. Núñez, R. (1999). Could the Future Taste Purple? In R. Núñez and W. Freeman, *Reclaiming cognition: The primacy of action, intention, and emotion*, pp. 41-60. Thorverton, UK: Imprint Academic.

22. Núñez, R. (2000). Mathematical idea analysis: What embodied cognitive science can say about the human nature of mathematics. Opening plenary address in *Proceedings of the 24th International Conference for the Psychology of Mathematics Education*, 1:3–22. Hiroshima, Japan.
23. Núñez, R. (2003). Fictive and metaphorical motion in technically idealized domains. *Proceedings of the 8ᵗʰ International Cognitive Linguistics Conference*, Logroño, Spain, July 20-25, p. 215.
24. Núñez, R.(in press). Creating Mathematical Infinities: The Beauty of Transfinite Cardinals. *Journal of Pragmatics*.
25. Núñez, R., Edwards, L., Matos, J.F. (1999). Embodied Cognition as grounding for situatedness and context in mathematics education. *Educational Studies in Mathematics*, *39*(1-3): 45-65.
26. Núñez, R. & G. Lakoff. (1998). What did Weierstrass really define? The cognitive structure of natural and ε-δ continuity. *Mathematical Cognition*, *4*(2): 85-101.
27. Núñez, R. & Lakoff, G. (in press). The Cognitive Foundations of Mathematics: The Role of Conceptual Metaphor. In J. Campbell (ed.) *Handbook of Mathematical Cognition*. New York: Psychology Press.
28. Núñez, R. & E. Sweetser, (2001). *Proceedings of the 7ᵗʰ International Cognitive Linguistics Conference*, Santa Barbara, USA, July 22-27, p. 249-250.
29. Sweetser, E. (1990). *From Etymology to Pragmatics: Metaphorical and Cultural Aspects of Semantic Structure*. New York: Cambridge University Press.
30. Sweetser, E. (1998). Regular metaphoricity in gesture: bodily-based models of speech interaction. In *Actes du 16ᵉ Congrès International des Linguistes*. Elsevier.
31. Talmy, L. (1996). Fictive motion in language and "ception." In P. Bloom, M. Peterson, L. Nadel, & M. Garrett (eds.), *Language and Space*. Cambridge: MIT Press.
32. Talmy, L. (1988). Force dynamics in language and cognition. *Cognitive Science*, 12: 49–100.
33. Talmy, L. (2003). *Toward a Cognitive Semantics. Volume 1: Concept Structuring Systems*. Cambridge: MIT Press.

Information-Theoretical Aspects of Embodied Artificial Intelligence

Olaf Sporns and Teresa K. Pegors

Department of Psychology, Indiana University, Bloomington IN 47405, USA
{osporns, tpegors}@indiana.edu
http://www.indiana.edu/~cortex/lab.html

Abstract. Embodied AI is a new approach to the design of autonomous intelligent systems. This chapter is about a new principle for the design of such systems that is deeply rooted in the notion of embodiment. Embodied action has causal effects on the nature and statistics of sensory inputs, which can in turn drive neural and cognitive processes. The statistics of sensory inputs can be captured by using methods from information theory, specifically measures of entropy, mutual information and complexity, on sensory data streams. Several such methods are outlined and their application to embodied AI systems is discussed.

1 Introduction

The creation of intelligent systems capable of autonomous behavior in complex environments represents one of the major challenges to science and engineering in the 21st century. In the past, the design of artificial intelligence (AI) revolved mainly around the implementation of appropriate rules and representations in a (disembodied) computational setting. The last decade or so has seen a radical paradigm shift towards "embodied AI", a new approach that explicitly incorporates aspects of body morphology, movement and plasticity into its theoretical framework. While this approach is still in its infancy, some general principles for the design of embodied AI systems are on the horizon. This chapter is about one such principle, focusing on the impact of embodiment on structuring sensory inputs. Structure, in this context, refers to statistical dependencies or relationships between receptors or sensing elements and can be measured using quantitative approaches from information theory. We will argue that structured sensory inputs can have a powerful influence on the information-processing capacity of the embodied system's control structure (e.g. its nervous system).

First we need to discuss some of the central design features that are shared by many kinds of natural and artificial embodied systems. We suggest that most embodied systems consist of three integrated components:

(a) A *control architecture*. This can be a biological nervous system or a simulated neural or cognitive model. Plasticity and adaptation (as the primary mechanisms of development and learning) primarily take place within this "internal" set of structures.

F. Iida et al. (Eds.): Embodied Artificial Intelligence, LNAI 3139, pp. 74–85, 2004.
© Springer-Verlag Berlin Heidelberg 2004

(b) A *body*. This can be either the body of an organism or that of an autonomous robot. The body has a specific morphology (arrangement of sensory surfaces, appendages, muscles etc.) as well as a movement repertoire. While we view the body as separate from the morphology of the control architecture itself, we note that the distinction between these two domains is rather less obvious in most biological organisms. Brain and body morphology have obviously evolved together and form a continuum rather than strictly separate domains.

(c) An *environment*. The environment normally contains various objects and events. In simulated models of embodied AI the environment is a part of the computer simulation. If a physical robot is used, the environment often consists of a specific real-world ecological niche, constructed for experimental purposes in a laboratory. Other embodied systems or organisms may form part of this econiche.

While each of these three components can be independently modeled and studied, embodied AI considers whole systems for which these three components are dynamically and reciprocally coupled. This dynamic coupling is essential for the design philosophy of embodied AI. Dynamic coupling, in this context, refers to continuous reciprocal interactions between brain, body and world, across multiple time scales. For example, it is obvious that neural signals (i.e. brain variables) can cause movements of the body and thus action in the environment. It is perhaps less obvious, but equally important to note that the effects of neural states on the environment can have an impact on the nature and on the statistics of sensory inputs reaching the nervous system. In other words, *an embodied system determines what its future inputs will be* and thus imposes structure on its own sensory input space [1,2]. Sensory inputs, in turn, have powerful roles to play in the development of neural structures and representations. The statistical structure of sensory inputs is therefore a crucial ingredient in learning and development.

We suggest that the generation of structure in input data is a fundamental principle of embodied AI. Only an embodied system which is coupled to its environment through sensorimotor interactions can actively structure its input space. In this chapter, we first briefly discuss the importance of information for brain function. Then, we outline ways to measure information in sensory inputs and explore the possibility of quantifying the contribution of embodied interactions with the environment towards the generation of statistical structure. At the end, we briefly discuss the potential relevance of these results for robotics and embodied AI.

2 Information and the Brain

Why is information important for the brain? As neurons respond to sensory stimuli they encode information about these stimuli in their firing patterns. Elevated firing rates or increased synchrony within neuronal populations are two main coding dimensions utilized by many neurons in different parts of the brain. Neural codes can be "read" by other neuronal populations and in turn affect their activity states and firing patterns. The concerted action of widely distributed neuronal populations in multiple brain areas presumably underlies all cognitive, perceptual and behavioral states. The importance of neural coding and information for brain function is now almost universally recognized.

Numerous theories of information-processing in the nervous system have been proposed over the past several decades. Some of these theories are attempting to formulate computational principles that relate the statistics of sensory stimuli to neural processing and representation. It is increasingly realized that biological nervous systems are embedded in often complex natural environments containing stimuli with specific statistical properties and that real-world constraints have an impact on the functional organization of the brain. We have argued previously [3] that there are (at least) two fundamental challenges in information processing that are faced by higher (mammalian) nervous systems.

(a)   Information about stimuli in the environment needs to be efficiently extracted and mapped to *functionally specialized* neurons in the brain.

(b)   The information then needs to be *functionally integrated* to allow the emergence of coherent brain states that can guide behavior.

Integration and segregation may be viewed, in some sense, as antagonistic principles. Functional segregation is consistent with the information-theoretical idea that neurons extract specialized information from input patterns by eliminating redundancy and maximizing information transfer. The idea that neurons perform highly effective (perhaps near-optimal) information extraction has also been called the efficient coding hypothesis [4], and its proponents have made significant efforts to characterize the information present in naturalistic sensory stimuli. In contrast to functional segregation, functional integration establishes statistical relationships (for example in the form of temporal correlations) between distinct and often remote cell populations and brain regions. This results in the generation of mutual information, a general measure of statistical "overlap" or dependence. By creating these mutual dependencies, local neuronal specialization may be degraded. Both, functional segregation and integration can have causal efficacy within the brain, in that the integrated action of specialized neurons can exert specific causal effects on other neurons that are located elsewhere.

We have hypothesized [5,6] that the degree to which a neural system combines functional specialization and functional integration is related to how well the system is adapted to its specific environment. If it extracts information well and then integrates it to generate coherent internal states, we may say that the system has a high degree of "matching" to its stimulus world: its internal connectivity has captured statistical regularities to a high degree. In previous computational work, we noted that high matching within a neural system is facilitated if the input data (that constitute the neural system's environment) contains high amounts of structure. This observation is central to our argument about a potential role of embodiment in shaping input statistics. We will now discuss various quantitative measures of information that can be used to analyze the role of embodiment in structuring sensory data.

3 Measures of Information

The entropy of a discrete random variable X occupying a finite set of states can be calculated from the variable's state distribution using Shannon's formula [7].

$$H(X) = -\Sigma_i p(i) \log p(i) . \tag{1}$$

If all possible states of the variable are equally likely, entropy is maximal. If only very few states are occupied, entropy is reduced. Given two random variables X_1 and X_2, their mutual information (MI) is

$$MI(X_1;X_2) = H(X_1) + H(X_2) - H(X_1;X_2) . \tag{2}$$

MI is high if the state of one variable provides information about the state of the other variable (i.e. if their joint entropy $H(X_1;X_2)$ is significantly less than the sum of their individual entropies). For discrete random variables, the value for MI can be obtained from the joint state probability matrix. MI usually refers to instantaneous statistical dependencies, without time-lag in the measurement of one variable over the other. Time-lagged mutual information expresses how much information the state of one variable provides about the state of the other variable, when both measurements are separated by a fixed time interval.

In a series of studies [3,5,6,8,9] aimed at characterizing information states in neural systems we introduced several multivariate statistical measures that were designed to capture global aspects of how much information (statistical dependence) is present within a given system of arbitrary size, and of how this information is distributed. A global estimate of the amount of statistical dependence within a given system or set of elements $X = \{x_1, x_2, \ldots x_n\}$ is provided by the difference between the individual entropies of the elements and the joint entropy of the entire set, called integration:

$$I(X) = \Sigma_i H(x_i) - H(X) . \tag{3}$$

Any amount of statistical dependence between the elements will express itself in a reduction of their joint entropy and thus in a positive value for I(X). If all elements are statistically independent their joint entropy is the sum of the element's individual entropies and I(X) = 0. Thus, integration quantifies the total amount of structure or statistical dependencies present within the system. Comparing Eq. 2 and 3 reveals that integration is the multivariate generalization of mutual information.

The interplay between segregation and integration within a given system is captured by the global structure of the system's covariance matrix (which captures all of its linear pair-wise interactions). Segregation and integration leave characteristic signatures in the pattern of statistical interactions and dependencies. As reviewed above, statistical dependencies between the elements of a system can be measured by estimating their entropy and mutual information. Systems that combine functional segregation and functional integration exhibit "interesting" structure that is present at different levels of scale, a hallmark of complexity [3]. Less complex systems contain no statistical structure (random systems) or contain structure only at one level (e.g. crystals) which simply repeats. To calculate the complexity for a given system, we derive the spectrum of average integration across all levels of scale. In general, if the system contains local structure (at a small spatial scale) as well as global structure (at a large scale) this measure will be high. Here, we use an equivalent measure of complexity that does not require the full spectrum of entropy and integration ($H(x_i|H−x_i)$ denotes the conditional entropy of one element, given the rest of the system):

$$C(X) = H(X) − \Sigma_i H(x_i|X−x_i) . \tag{4}$$

We have shown previously [6] that complexity is high for systems that effectively combine functional segregation and integration, e.g. by incorporating specialized elements that are capable of global (system-wide) interactions. On the other hand, complexity is low for random systems, or for systems that are highly uniform (or, in other words, systems that lack either global integration or local specialization).

In neuronal networks, there is a strong relationship between complexity (as determined from the pattern of statistical interactions among elements of a system) and the pattern of anatomical or structural connections [6]. Complexity (and other informational measures, such as entropy or integration) can be used as cost functions in simulations designed to optimize network architectures. We found that networks optimized for high complexity showed structural motifs that are very similar to those observed in real cortical connection matrices [6,9], in particular a tendency to form clusters, short characteristic path lengths and short wiring lengths. Other informational measures produced networks with strikingly different structural characteristics.

Given that the simultaneous generation and integration of information within the brain is such a challenging task, what impact does it have for the brain to be embodied? Does embodiment add to the difficulty of the information-processing challenges faced by real brains or does it help in solving them? In order to approach this question, we turn to a very simple demonstration of how measures of information might be used to quantify structure in sensory data, and of how movement strategies that involve a high degree of coupling between an agent and an environment can result in generating useful structure in sensory inputs.

4   Structuring Sensory Data: Two Examples

These measures of information, integration and complexity can be applied to time series of random variables irrespective of their origin, whether they are neural or non-neural. Specifically, we can apply these measures to time series of sensory inputs to gain insight into their informational content. The first application of this kind was carried out by Lungarella and Pfeifer [10] and the examples we introduce in this chapter are based on their initial study.

4.1   Simulation of Visual Tracking

Figure 1 (left) shows the basic layout of a computer simulation of the sampling of visual data. The environment consists of an array of 100×100 pixels with randomly

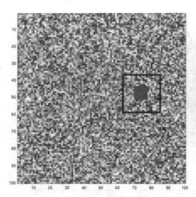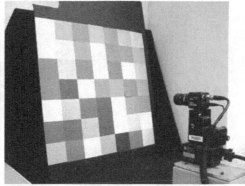

Fig 1. (Left) Simulation environment (100×100 pixels) with the agent's window (20×20 pixels) centered on the object (red square, 8×8 pixels).   (Right) Pan/tilt camera platform.

assigned color values (RGB, 8-bit resolution). Color values of all pixels are updated once every time step, resulting in independent discrete (8-bit) time series for each of the pixels. The environment also contains a single object, a red square with a size of 8×8 pixels. This object is moving through the environment in a random path. The environment houses a single agent, viewing a portion of the environment through a window (20×20 pixels), which represents its visual field. The agent can move about the environment, by displacing the window, which results in a time series of visual images (input). We implemented two distinct movement strategies for the window:

- The first strategy ("random") involved random movements where new locations were chosen independent of the visual image itself.
- The second strategy ("tracking") involved the selection of movements depending on sensory visual inputs. The agent's neural architecture consisted of topographic maps of receptors that were sensitive to red, green and blue pixel values. A scaled linear combination of these receptor maps was generated. The scaling factors were set such that the color red was preferentially detected. Then, the output of this operation was passed through a spatial filter chosen to enhance regions of coherent color values, similar to an attentional "saliency map". In this map, the spatial location with the maximal activation value was labeled and a saccadic movement of the window to that location was generated.

What distinguishes these two movement strategies is that the random strategy does not involve coupling of the agent's actions to the environment, while the tracking strategy does involve such coupling. For both movement strategies, the sensory data within the moving window are recorded for later analysis. A typical simulation run lasts for 100,000 time steps. Movies of the "random" and "tracking" conditions can be downloaded at www.indiana.edu/~cortex/lab.htm. At the end of a typical run, the recorded sensory data consists of a single matrix of 400 random variables sampled at 8-bit resolution for 100,000 time steps, for each of the color channels red, green, and blue. Only the red channel is used in the present analysis.

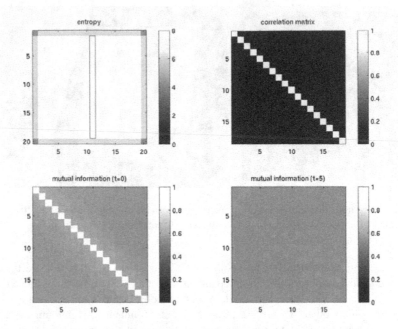

Fig. 2. Movement strategy "random". Plots of entropy (upper left panel), instantaneous mutual information (lower left panel), time-lagged mutual information (lower right panel) and correlation matrix (upper right panel). Entropy is shown in a topographic map over the 20×20 pixels of the window. A strip of 18 pixels along the central vertical axis is marked (rectangular box). Data from these pixels is used for the analyses shown in the other panels.

Fig. 2 and 3 show representative data sets obtained from two individual simulations, with movement strategies "random" and "tracking", respectively. Each figure shows data for entropy and mutual information as well as the correlation matrix for visual inputs.

For the "random" movement strategy, the entropy of the visual inputs at all locations within the moving window is homogeneous and near-maximal (8 bits, for a state space with 2^8 states). The instantaneous mutual information is uniformly low (less than one bit). Similarly, the time-lagged mutual information (time delay = 5 time steps) is uniform and approximates zero. The residual mutual information seen in both plots is due to the incomplete coverage of the joint state space (2^{16} bins). Discretizing the data using lower resolution (e.g. 5-bit) eliminates this residual MI completely (data not shown). Not surprisingly, given the absence of any statistical relationships between any of the visual inputs, the correlation matrix is flat with correlations very close to zero throughout.

For the "tracking" strategy, the entropy of the visual inputs at or near the foveal part of the visual field is markedly reduced to around 3 bits. In addition to lowered

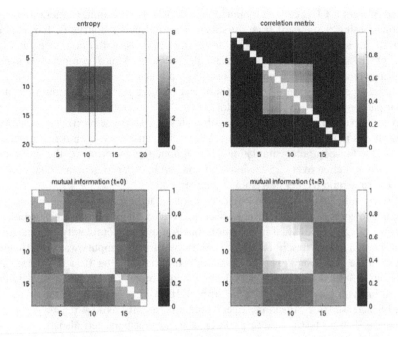

Fig. 3. Movement strategy "tracking". Plots of entropy (upper left panel), instantaneous mutual information (lower left panel), time-lagged mutual information (lower right panel) and correlation matrix (upper right panel).

entropy, foveal visual inputs have high mutual information, in excess of 1 bit. Time-lagged mutual information is also relatively high for visual inputs at or near the foveal region, indicating that these input states remain correlated over several input frames. While extra-foveal visual inputs do not show any consistent cross-correlation, the foveal inputs exhibit significant cross-correlations of around 0.6-0.8.

The correlation matrices for "random" and "tracking" simulations can be used to calculate the overall complexity of the visual data. In order to calculate the joint entropies and mutual information directly from the correlation matrix we transformed each of the time series of visual inputs into an equivalent time series with a Gaussian amplitude profile, which maintains (approximately) the same linear and non-linear interactions. Then, integration and complexity are calculated using equations 3 and 4, following standard formulae for Gaussian multivariate processes [7]. We obtain values of $I(X) = 0.0130$ and $C(X) = 0.0011$ for the "random" condition and $I(X) = 2.4402$ and $C(X) = 0.0961$ for the "tracking" condition.

4.2 A Simple Robot Experiment

In addition to the computer simulation discussed above, we also constructed a robotic platform designed to sample visual inputs. The robotic platform (Fig. 1, right) consisted of a color CCD camera mounted on a 2 DOF pan-tilt unit. Camera images were captured using a standard frame grabber, acquired under constant illumination and spatially averaged to yield a resolution of 16×12 pixels, with one image each for the red, green and blue channels of the color camera. Motor commands moved the pan/tilt servos to specified positions resulting in image displacement. The stimuli used in these experiments were "color Mondrians", i.e. large collages composed of small color patches of a broad range of colors.,

We compared two different motor strategies (cf. above section): 1) "random": the camera was moved at random; 2) "tracking": the camera was controlled as described above for the computer simulation. Throughout the experiment, the stimulus was manually switched every 25 seconds (50 images) in order to generate changing visual scenes that would trigger new tracking movements.

Image time series acquired for two representative experiments using these two motor strategies were normalized, discretized to 5-bit resolution and examined for patterns of entropy, mutual information, integration and complexity. All calculations were carried out as described above.   Entropy of visual inputs was significantly reduced in the "tracking" over the "random" condition (3.39 ± 0.31 bits versus 3.72 ± 0.29 bits, $t(382) = 10.37$, $p<0.001$). Instantaneous mutual information for neighboring visual inputs located along a central strip of 10 pixels (as in Fig. 1, left) was significantly increased for "tracking" over "random" conditions (1.46 ± 0.24 bits versus 1.18 ± 0.19 bits, $t(16) = 2.76$, $p<0.01$), a trend that persisted also for time-lagged mutual information (lag = 5 time steps, data not shown).   Integration and complexity for these inputs were $I(X) = 0.9858$ and $C(X) = 0.0985$ for "random" and $I(X) = 2.2296$ and $C(X) = 0.1931$ for "tracking". Corresponding correlation matrices showing elevated cross-correlations between neighboring visual inputs for the "tracking" condition are shown in Fig. 4.

4.3 Summary of Results

These examples serve the purpose of illustrating the potential use of quantitative measures of information in the context of robotics. The interpretation of the computational results is relatively straightforward.   The "tracking" movement strategy involves the coupling of the agent's sensory surface to patterns and changes in the environment.   No such coupling occurs in the "random" strategy. As a result of the propensity of the agent/camera to track moving red objects, visual inputs near the fovea tend to sample pixels with high values for the color red, thus altering their overall state space distributions and reducing the input entropy. The spatial extent of red objects (covering approximately one-third of the total visual angle) generates correlations in the states of neighboring visual inputs at or near the foveal region of the visual field. Thus, these inputs exhibit not only reduced entropy, but also increased mutual information and cross-correlation. The temporally continuous nature of tracking movements results in elevated mutual information "across time", i.e. mutual information between neighboring visual inputs that persists over an extended series of input

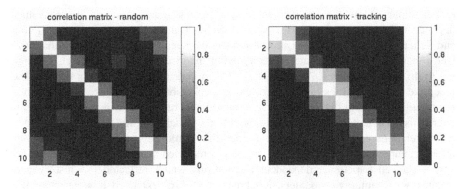

Fig. 4. Comparison of correlation matrices obtained from data collected with the pan/tilt camera platform using movement strategies "random" (left) and "tracking" (right).

frames. The overall pattern of inputs in the "tracking" condition contains more statistical dependencies and is significantly more complex than for the "random" strategy, with a subset of inputs exhibiting high correlations ("local structure").

5 Application to Embodied Artificial Intelligence

The above examples were deliberately simplified and idealized to illustrate how informational patterns in inputs may depend on the coupling between an agent and its environment. Our results are in agreement with those of other studies that have shown that simple sensorimotor functions like gaze direction and foveation can generate high mutual information and complexity in visual inputs [10,11]. Other examples of (in Rolf Pfeifer's words) "generating good structure" in sensory inputs exist, but have so far not been subjected to a rigorous information-theoretical analysis. Fitzpatrick and Metta, for example, have described a robot that can push and physically displace objects in a visual scene [12]. The action of the robot causes the sudden appearance of spatially and temporally correlated movement in the visual array, which can be used to segment the object from a background. This is an important demonstration of how the embodiment of a robot can simplify an otherwise very difficult visual task, that of segmenting the image into discrete objects. Another sensorimotor function with very dramatic effects on the statistics of visual inputs is attention-driven saccades (camera/eye movements) which determine the direction of gaze [13]. Depending on the state of the attentional system different kinds of objects are preferentially selected and placed in the central region of the visual array where they may be subjected to closer visual analysis. The use of systems directing the gaze of a robot clearly shows that action and perception of neuro-robotic systems form closely coupled dynamical loops and are often inseparably linked.

Other examples of how embodiment helps in structuring input spaces have hinted at the potential importance of this effect in development and learning. There are several examples of how the development of specific neural and cognitive functions may depend on the embodied actions of a robot. Scheier and Lambrinos [14] and Pfeifer and Scheier [2] have developed several models of perceptual categorization that utilized robot behavior to generate inputs allowing the discrimination of objects belonging to different categories. Almassy et al. [15] modeled the development of complex receptive field properties in visual cortex in a robot behaving in a real-world environment. They found that self-generated movements of the robot resulted in smooth lateral displacements of objects within the robot's visual field, thus generating temporal correlations over multiple image frames that could be exploited by neurons in inferior temporal cortex. In addition, the map of modeled neurons in inferior temporal cortex showed experience-dependent fluctuations which reflected the history and frequency of stimulus encounters. Krichmar et al. [16] extended this aspect of the model and recorded systematic changes in receptive field properties of object-selective visual neurons with object composition of the environment, demonstrating the experience-dependence of perceptual categorization.

In a model of reward conditioning in an autonomous robot, Alexander and Sporns [17,18] found evidence for complex interactions between behavior and neural states that can significantly influence developmental trajectories and learning patterns. They found a surprising degree of coupling between neural and behavioral variables even when examining relatively simple environments and neural structures. The robot's actions altered the spatial distribution of rewarding objects, which in turn impacted on the timing of rewarding stimulus encounters. The difference in the timing of the rewarding events led to differences in the development of synaptic patterns representing predictions about future rewards.

As these studies of embodied systems show, robots and organisms do not passively absorb information from their surrounding environment, but their actions *on* the environment select and shape this information. Informational patterns in sensory inputs can be exploited by neural circuits and promote the stabilization of matching neural connections that incorporate recurrent statistical features. There is abundant evidence that the statistics of sensory inputs are of great importance in learning and development. The sensitivity of brain tissue to patterns in sensory inputs gives special significance to the causal influence of embodiment on input statistics. The examples presented in this chapter only provide a very preliminary glimpse of the fundamental role of embodiment in shaping the statistics of sensory inputs. Much more work remains to be done in this emerging area of artificial intelligence, by investigating the causal role of motor actions and behavior in selecting inputs and generating statistical regularities that can be exploited by neural mechanisms. These studies may lead to a new set of principles and quantitative measures that may help guide robot design to create devices whose motor capabilities match their internal processing power.

Acknowledgements. Supported by NIH/NIDA grant 1R21DA15647-01 to O.S. T.K.P. was supported through Indiana University's Science, Technology, and Research Scholar's Program.

References

1. Nolfi, S., Parisi, D.: Self-Selection of Input Stimuli for Improving Performance. In: Bekey, G.A. and Golberg, K.Y. (Eds.) Neural Networks in Robotics, pp. 403-418 (1993)
2. Pfeifer, R., Scheier, C.: Understanding Intelligence. MIT Press, Cambridge, MA (1999)
3. Tononi, G., Edelman, G.M., Sporns, O.: Complexity and Coherency: Integrating Information in the Brain. Trends Cogn. Sci. 2 (1998) 474-484
4. Simoncelli, E.P. and Olshausen, B.A.: Natural Image Statistics and Neural Representation. Annu. Rev. Neurosci. 24 (2001) 1193-1216
5. Tononi, G., Sporns, O., Edelman, G.M.: A Complexity Measure for Selective Matching of Signals by the Brain. Proc. Natl. Acad. Sci. USA 93 (1996) 3422-3427
6. Sporns, O., Tononi, G., Edelman, G.M.: Theoretical Neuroanatomy: Relating Anatomical and Functional Connectivity in Graphs and Cortical Connection Matrices. Cerebral Cortex 10 (2000) 127-141
7. Cover, T.M. Thomas, J.A.: Elements of Information Theory. Wiley, New York (1991)
8. Tononi, G., Sporns, O., Edelman, G.M.: A Measure for Brain Complexity: Relating Functional Segregation and Integration in the Nervous System. Proc. Natl. Acad. Sci. USA 91 (1994) 5033-5037
9. Sporns, O., Tononi, G.: Classes of network connectivity and dynamics. Complexity 7 (2002) 28-38
10. Lungarella, M., Pfeifer, R.: Robots as Cognitive Tools: Information-Theoretic Analysis of Sensory-Motor Data. Proc. 2001 IEEE-RAS Intern. Conf. Humanoid Robots, pp. 245-252 (2001).
11. Sporns, O. Pegors, T.: Generating Structure in Sensory Data through Coordinated Motor Activity. Proceedings IJCNN 2003 (2003) 2796
12. Fitzpatrick, P., Metta, G.: Grounding Vision through Experimental Manipulation. Phil. Trans. R. Soc. Lond. A 361 (2003) 2615-2625.
13. Breazeal, C., Edsinger, A., Fitzpatrick, P., Scassellati, B.: Active Vision for Sociable Robots. IEEE Trans. Man Cybernetics Systems 31 (2001) 443-453
14. Scheier, C., Lambrinos, D.: Categorization in a Real-World Agent using Haptic Exploration and Active Perception. Proc. SAB96 (1996) 65-74
15. Almassy, N., Edelman, G.M., Sporns, O.: Behavioral Constraints in the Development of Neuronal Properties: A Cortical Model Embedded in a Real World Device. Cereb. Cortex 8 (1998) 346-361
16. Krichmar, J.L., Snook, J.A., Edelman, G.M., Sporns, O.: Experience-Dependent Perceptual Categorization in a Behaving Real-World Device. In: Animals to Animats 6, Meyer, J.A.; Berthoz, A.; Floreano, D.; Roitblat, H.; Wilson, S.W., (eds.), MIT Press, Cambridge (2000), pg. 41-50.
17. Sporns, O., Alexander, W.H.: Neuromodulation and Plasticity in an Autonomous Robot. Neural Netw. 15 (2002) 761-774
18. Alexander, W.H., Sporns, O.: An Embodied Model of Learning, Plasticity and Reward. Adapt. Beh. 10 (2003) 141-159

Robot Bouncing: On the Synergy Between Neural and Body-Environment Dynamics

Max Lungarella* and Luc Berthouze

Neuroscience Research Institute
Tsukuba AIST Central 2, Japan
maxl@isi.imi.i.u-tokyo.ac.jp, luc.berthouze@aist.go.jp

Abstract. The study of how infants strapped in a Jolly Jumper learn to bounce can help clarify how they explore different ways of exploiting the dynamics of their movements. In this paper, we describe and discuss a set of preliminary experiments performed with a bouncing humanoid robot and aimed at instantiating a few computational principles thought to underlie the development of motor skills. Our experiments show that a suitable choice of the coupling constants between hip, knee, and ankle joints, as well as of the strength of the sensory feedback, induces a reduction of movement variability, and leads to an increase in bouncing amplitude and movement stability. This result is attributed to the synergy between neural and body-environment dynamics.

1 Introduction

Despite the availability of many descriptive accounts of infant development, modeling how motor abilities unfold over time has proven to be a hard problem [1,2,3]. Existing models are based on general principles and specific mechanisms which are assumed to underlie the changes in early motor development.

One such mechanism is self-exploration through spontaneous activity. An important precursor of later motor control [4,5,6], its main role seems to be the exploration of various musculo-skeletal organizations in the context of multiple constraints such as environment, task, architecture of nervous system, muscle strength, mass of the limbs, and so on. A growing number of developmental psychologists has started to advocate the view that self-exploration through spontaneous movements helps infants bootstrap new forms of motor activity, as well as discover more effective ways of exploiting the dynamics generated by their bodily activities [7,1,8,2,9]. It has been suggested that through movements that garner information specific to stable regions in the high-dimensional space of possible motor activations, self-exploration can lead to a state of awareness about body and environment [1]. In fact, fetuses (as early as 8 to 10 weeks after conception) as well as newborn infants display a large variety of transient and spontaneous movement patterns such as infant stepping and kicking [3], spontaneous arm movements [10], and general movements and sucking movements [11]. Infants probably learn about their body by performing movements over and over again, and by

---

* New affiliation: Department of Mechano-Informatics, School of Information Science and Technology, The University of Tokyo, Japan.

F. Iida et al. (Eds.): Embodied Artificial Intelligence, LNAI 3139, pp. 86–97, 2004.

exploiting the continuous flow of sensory information from multiple sensory modalities. In doing so, they explore, discover, and eventually select – among the myriad of available solutions – those that are more adaptive and effective [7].

The control of exploratory movements has been traditionally attributed to neural mechanisms alone. Prechtl, for instance, linked the production and regulation of spontaneous motility in infancy "exclusively" to endogenous neural mechanisms, such as central pattern generators [11]. This claim is somewhat substantiated by the fact that in many vertebrate species, central pattern generators appear to generate the rhythm and form of the bursts of motoneurons [12], or to govern innate movement behaviors altogether [4].

In the last two decades, however, new evidence has pushed forward an alternative and multi-causal explanation theoretically grounded into dynamic systems theory [3]. According to this view, coordinated motor behavior is also the result of a tight coupling between the neural and biomechanical aspects of movement, and the environmental context in which the movement occurs [1,13,14,3]. Spontaneous movements are not mere random movements, but are organized (or better, self-organize), right from the very start, into recognizable patterns involving various parts of the body, such as head, trunk, arms, and legs. Spontaneous kicks in the first few months of life, for instance, appear to be particularly well-coordinated movements characterized by a tight coupling [6, 3], and by short phase lags between the hip, knee and ankle joints [5]. Rigid phase-locked movements can be interpreted as a "freezing" of a number of degrees of freedom that must be controlled by the nervous system, thus resulting in a reduction of the movement variability and complexity, and in a faster learning process [15,16]. During development, the strong synchrony is weakened, and the degrees of freedom are gradually "released" [5,6,17]. The ability to change the patterns of coordination between various joints to accomplish a task is an important aspect of infants' motor development [7]. It has been shown that tight interjoint coupling persisting beyond the first few months of life may lead to poor motor development, or may even be associated with abnormal development [17].

In a previous paper, we examined the effects of "freezing and freeing of degrees of freedom" [18] in a swinging biped robot. The study showed that by freezing (that is, rigidly coupling) and by subsequently freeing the mechanical degrees of freedom, the sensorimotor space was more efficiently explored, and the likelihood of a mutual regulation of body-environment and neural dynamics (that is, entrainment) was increased. The aim of this chapter is to further our understanding of the role played by the coupling (a) between joints, and (b) between the sensory apparatus and the neural structure for the acquisition of motor skills. To achieve this goal, we embedded a pattern generating neural structure in a biped robot, and by manually altering various coupling constants, we systematically studied their interaction with the body-environment dynamics in the context of a real task (bouncing).

2   Hypotheses on Infant Bouncing Learning

Goldfield et al. [19] performed a longitudinal study in which eight six-months old infants strapped in a "Jolly Jumper" (i.e., a harness attached to a spring) were observed once a week, for a period of several weeks, while learning to bounce. They concluded that in the course of learning, the infants' motor activity could be decomposed into an initial "assembly phase", during which kicking was irregular and variable in period, followed by a "tuning phase" characterized by bursts of more periodic kicking and long bouts of sustained bouncing, during which infants seemed to refine and adapt the movement to the particular conditions of the task. A third phase was initiated by a sudden doubling of the bout length, and was characterized by oscillations of the mass-spring system at its resonant frequency, a sensible rise of amplitude, and a decrease of the variability of the period of the oscillations.

A few principles can be derived from this study. First, there is no need to postulate a set of preprogrammed instructions or predefined motor behaviors. It is by means of a process of self-organization and self-discovery, and through various spontaneous (seemingly random) movements that infants explored their action space and eventually discovered that kicks against the floor had "interesting" consequences [19]. After an initial exploratory phase (assembly), the infants selected particular behaviors and began to exploit the physical characteristics of the mass-spring system. Goldfield and collaborators advanced the hypothesis that, in general, infants learning a task may try out different musculo-skeletal organizations by exploring the corresponding parameter space, driven by the dynamics of the task as well as by the existing repertoire of skills and reflexes.

Second, to achieve effective and continuous bouncing, i.e., bouncing characterized by simultaneous leg extensions, the infants had to learn patterns of intersegmental coordination. Thus, the infants had to explore different force and timing combinations for the control of their movements, and to integrate the environmental information impinging on various sensory modalities, i.e., visual, vestibular, and cutaneous. Unfortunately, the study performed by Goldfield et al. did not provide any kinematic or kinetic analysis of the development of the infants' movement patterns. In line with the findings reported in [18,6,17], we hypothesize that in order to reduce movement complexity, the initial movements had to be performed under tight intersegmental coupling. As development and learning progressed, the couplings were weakened, and more complex movement patterns could be explored. Thelen and colleagues put forward evidence showing that in infants the loosening of the tight joint coupling may not necessarily be a consequence of maturation of the nervous system alone [3], but instead may be also ascribed to changes in muscle mass, body composition, and body proportion.

Third, the rhythmic nature of the task (bouncing) can be interpreted as a particular instance of Piagetian circular reaction[1]. Rhythmic (not necessarily task-oriented) activity is highly characteristic of emerging skills during the first year of life. Thelen and Smith suggested that oscillatory movements are the by-product of a motor system under emergent control, that is, when infants are in the process of attaining some degree of intentional control of their limbs or body postures, but when their movements are not fully goal-corrected [3].

Finally, this study highlighted the necessity of a value system to evaluate the consequences of the movements performed, and to drive the exploratory process. Value

---

[1] Circular reactions represent an essential sensorimotor stage of Piaget's developmental schedule [20], which refer to the repetition of an activity in which the body starts in one configuration, goes through a series of intermediate stages, and eventually returns to the initial configuration.

systems are known to mediate plasticity and to modulate learning in an unsupervised and self-organized manner, allowing organisms to be adaptive, and to learn on their own via self-generated and spontanenous activity. They also create the necessary conditions for the self-organization of dynamic sensory-motor categories, that is, movement patterns.

3 Experimental Setup

To test our computational hypotheses, we decided to replicate Goldfield et al.'s experiments using a small-sized humanoid robot with 12 mechanical degrees of freedom (Fig. 1). The robot was suspended in a leather harness attached to two springs. Each leg of the robot had three segments (thigh, shank, and foot) and five joints, but only three of the latter (i.e., hip, knee and ankle) were used. Each joint was actuated by a high-torque RC-servo module. These modules are high-gain positional open-loop control devices and do not provide any feedback on the position of the corresponding joint. In fact, there was no need to measure the anatomical angles of hip, knee and ankle, since these values were available as the set positions of the RC-servo modules. Exteroceptive and proprioceptive information were also taken into account. Ground reaction forces were measured by means of force sensitive resistors placed under the feet of the robot (two per foot). To reduce impact forces in the joints of the robot and to add some passive compliance, the soles of the robot's feet were covered with soft rubber. Torsional movements around the z-axis were measured with a single-axis solid-state gyroscope. Linear accelerations in the sagittal plane were estimated by a dual-axis accelerometer (Fig. 1 right).

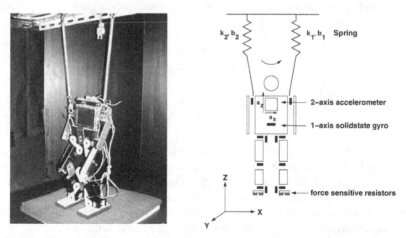

Fig. 1. Left: Humanoid robot used in our experiments. Right: Schematic representation of the robotic setup.

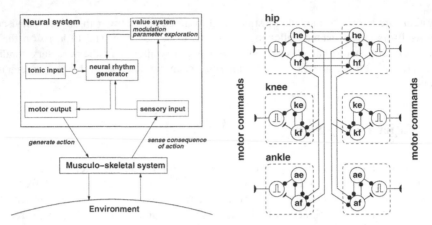

Fig. 2. Left: Basic structure of the neuro-musculo-skeletal system. The arrows in the model show the information flow. Right: Neural rhythm generator composed of six neural oscillators. The solid circles represent inhibitory, and the half-circles are excitatory connections. Abbreviations: he=hip extensor, hf=hip flexor, ke=knee extensor, kf=knee flexor, ae=ankle extensor, af=ankle flexor. Not shown are proprioceptive feedback connections and tonic excitations.

3.1   Neural Rhythm Generator

Figure 2 (right) depicts a schematic representation of the neuro-musculo-skeletal system inspired by [14]. The neural rhythm generator or central pattern generator [12] was constructed by using six neural oscillators, each of which was responsible for a single joint (Fig. 2 right). We modeled the individual neural oscillators according to the following set of nonlinear differential equations [21]:

$$\tau_u \, \dot{u}_f = -u_f - \beta \, v_f - \omega_c \, g(u_e) - \omega_p \, g(F_{eed}) + te$$
$$\tau_u \, \dot{u}_e = -u_e - \beta \, v_e - \omega_c \, g(u_f) - \omega_p \, g(-F_{eed}) + te$$
$$\tau_v \, \dot{v}_f = -v_f + g(u_f)$$
$$\tau_v \, \dot{v}_e = -v_e + g(u_e)$$
$$y_{out} = u_f - u_e$$

where u_e and u_f are the inner states of neurons e (extensor) and f (flexor), v_e and v_f are variables representing the degree of adaptation or self-inhibition of the extensor and flexor neurons. The external tonic excitation signal te determines the amplitude of the oscillation. β is an adaptation constant, ω_c is a coupling constant controlling the mutual inhibition of neurons e and f, τ_u and τ_v are time constants, and determine the strength of the adaptation effect. The operator $g(x) = max(0, x)$ returns the positive part of x. The difference of the output of the extensor and the flexor neuron of each unit oscillator was fed to a pulse generator. Its output y_{out} was the angle of the RC-servo associated with the corresponding unit oscillator. Sensory feedback to the pattern generator F_{eed} occurred through four the pressure sensors located under the robot's feet. The value of the afferent feedback was computed as the sum of the sensed ground reaction forces, weighted by the variable ω_p. Appropriate joint synergies among ipsilateral joints, i.e., appropriate phase relationships between the corresponding neural oscillators, were produced by feeding the flexor unit of one oscillator with a combination of the output of the extensor and flexor units of the other oscillator. As shown by Fig. 2, reciprocal inhibitory connections between corresponding flexor and extensor neurons of the left and right hip joint were also implemented.

3.2 Selection of the Neural Control Parameters

The adaptation constant β and the degree of mutual inhibition between extensor and flexor neuron of a single neural oscillator were fixed throughout the whole study to $\beta = 2.5$ and $\omega_c = 1.0$. The tonic excitation was fixed to $te = 1.0$, and the intersegmental coupling constant to $\omega_s = 0.75$. The high value of the latter constant induced kicking patterns with a tight joint coupling. According to Williamson [22], the time constants τ_u and τ_v determine the shape and the speed of the oscillator output. In order to guarantee stable oscillations, the ratio $r = \tau_u/\tau_v$ should be kept in the interval $[0.1, 0.5]$. In all experiments, we fixed the ratio r to 0.5. The sensory feedback coefficient ω_p was variable, and was set as specified in each sub-section.

4   Experiments and Discussion

To model and analyze our experimental results, we assumed an ideal mass-spring-damper system. This model represents a first attempt to identify a relationship between oscillation frequency, amplitude of the oscillation, and other parameters. The differential equation governing the free oscillation of the mass-spring-damper system is $m\,\ddot{x}(t) + b\,\dot{x}(t) + k\,x(t) = 0$. In our case, m is the mass of the robot, b is the damping coefficient of the spring and k its spring constant. The equation has solutions of the form: $x(t) = A\,e^{-bt/2m}\cos(\omega_d t + \phi)$, where A (amplitude of the oscillation) and ϕ (phase) are determined by the initial displacement and velocity of the robot. $\omega_n = \sqrt{k/m}$ is defined as the undamped natural frequency of the mass-spring-damper system and $\omega_d = \sqrt{\omega_n^2 - (b/2m)^2} < \omega_n$ is its damped natural frequency. The mass of the robot (fixed throughout all experiments) was $m = 1.33kg$. The estimated spring constant was $k_1 = k_2 = 25.5N/m$, and the damping coefficient was $b = 0.065kg/sec$ for both springs (Fig. 1 left). For the computation of b, we assumed a viscous frictional force, proportional to the velocity of the oscillation.

In all experiments, we recorded the system's movements by tracking the position (relative to an earth-fixed frame of reference) of colored markers placed on the robot's hip, knee and ankle. The experiments were organized according to the complexity of their environmental interaction (with/without ground contact, with/without sensory feedback).

4.1   Scenario 1 – Free Oscillations

This scenario served to assess the basic properties of the real system and of the corresponding mass-spring-damper model needed to qualify oscillatory behaviors (and materialize the presence of entrainment). The robot's joints were not actuated, and the robot was set so that its feet could not touch the ground no matter the amplitude of the vertical oscillations. At the onset of the experiment, the robot was lifted by an arbitrarily chosen height, and then let oscillate freely. The resulting motion was harmonic and underdamped, with an exponentially decreasing amplitude of the form $e^{-\alpha t} sin(2\pi t/T)$, a decay coefficient $\alpha = 0.124/sec$, and a period $T = 1.01 sec$. Hence, the resonance frequency of the system could be estimated to be $f_R = 1/T = 0.99 Hz \approx \omega_d/2\pi$. The effective spring constant of the system was $K_{eff} = 50.5\ N/m$, which is almost twice the spring constant of each spring. From our measurements, we estimated the effective damping coefficient to be approximately $B_{eff} = 0.33 N\ sec/m$. Note that B_{eff} is not twice the damping coefficient of a single linear spring, as might be inferred by the value of K_{eff}. This clearly shows that the system is not a close-to-ideal mass-spring system, and that a more rigorous approach would have to consider a better model for the damping force. For instance, viscous frictional forces proportional to the square of the velocity of the mass should be taken into account.

4.2   Scenario 2 – Forced Oscillations Without Ground Contact

In this experiment, the robot's joints were actuated such that the equation describing the motion of the robot was $m\ddot{x}(t) + b\dot{x}(t) + kx(t) = F(t)$, where driving force $F(t)$ is a function of the paramter settings of the neural oscillators and of the amplitude of the robot's limb movements (as suggested by Goldfield [19]). In other words, the movement of the robot can be modeled as a forced mass-spring system, with the robot's kicking movements representing the driving force. As in scenario 1, the robot could not reach the ground with its feet. After an initial transient, the system converged to a steady state, a forced harmonic oscillation. Vertical resonance was achieved for the parameter setting $(\tau_u = 0.108, \tau_v = 0.216)$, and resulted in an average vertical displacement from the rest position of $10.6 cm$, and a peak displacements exceeding $17 cm$. The dominant frequency of the oscillation, estimated via a spectral analysis of the vertical component of the hip marker position, was $f_{Hip} = 1.01 Hz$, which was very close to the previously estimated resonant frequency of the system $f_R = 0.99 Hz$. Interestingly, the system displayed at least three oscillatory modes. This behavior is akin to spontaneous activity in infants, who enter preferred stable states and exhibit abrupt phase transitions between states [1]. Parameter settings close to $(\tau_u, \tau_v) = (0.066, 0.132)$ led to a strong horizontal oscillatory motion, whereas for $\tau_u > 0.150$ and $\tau_v > 0.300$, there was an evident torsional movement. For $\tau_u < 0.06$, vertical oscillations were essentially unexistent.

4.3   Scenario 3 – Forced Oscillations with Ground Contact ($\omega_p = 0$)

The goal of this set of experiments was to assess the effect of ground contact on the oscillatory movement observed in scenario 2, in the absence of afferent feedback from the touch sensors (i.e., $\omega_p = 0$). At the onset of each experimental run, we made sure that the robot's feet could touch the ground. To correct for the lack of compliance in the robot's joints, the ground was covered with soft material. The introduction of this additional nonlinear perturbation led (given appropriate neural control parameters) to the emergence of a new behavior: bouncing. Figure 3 shows the result of three different parameter configurations. A suitable model of the movement of the robot's center of mass needs also to take into account the nonlinear interaction with the ground, and the stiffness and damping characteristics of the floor and the feet. We propose the following linear model (see also [19]): $m\,\ddot{x}(t) + B_{eff}\,\dot{x}(t) + K_{eff}\,x(t) = F(t)$, where $F(t) = 0$ when the feet are off the ground and $F(t) = F_0 - F_0 sin(2\pi f t)$, $F_0 > 0$, when the feet are on the ground, with K_{eff} (effective spring constant) and B_{eff} (effective damping coefficient) incorporating the effect of springs, feet and floor.

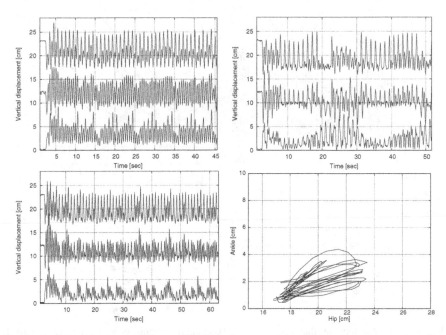

Fig. 3. Forced harmonic oscillations with ground contact (bouncing) in the absence of sensory feedback ($\omega_p = 0$). Top: $\tau_u = 0.108, \tau_v = 0.216$ and $\tau_u = 0.140, \tau_v = 0.280$, bottom: $\tau_u = 0.114, \tau_v = 0.228$ (phase plot on the right). In all graphs, the three curves represent the vertical displacement of the ankle, knee and hip marker in cm.

4.4 Scenario 4 – Forced Oscillations with Ground Contact ($\omega_p > 0$)

Afferent sensory feedback and contact with the ground induced a "haptic closure" of the sensory-motor loop, which turned the linear and externally driven mass-spring system of experiments 2 and 3 into an autonomous limit-cycle system with the intrinsic timing determined by the moment of foot contact with the ground and by the gain of the feedback connection ω_p. In other words, the kicking frequency (implicitly timed by the neural oscillators) and its phase relationship with the bouncing was regulated by haptic information, and resulted in entrainment between time of ground contact and period of the neural oscillators. A positive ω_p had at least two advantages: (a) it led to a stabilized and sustained bouncing, and (b) to an increase of its amplitude (measured as the difference between successive maxima and minima of the vertical displacement). These effects are visualized in Figure 4 top-left, in which the parameters were $(\tau_u, \tau_v) = (0.114, 0.228)$ and $\omega_p = 0.5$. The phase plot of the same time series is depicted in Figure 4 (top-right). The phase plots in figures 3 and 4 clearly demonstrate the stabilizing effects of

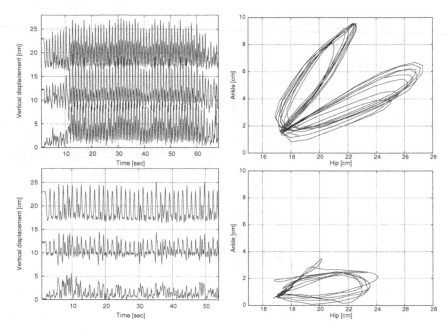

Fig. 4. Forced harmonic oscillations with ground contact (bouncing) in presence of sensory feedback ($\omega_p > 0$). Top row: $\omega_p = 0.5, \tau_u = 0.114, \tau_v = 0.228$, bottom row: $\omega_p = 0.75, \tau_u = 0.140, \tau_v = 0.280$.

sensory feedback. In Fig. 4 (top-right), the parameters were ($\tau_u = 0.140, \tau_v = 0.280$) and $\omega_p = 0.75$, and the bouncing was stable and sustained. For $\omega_p = 0$, however, the bouncing suddenly collapsed and exhibited more variability (Fig. 3 top-right).

The influence of sensory feedback on the bouncing amplitude is evident by comparing Fig. 3 (bottom) with Fig. 4 (top). In the latter case, the maximum vertical displacement of the hip relative to the initial position of the ankle marker was $27.3cm$, and its maximum vertical displacement relative to the initial position of the hip marker was $4.4cm$. The dominant frequency of the vertical oscillation (determined via a spectral analysis of the hip marker) was $f_{Hip} = 0.93Hz$, whereas $f_{Hip} = 0.95Hz$ for the same parameter configuration but with $\omega_p = 0$. Thus, the sensor feedback also affected the frequency of the oscillation. After a short initial transient, the robot settled into a stable oscillatory movement but did not bounce.

In this scenario, the model is more complicated and has to take into account the change of phase and timing due to the sensory feedback. This is realized by introducing a new variable ϕ such that $m\,\ddot{x}(t) + B_{eff}\,\dot{x}(t) + K_{eff}\,x(t) = F(t, \phi)$.

5 Discussion and Conclusion

The question of how sensory feedback interacts with the central pattern generator is still open [14]. As a demonstration that sensory feedback is not necessary for the generation and coordination of rhythmic activity, experiments in completely isolated spinal cords and in deafferented animals (i.e., without sensory feedback) have shown that the patterns generated by these type of structures are very similar to those recorded in intact animals [23]. What emerged from our study is that a suitable choice of the intersegmental coupling constant, as well as of the gain of the sensory feedback reduces movement variability, increases bouncing amplitude, and leads to stability. We attribute this result to the entrainment of neural and body-environment interaction dynamics. In other words, the neural system of our model is designed to produce a basic pattern of muscle activation established not only by the connections between the neural oscillators, but also by the input of sensory signals representing body movements and the coupling with the environment. Through a recurrent interaction in the sensorimotor loop, the variability and instability of the movements are stabilized into a limit cycle. In the sense that such a coupling produces an effect greater than the sum of the individual components, it is a synergistic coupling. A similar finding, in the case of biped walking, was reported by Taga [14].

Goldfield et al. [19] suggested that the developmental transformation of spontaneous motor activity into task-specific movements consists of two phases, which they called assembly and tuning phase. While assembly refers to the self-organization of relationships between the components of the system, tuning is concerned with the adaptation of the system parameters to particular conditions. In this paper, we have primarily focused on the tuning phase by making the premise that the assembly phase results in a positive intersegmental coupling between hip, knee and ankle. It is interesting to consider the issue of the mechanisms underlying the assembly phase. Although bouncing is intrinsically a rhythmic activity for which central pattern generators represent suitable neural structures, there is no evidence that newborn infants move their limbs in a manner consistent with the output of central pattern generators, and indeed, sporadic kicking movements are more plausible candidates. Given that neural oscillators are usually modeled as a set of mutually inhibitory neurons, the assembly phase could be a process during which the topology of a vanilla-type cell assembly changes, driven by feedback from the environment, and by a value system (based on the amplitude of the oscillations, for instance).

With respect to the tuning phase, there is still much to do. In some sense, tuning refers to the non-stationary regime which occurs before stabilization of movement patterns. In other words, it is the by-product of the entrainment between neural control structure and environment – when sensory feedback turns the system into an autonomous limit-cycle system. At a lower level of control, tuning could also be implemented as changes in gain or time-constants of the neural oscillators. An autonomous implementation of such parameter tuning could be realized via a mechanism of Boltzmann exploration driven by a value system (Fig. 1 right). The authors have successfully used this combination in a pendulating humanoid robot [18].

Yet, all this may not be sufficient to hypothesize a valid model of child motor development as there is evidence that kicking behaviors display spatio-temporal patterns. In particular, Taga et al. [24] recently discussed the chaotic dynamics of spontaneous movements in human infants. Thus, formulating the development of those skills in a dynamical systems framework would be highly desirable so that an appropriate set of adaptive mechanisms could be implemented and tested against human data.

References

1. E.C. Goldfield. *Emergent Forms: Origins and Early Development of Human Action and Perception*. Oxford University Press: New York, 1995.
2. O. Sporns and G.M. Edelman. Solving bernstein's problem: a proposal for the development of coordinated movement by selection. *Child Development*, 64:960–981, 1993.
3. E. Thelen and L. Smith. *A Dynamic Systems Approach to the Development of Cognition and Action*. MIT Press: Cambridge, MA. A Bradford Book, 1994.
4. H. Forssberg. Neural control of human motor development. *Current Opinion in Neurobiology*, 9:676–682, 1999.
5. J.P. Piek. Is a quantitative approach useful in the comparison of spontaneous movements in fullterm and preterm infants? *Human Movement Science*, 20:717–736, 2001.
6. E. Thelen and D. Fischer. The organization of spontaneous leg movements in newborn infants. *Journal of Motor Behavior*, 15:353–377, 1983.
7. R.M. Angulo-Kinzler. Exploration and selection of intralimb coordination patterns in 3-month-old infants. *J. of Motor Behavior*, 33(4):363–376, 2001.
8. K. Schneider, R.F. Zernicke, B. Ulrich, J. Jensen, and E. Thelen. Understanding movement control in infants through the analysis of limb intersegmental dynamics. *Journal of Motor Behavior*, 22:493–520, 1990.
9. C. Von Hofsten. Prospective control: A basic aspect of action development. *Human Development*, (36):253–270, 1991.
10. J.P. Piek and R. Carman. Developmental profiles of spontaneous movements in infants. *Early Human Development*, 39:109–126, 1994.
11. H.F.R. Prechtl. The importance of fetal movements. In K.J.Connolly and H.Forssberg, editors, *Neurophysiology and Neuropsychology of Motor Development*, pages 42–53. Mac Keith Press, 1997.
12. S. Grillner. Neurobiological bases on rhythmic motor acts in vertebrates. *Science*, 228:143–149, 1985.
13. S.J.A. Kelso. *Dynamic Patterns*. MIT Press: Cambridge, MA. A Bradford Book, 1995.
14. G. Taga. A model of the neuro-musculo-skeletal system for human locomotion: Emergence of basic gait. *Biological Cybernetics*, 73:97–111, 1995.
15. N. Bernstein. *The Co-ordination and Regulation of Movements*. Pergamon: London, UK, 1967.
16. M.T. Turvey and P. Fitzpatrick. Commentary: Development of perception-action systems and general principles of pattern formation. *Child Development*, (64):1175–1190, 1993.
17. J. Vaal, A.J. van Soest1, B. Hopkins, L.T.L. Sie, and M.S. van der Knaap. Development of spontaneous leg movements in infants with and without periventricular leukomalacia. *Experimental Brain Research*, 135:94–105, 2001.

18. M. Lungarella and L. Berthouze. On the interplay between morphological, neural and environmental dynamics: A robotic case-study. *Adaptive Behavior*, 10(3/4):223–241, 2002.
19. E.C. Goldfield, B.A. Kay, and W.H. Warren. Infant bouncing: the assembly and tuning of an action system. *Child Development*, 64:1128–1142, 1993.
20. J. Piaget. *The Origins of Intelligence*. Routledge: New York, USA, 1953.
21. K. Matsuoka. Sustained oscillations generated by mutually inhibiting neurons with adaptation. *Biological Cybernetics*, 52:367–376, 1985.
22. M.M. Williamson. Neural control of rhythmic arm movements. *Neural Networks*, 11(7/8):1379–1394, 1998.
23. A. Ijspeert. Vertebrate locomotion. In M.Arbib, editor, *The Handbook of Brain Theory and Neural Networks*. MIT Press: Cambridge, MA. Bradford Book, 2002.
24. G. Taga, R. Takaya, and Y. Konishi. Analysis of general movements of infants towards understanding of developmental principle for motor control. In *Proc. of 1999 IEEE Int. Conf. on Systems, Man, and Cybernetics*, pages 678–683, 1999.

The Need to Adapt and Its Implications for Embodiment

Lukas Lichtensteiger

Artificial Intelligence Lab, Computer Science Department, University of Zurich,
Andreasstrasse 15, CH-8050 Zurich, Switzerland,
llicht@ifi.unizh.ch,
http://www.ifi.unizh.ch/ailab/people/llicht/

Abstract. We present the hypothesis that an important factor for the
choice of a particular embodiment for a natural or artificial agent is the
effect of the embodiment on the agent's ability to adapt to changes in
the environment. To support this hypothesis, we discuss recent empirical results where sensor morphology was found to significantly affect the
time needed for learning a given task. Also, we discuss other recent experiments where a unique optimal sensor morphology could be evolved
simply by requiring that the agent had to learn its task as quickly as
possible. Both these findings are explained by the recently discovered
"Principle of Unique Local Gain Factors for Optimal Adaptation" which
provides a first step towards a general mathematical setting for understanding the interdependence between an agent's embodiment and its
learning performance.

1 Introduction

Although the importance of embodiment for intelligent behavior in both animals
and robots has been realized already more than a decade ago [1,2] the interdependence between an agent's body (morphology, materials, etc), its brain and its
task environment is still not understood very well. For example, animals show
an abundance of different sensor morphologies and it is believed that these differences relate to differences in the respective task environments of the animals
(e.g., facet density distributions in arthropod compound eyes can vary strongly
depending on species, sex and habitat [3]). However, so far very little is known
about this correspondence. A number of qualitative guidelines such as design
principles [4,5,6] have been proposed but to-date a more quantitative theory is
still lacking. In this paper we discuss implications on embodiment caused by an
agent's need to optimally adapt to its task environment. We review recent experimental results relating learning speed to particular sensor morphologies and
discuss them in the light of a recently discovered general mathematical framework describing the interdependence between an agent's body morphology and
its learning performance.

F. Iida et al. (Eds.): Embodied Artificial Intelligence, LNAI 3139, pp. 98–106, 2004.
© Springer-Verlag Berlin Heidelberg 2004

2 Modulating Agent-Environment Interaction by Adapting the Embodiment

Consider the general setting of a (natural or artificial) agent interacting with a real world environment. It is clear that this always requires the agent to be physically embodied: the agent and its environment can only influence each other through physical interaction. All interaction happens as a consequence of physical laws and is continuously taking place (it cannot be "switched off"). Even a completely "dumb" body interacts with the environment in this way and it can already perform certain (very simple) "tasks" like heating itself up to the environment's temperature, rolling down an incline, etc. We can view the physical interaction between an agent's body and the environment as the fastest (innermost) control loop of the agent's behavior (see figure 1, middle left, where the thickness of the arrows symbolizes the "bandwidth" of the interaction). However, even this simplest type of agent can be adapted to its task and environment by optimizing its body (shape, materials, etc). This is symbolized in the lower left quadrant of figure 1 by the "evolutionary fitness" feedback loop acting on the agent's body ("phylogenetic adaptation").

The next more complex type of agent still only consists of a body without any internal controller whatsoever, but this time the body itself is (continuously) adaptive, i.e., its shape (or other properties) can be dynamically modified by the interaction with the environment. This already allows for more complex tasks: For example, a passive dynamic walker can "walk" in a "human-like" way down an incline simply by exploiting the complex dynamics of the physical interaction between its moving limbs and the environment [7]. On a longer time-scale also for this type of agent the body can be optimized for a given task environment, for example by evolving optimal limb mass distributions (same feedback loop in the lower left quadrant of figure 1).

In order to cope with more complex tasks and environments agents with internal controllers ("brains") are needed. Their controller in a way "modulates" the basic physical agent-environment interaction by dynamically modifying the agent's body (its shape or other properties) in a specific way, depending on information obtained from the environment through sensory inputs (outer control loop in the top right quadrant of figure 1, from "sensory inputs" to "effector outputs"). By being able to adapt its own body morphology to a given task environment this type of agent can be seen as a kind of Morpho-functional machine [8]. Sometimes different time-scales can be discriminated for this adaptation: Short-term body control like muscle activity, and medium-term modifications like body development, self-assembly, self-repair, etc. Depending on the speed required for the respective control loop, the controllers could also be implemented using different types of substrates, e.g. electrical signals (muscle activity) or biochemical metabolism (growth control, self-repair). Since typically the control loop of the "brain-body interaction" is much slower than the physical interaction between body and environment it is also useful for this type of agent to exploit as much as possible the specific details of the physical interaction with the environment by optimizing its body morphology (e.g., by evolving limb mass distributions) and only using the controller for modulating the intrinsic dynamics of the agent-environment interaction (see e.g. [9] for an interesting application of this idea). Again, this long-term adaptation is done through the "phylogenetic adaptation" feedback loop in the lower left quadrant of figure 1).

However, for this type of agents adapting the body not only has consequences for the body-environment interaction, but it influences the body-brain interaction as well: Since the controller cannot interact directly with the physical world and can only obtain its information about the environment through the agent's body (by means of sensors) this means that the actual state of the body (shape, dynamical configuration, position in the environment, etc) influences directly what the sensors deliver to the controller. In order to successfully perform its task in the environment therefore the agent not only needs to control its body to modulate the physical body-environment interaction in the desired way, but it also has to do it in a way as to generate optimal sensory inflow for the controller in terms of information content needed. On short time-scales this amounts to controlling the body's actuator activity in such a way that it can collect the best information possible, e.g., by using sensory-motor coordination [5]. On medium and long time-scales the same goal can be achieved by creating a body shape (e.g., a specific sensor morphology) that by its intrinsic physical properties predominantly extracts the most relevant information from the environment. This long term adaptation can both be done ontogenetically (through development) or phylogenetically (through evolution). In a series of earlier experiments on an adaptive artificial compound eye we were able to show that it is possible to evolve optimal sensor morphologies for specific tasks [10,11,12,13].

3   Optimal Embodiment for Adaptive Controllers

Throughout the previous section we have been focusing on the adaptation of the agent's body, either through "external" optimization (like evolution) or by controlling its adaptation with an internal "brain". However, for many task environments it is advantageous to be able to adapt the controller itself as well. While for long-term adaptation this can again be done by an "external" process like evolution (feedback loop in the lower right quadrant of figure 1) for short and medium term adaptation the controller has to be able to adapt itself, i.e., it has to be able to learn (inner feedback loop in the top right quadrant in figure 1, where we have abstracted all types of learning processes as adaptation of the controller induced by some "value signal": In the same way that an effector output can trigger an adaptation of the agent's body an active "value signal" can induce a modification inside the agent's brain). Depending on the controller's substrate this adaptation could manifest itself for example as neural plasticity (modification of neural connectivity and synaptic weights) or also biochemical plasticity (e.g., immune system "learning"). Regardless of the actual implementation, it is important to realize that learning is controlled by the controller itself and not directly by the environment and therefore ultimately also depends on the information that is provided by the agent's sensors. Consequently, in addition to all

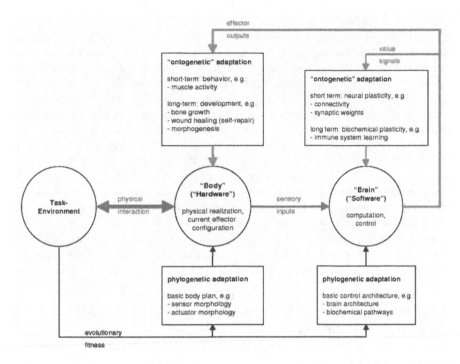

Fig.1. Schematic view of general agent-environment interaction showing the interdependence of a (task) environment, an agent's body and its internal controller ("brain"). Only the agent's body can directly interact with the environment. The processes described in the rectangular boxes can adapt the "body" and/or the "brain" of the agent on different time-scales and in this way modulate the interaction between task environment and body (physical interaction) and between body and brain (sensory inputs, effector outputs). The thickness of the arrows denotes the "bandwidth" of the interaction: thick arrows signify frequent interaction and fast control loops. Details see text.

the other interdependencies between body and brain described in the previous section, for the case of learning the agent's body also affects the adaptation of the brain. In general, it seems natural to ask if the shape of an agent's body - especially the sensor morphology - can influence the learning performance of its controller. Indeed, it has been found in simulation experiments on an adaptive artificial compound eye that the time needed to learn a given task can vary by orders of magnitude depending on the particular sensor morphology employed by the agent [14].

These results suggest to optimize the shape of the body (and in particular the sensor morphology) for extracting sensory information from the environment in a way that allows the controller to learn as quickly as possible. In principle this adaptation of the body for learning could also be done by short-term or medium-term processes like muscle activity or body development. However, since learning itself is necessarily a medium-term process it is much better to optimize it by a process on an even longer timescale (in an "outer control loop"), e.g., through phylogenetic adaptation by an evolutionary process. In a series of recent experiments we have been evolving body shape (sensor morphology) of simulated agents with the goal of optimizing learning performance (adaptivity) of their controller [15]. Sensory input from an adaptive artificial compound eye was fed into a standard two-layer feed-forward network that had to learn a given task using backpropagation. The time the neural network required for learning was then directly used as fitness value for evolving the morphology of the compound eye. It turned out that for a given task environment there was always a unique sensor morphology which allowed the controller to learn fastest. In other words: Only requiring from the agent that it had to learn the task as quickly as possible (without any other constraints on its fitness) was enough to always evolve the same unique, distinct sensor morphology specific for the given task environment. Interestingly, this special morphology that turned out to be optimal for learning is qualitatively very similar to the morphologies found in some biological compound eyes (e.g., in flies and bees) [3].

4   The Principle of Unique Local Gain Factors for Optimal Adaptation

Along a theoretical line of research we were able to show recently that the results described in the previous paragraph can actually be understood as a consequence of a general principle that we call the principle of Unique Local Gain factors for optimal Adaptation (ULGA). By *local gain factors* we simply mean dedicated constant parameters that are multiplied to the adaptive weights in a controller that has to learn a given problem. The ULGA principle then states that for any given task environment where the learning problem can be described by a non-degenerated parabolic cost function and for standard gradient descent learning there always exists a *unique* optimal set of non-negative local gain factors such that the system can learn the task in minimal time. Note that we require that the minimum of the given cost function is non-degenerate (i.e., it consists only of a single point in weight space), and that its location remains constant over time (otherwise the optimal local gain factors would change over time as well). So far the ULGA principle has only been proved for standard gradient descent learning [16]; however, we believe that it may also hold for other (maybe more biologically plausible) learning schemes.

A consequence of the ULGA principle for the evolution of sensor morphologies is the following: Assume that the effect of using a specific sensor morphology is to scale each individual sensory input channel by a specific factor. Then, for the case of controllers with linearly weighted inputs (which include most of the commonly used neural network controllers, e.g., all multi-layer perceptrons), this is the same as multiplying each input weight with the corresponding factor instead. If this is done exactly in the right way, i.e., by effectively multiplying each input channel by its optimal local gain factor given by the ULGA principle, then the controller will be able to learn the task in the shortest possible time and therefore the corresponding agent will have a very high performance in its task environment and consequently a very high fitness.

Of course, choosing a particular morphology usually also has other effects besides influencing learning time which can also affect the agent's overall fitness rating (for example, the agent could become too bulky for a certain sensor morphology, the development of a sophisticated morphology becomes too expensive, etc). However, we believe that the ability to learn a task quickly and to rapidly adapt the controller to small changes in the task environment is a very important factor for an agent behaving in the real world, and consequently, we think it is very important to take the effects on learning speed into account when designing or evolving sensor morphologies.

In the simulation experiments described in the previous section simply using learning speed as fitness value already provided enough constraints to evolve a unique optimal sensor morphology. The reason for this direct correspondence is the fact that for sensing optical flow using a compound eye the local gain factors are directly proportional to the facet density which in turn determines the morphology of the eye.

We would like to add the following technical remark: Although the existence of a unique set of optimal local gain factors is theoretically guaranteed by the ULGA principle finding the actual optimal values for a given problem is usually highly non-trivial and can in general only be done numerically using some optimization algorithm like artificial evolution. However, since the existence of a unique global minimum (without any additional local minima!) is theoretically guaranteed finding the actual optimal values should always be possible.

5   Further Implications of the ULGA Principle

Actually, the ULGA principle is not restricted to weights associated with sensory inputs, but it can be applied to any adaptive weight in the controller. Therefore, the principle is also applicable for example to purely internal connections in a neural network. In this situation the existence of a unique ensemble of optimal local gain factors can be seen as another example of optimizing morphology for learning speed, except that in this case not the sensor morphology is adapted but instead part of the "morphology" of the brain - the "gains" of the neural connections - is optimized (symbolized by the "phylogenetic adaptation" feedback loop in the lower right quadrant of figure 1).

Taking this argument even further, we can consider the whole ensemble "environment plus agent" as an adaptive system governed by a certain cost function. In this way we are basically "including the environment in the learning loop". Then we can apply the ULGA principle also on the effector side: If there are a number of controller outputs where each output is multiplied by a dedicated constant parameter (a local gain factor) and an adaptive weight (such that the system only depends on the product of these three factors), then there exists a unique set of optimal values for these local gain factors such that the system as a whole (including agent and environment) can learn the corresponding weights in a minimum of time.

The significance of this is as follows: Assume, for example, that a controller is driving some actuators in a way that the force delivered by each actuator is proportional to the value of the corresponding controller output multiplied by a dedicated weight for each output (this can easily be achieved for example with electric motors). Assume further that the task of the agent only depends on the kinematics of the actuators but not on their dynamics, i.e., that not the forces itself are relevant for the system but only the acceleration of individual body parts, i.e., the actuating force divided by actuated mass. (This can be true for example for a robot arm that just has to follow a certain trajectory with prescribed speed). Under these conditions the ULGA principle can be applied where the local gain factors are now simply the reciprocals of the masses of the actuated body parts. The principle then asserts that for each actuator there exists a unique optimal mass for the actuated body part (e.g., for the corresponding limb) such that the system can learn its task in a minimum of time. In this way the ULGA principle allows (at least for some cases) to relate not only the morphology of an agent's sensors but also the morphology of its actuators and controller directly to its need to adapt as quickly as possible.

6   Conclusions

In this paper we have been studying the interdependence between an agent's embodiment, its controller, and its task environment, with a special focus on adaptation. For different levels of complexity of agents we have been discussing different mechanisms (acting on different time-scales) for how an agent can adapt to optimally perform its task in a given (partially unknown and possibly changing) environment. Throughout the paper we have been presenting examples that suggest an important role of the agent's embodiment for adaptation on all time-scales and levels of complexity.

Furthermore, we believe that a large part of the highly specific embodiments found in animals can actually be explained by the agent's continuous need to adapt: An agent's embodiment can be adapted in order to directly optimize the dynamics of the physical body-environment interaction, it can be adapted to optimize the information content delivered by the sensors to the agent's brain, and it can be optimized to allow the brain to learn a task as quickly as possible.

We believe that especially the third strategy is very important: Since for most sufficiently complex tasks the controller needs to be able to constantly cope with small changes in the task and/or in the environment it is possibly even more important to have a body that is optimized for fast learning instead of simply being an optimal solution for one particular environmental situation: An animal that can adapt its behavior faster to small changes in the environment will have a better lifetime performance in its ecological niche.

This hypothesis is supported both by empirical results from simulation experiments on an artificial compound eye as well as by a novel theoretical framework:

The principle of unique local gain factors for optimal adaptation (ULGA). This principle can predict the experimental results very well (evolution of a unique morphology that is optimal for learning) and it can also be applied to a large number of other problems in the study of the interdependence between an agent's morphology (sensors, actuators and controller) and its adaptivity.

References

1. Brooks, R.A.: Intelligence without reason. In Mylopoulos, J., Reiter, R., eds.: Proceedings of the 12th International Conference on Artificial Intelligence (IJCAI-91), San Mateo, CA, Morgan Kaufmann (1991) 569–595

2. Brooks, R.A.: Intelligence without representation. Artificial Intelligence 47 (1991) 139–159

3. Land, M.: Variations in the structure and design of compound eyes. In Stavenga, D., Hardie, R., eds.: Facets of Vision. Springer, Berlin, Germany (1989) 90–111

4. Pfeifer, R.: Building fungus eaters: Design principles of autonomous agents. In Maes, P., Mataric, M.J., Meyer, J.A., Pollack, J., Wilson, S.W., eds.: From animals to animats 4: Proceedings of the fourth international conference on simulation of adaptive behavior, Cambridge, MA, MIT Press (1996) 3–12

5. Pfeifer, R., Scheier, C.: Understanding Intelligence. MIT Press, Cambridge, MA (1999)

6. Pfeifer, R.: On the role of morphology and materials in adaptive behavior. In Berthoz, A., Floreano, D., Meyer, J.A., Roitblat, H., Wilson, S., eds.: From Animals to Animats 6: Proceedings of the Sixth International Conference on the Simulation of Adaptive Behaviour, Cambridge, MA, MIT Press (2000) 23–32

7. McGeer, T.: Passive dynamic walking. Int. J. Robotics Research 9 (2) (1990) 62–82

8. Hara, F., Pfeifer, R., eds.: Morpho-functional machines: the new species; designing embodied intelligence. Springer, Tokyo, Japan (2003)

9. Wisse, M., van Frankenhuyzen, J.: Design and construction of mike: a 2d autonomous biped based on passive dynamic walking. In: Proceedings of the 2nd International Symposium on Adaptive Motion of Animals and Machines AMAM2003, Kyoto, Japan (2003)

10. Lichtensteiger, L., Eggenberger, P.: Evolving the morphology of a compound eye on a robot. In: Proceedings of the 3rd European Workshop on Advanced Mobile Robots Eurobot'99, Piscataway, NJ, IEEE Press (1999) 127–134

11. Lichtensteiger, L.: Towards optimal sensor morphology for specific tasks: Evolution of an artificial compound eye for estimating time to contact. In McKee, G.T., Schenker, P.S., eds.: Sensor Fusion and Decentralized Control in Robotic Systems III, Proceedings of SPIE Vol. 4196, Bellingham, WA, Society of Photo-Optical Instrumentation Engineers (SPIE) (2000) 138–146

12. Lichtensteiger, L., Salomon, R.: The evolution of an artificial compound eye by using adaptive hardware. In: Proceedings of the 2000 Congress on Evolutionary Computation (CEC2000), San Diego, CA, IEEE Computer Society Press (2000) 1144–1151

13. Lichtensteiger, L.: Evolving task specific optimal morphologies for an artificial insect eye. In Hara, F., Pfeifer, R., eds.: Morpho-functional machines: the new species; designing embodied intelligence. Springer, Tokyo, Japan (2003) 41–57

14. Lichtensteiger, L., Pfeifer, R.: An optimal sensor morphology improves adaptability of neural network controllers. In Dorronsoro, J.R., ed.: Proceedings of the International Conference on Artificial Neural Networks (ICANN 2002), Lecture Notes in Computer Science LNCS 2415, Heidelberg, Germany, Springer (2002) 850–855
15. Lichtensteiger, L.: How the need to learn fast can shape bodies. submitted (2004)
16. Lichtensteiger, L.: Existence and uniqueness of optimal local gain factors for gradient descent learning. submitted (2004)

How Should Control and Body Systems Be Coupled?
A Robotic Case Study

Akio Ishiguro[1] and Toshihiro Kawakatsu[2]

[1] Dept. of Computational Science and Engineering, Nagoya University
Furo-cho, Chikusa-ku, Nagoya 464-8603, Japan
ishiguro@cse.nagoya-u.ac.jp
[2] Dept. of Physics, Tohoku University
Aoba, Aramaki, Aoba-ku, Sendai 980–8578, Japan
kawakatu@cmpt.phys.tohoku.ac.jp

Abstract. This study is intended to deal with the interdependency between control and body systems, and to discuss the "relationship as it should be" between these two systems. To this end, a decentralized control of a multi-legged robot is employed as a practical example. The result derived indicates that the convergence of decentralized gait control can be significantly ameliorated by modifying its interaction dynamics between the control system and its body system to be implemented. We also discuss a property expected to emerge under the "well-balanced coupling" particularly from the viewpoint of learning, by borrowing the idea from the "protein folding problem".

1 Introduction

In robotics, traditionally, a so-called *hardware first, software last* based design approach has been employed, which seems to be still dominant. Recently, however, it has been widely accepted that the emergence of intelligence is strongly influenced by not only control systems but also their embodiments, that is the physical properties of a robots' body[1]. In other words, the intelligence emerges through the interaction dynamics among the control systems (*i.e.* brain-nervous systems), the embodiments (*i.e.* musculo-skeletal systems), and their environment (*i.e.* ecological niche). In sum, control dynamics and its body (*i.e.* mechanical) dynamics cannot be designed *separately* due to their tight interdependency. This leads to the following suggestions: (1) there should be a "well-balanced coupling" between control and body dynamics, and (2) one can expect that quite interesting phenomena will emerge under such well-balanced coupling.

On the other hand, since the seminal works of Sims[2][3], so far various methods have been intensively investigated in the field of Evolutionary Robotics by exploiting concepts such as *co-evolution*, in the hope that they allow us to simultaneously design control and body systems[4][5]. Most of them, however, have mainly focused on automatically creating both control and body systems, and

F. Iida et al. (Eds.): Embodied Artificial Intelligence, LNAI 3139, pp. 107–118, 2004.

thus have paid less attention to gain an understanding of well-balanced coupling between the two systems. To our knowledge, still very few studies have explicitly investigated this point (*i.e.* appropriate coupling)[1].

In light of these facts, this study is intended to deal with the interaction dynamics between control and body systems, and to analytically and synthetically discuss a well-balanced relationship between the dynamics of these two systems. More specifically, the aim of this study is to clearly answer the following questions:

- how should these two dynamics be coupled?
- what sort of phenomena will emerge under the well-balanced coupling?

Since there are virtually no studies in existence which clearly discuss what a well-balanced coupling is, it is of great worth to accumulate various case studies at present. Based on this consideration, a decentralized control of a multi-legged robot consisting of several identical body segments is employed as a practical example. The derived result indicates that the convergence of decentralized gait control can be significantly ameliorated by modifying both control dynamics (*e.g.* information pathways among the body segments) and body dynamics (*e.g.* stiffness of the spine). We also discuss an idea regarding an emergent property expected to be observed under the "well-balanced coupling" from the viewpoint of learning, inspired by the *protein folding problem*.

2   Lessons from Biological Findings

Before explaining our approach, it is worthwhile to look at some biological findings. Beautiful instantiations of well-balanced couplings between nervous and body systems can be found particularly in insects. In what follows, let us briefly illustrate some of these instantiations.

Compound eyes of some insects such as houseflies show special *facet* (*i.e.* vision segment) distributions; the facets are densely spaced toward the front whilst widely on the side. Franceschini *et al.*[6] demonstrated with a real physical robot[2] that this non-uniform layout significantly contributes to detect easily and precisely the movement of an object without increasing the complexity of neural circuitry.

Another elegant instantiation can be observed in insects' wing design[9][10]. As shown in Fig.1(a), very roughly speaking, insects' wings are composed of hard and soft materials. It should be noted that the hard material is distributed asymmetrically along the moving direction. Due to this material configuration, insects' wings show complicated behavior during each stroke cycle, *i.e.*, twist

---

[1] Pfeifer introduced several useful design principles for constructing autonomous agents[1]. Among them *the principle of ecological balance* does closely relate to this point, which states that control systems, body systems and their material to be implemented should be balanced. However, there still remains much to be understood about how these systems should be coupled.

[2] Another interesting robot can be found in [7][8].

and oscillation. This allows them to create useful aerodynamic force, and thus they can realize agile flying. If they had symmetrical material configuration as shown in Fig.1(b), the complexity of neural circuitry responsible for flapping control would be significantly increased.

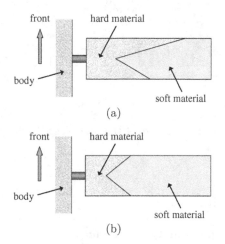

(a)

(b)

F ig.1. Material configuration in insects' wings.

3   The Model

In order to investigate *well-balanced coupling as it should be* between control and body systems, a decentralized control of a multi-legged robot is taken as a case study. Fig. 2 schematically illustrates the structure of the multi-legged robot. As shown in the figure, this robot consists of several identical body segments, each of which has two legs, *i.e.*, right and left legs. For simplicity, the right and left legs of each body segment are allowed to move in phase. In addition, the duty factor[3], trajectory and period of all the leg movement are assumed to be identical, which have to be prespecified before actually moving the robot. For convenience, hereafter the phase of the leg movement of the ith body segment is denoted as θ_i $(i = 1, 2, \cdots, n)$. Thus, the control parameters in this model end up to be the set of the phases $\theta_1, \theta_2, \cdots, \theta_n$.

The task of this robot is to realize rapid gait convergence which leads to a gait with minimum energy consumption rate from arbitrary initial relative-phase conditions. Note that each body segment controls the phase of its own legs in a decentralized manner, which will be explained in more detail in the following section.

---

[3] The duty factor is defined by the fraction of the cycle for which a given leg is in contact with the ground.

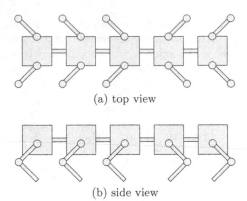

(a) top view

(b) side view

F ig.2. A Schematic of the structure of the multi-legged robot employed in this case study.

4   Proposed Method

4.1   Analysis of the Gait Convergence

Based on the above arrangements, this section analytically discusses how the control and body dynamics influence the gait convergence. Let P be the total energy consumption rate of this robot, then P can be expressed as a function of the phases as:

$$P = P(\boldsymbol{\theta}), \tag{1}$$

$$\boldsymbol{\theta} = (\theta_1, \theta_2, \cdots, \theta_n)^T. \tag{2}$$

Here, for purposes of simplified analysis, a simple learning scheme based on a *gradient method* is employed. It is denoted by

$$\Delta\boldsymbol{\theta}^{(k)} = -\boldsymbol{\eta}\frac{\partial P(\boldsymbol{\theta})}{\partial\boldsymbol{\theta}}\bigg|_{\boldsymbol{\theta}^{(k)}}, \tag{3}$$

where $\Delta\boldsymbol{\theta}^{(k)}$ is the phase modification at time step k, $\boldsymbol{\eta}$ is an $n \times n$ matrix which specifies how a body segment will exploit the information about phase modification done in other body segments in its determination of the phase modification. Based on Equation (3), the set of the phases at time step k is expressed in the following form:

$$\boldsymbol{\theta}^{(k+1)} = \boldsymbol{\theta}^{(k)} + \Delta\boldsymbol{\theta}^{(k)} = \boldsymbol{\theta}^{(k)} - \boldsymbol{\eta}\frac{\partial P(\boldsymbol{\theta})}{\partial\boldsymbol{\theta}}\bigg|_{\boldsymbol{\theta}^{(k)}}. \tag{4}$$

Let $\boldsymbol{\theta}^{(\infty)}$ be a set of converged phases. By performing the Taylor series expansion around $\boldsymbol{\theta}^{(\infty)}$, the partial differentiation of $P(\boldsymbol{\theta})$ with respect to $\boldsymbol{\theta}$ is:

$$\frac{\partial P(\boldsymbol{\theta})}{\partial \boldsymbol{\theta}} \simeq C(\boldsymbol{\theta} - \boldsymbol{\theta}^{(\infty)}), \tag{5}$$

$$C = \frac{\partial^2 P(\boldsymbol{\theta})}{\partial \boldsymbol{\theta} \partial \boldsymbol{\theta}} \bigg|_{\boldsymbol{\theta}^{(\infty)}}, \tag{6}$$

where C is an $n \times n$ *Hesse* matrix. Hence, the substitution of Equation (5) into Equation (4) yields:

$$\boldsymbol{\theta}^{(k+1)} = \boldsymbol{\theta}^{(k)} - \eta C(\boldsymbol{\theta}^{(k)} - \boldsymbol{\theta}^{(\infty)}). \tag{7}$$

For the sake of the following discussion, a *residual vector* $e^{(k)}$ is introduced, which is equivalent to $\boldsymbol{\theta}^{(k)} - \boldsymbol{\theta}^{(\infty)}$. Then, Equation (7) can be rewriten as:

$$e^{(k+1)} = A e^{(k)}, \tag{8}$$

$$A = I - \eta C, \tag{9}$$

where I is an $n \times n$ unit matrix.

4.2   Physical Meaning of η and C

A in Equation (8) is a matrix which characterizes the property of gait convergence. This will automatically lead to the following fact: for rapid convergence, the spectral radius of A should be less than 1.0.

What should be stressed here is the fact that as shown in Equation (9) the matrix A is composed of the two matrices: η and C. As has been already explained, the matrix η specifies the information pathways (or neuronal/axonal interconnectivity) among the body segments, which will be used to calculate the phase modification. This implies that the matrix η does relate to the design of the control dynamics.

On the other hand, it is obvious from the definition (see Equation (6)) that C is a matrix whose nondiagonal elements will be salient as the *long-distance interaction* between the body segments through the physical connections (*i.e.* the spine of the robot) becomes significant. This strongly suggests that the property of this matrix is remarkably influenced by the design of the body dynamics.

4.3   An Effective Design of the Body Dynamics

The design of the control dynamics can be easily done by tuning the elements of the matrix η. In contrast, much attention has to be paid to the design of the body dynamics. This is simply because one cannot *directly* access the elements of the matrix C nor tune them unlike in the case of the matrix η.

Before introducing our proposed method, let us briefly conduct a simple yet instructive thought experiment. Imagine a multi-legged robot in which its body segments are tightly connected via a *rigid* spine. In such a case, the phase modification of a certain leg will significantly affect the energy consumption rate of distant legs due to the effect of the *long-distance* interaction.

As has been demonstrated in this thought experiment, the *stiffness* of the spine generates considerable influence on the property of C, particularty the values of its nondiagonal elements. Therefore, it seems to be reasonable to connect the body segments via a *springy* joint. This idea is schematically illustrated in Fig. 3, where we only show the two body segments for clarity.

Based on the above consideration, a well-balanced design is investigated by tuning the parameters in the matrix η and the ones of the springs inserted between the body segments, which will lead to a reasonable gait convergence.

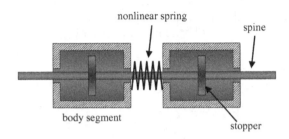

F ig.3. An effective structure for adjusting the body dynamics. Each body segment can move passively along the spine owing to the springs inserted between the body segments.

5   Preliminary Simulation Results

In order to efficiently investigate the well-balanced coupling, a simulator has been developed. The following simulations have been conducted with the use of a physics-based, three-dimensional simulation environment[13]. A view of the simulator is shown in Fig. 4. This system simulates both the internal and external forces acting on the agent and objects in its environment, as well as various other physical properties such as contact between the agent and the ground, and torque applied by the motors to the joints.

Before carrying out a thorough search of the design parameters, a preliminary experiment has been done to understand the influence of the two dynamics on the gait convergence. In this experiment, the property of the spring inserted between the body segments is assumed to be expressed as:

$$f = -k(\Delta x)^{\alpha}, \tag{10}$$

where f is the resultant force, k is a spring constant, α controls the degree of the nonlinearity of the spring, and Δx is a displacement.

Shown in Fig. 5 are data resulting from this experiment; the vertical axis denotes the total energy consumption rate whilst the horizontal axis depicts the number of modification of the phases conducted, *i.e.*, the number of learning

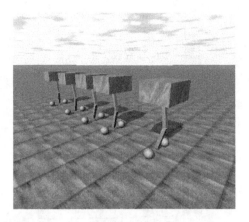

F ig . 4 . A view of the developed simulator.

steps. Note that each graph was obtained by averaging over 10 different initial relative-phase conditions. As a rudimentary stage of the investigation, only α was varied under the following conditions: the number of the body segments was 5; duty factor 0.65; k 1.0; and η set to

$$
\begin{pmatrix}
0.005 & 0.0 & 0.0 & 0.0 & 0.0 \\
0.0 & 0.005 & 0.0 & 0.0 & 0.0 \\
0.0 & 0.0 & 0.005 & 0.0 & 0.0 \\
0.0 & 0.0 & 0.0 & 0.005 & 0.0 \\
0.0 & 0.0 & 0.0 & 0.0 & 0.005
\end{pmatrix}.
$$

As shown in Fig. 5, the gait convergence is highly influenced by the parameter α. This is due to the fact that the long-distance interaction among the body segments depends on α, which leads to varying the property of the matrix C. In spite of the simplicity, these results strongly support the conclusion that the body dynamics imposes significant influence on the gait convergence.

6   Discussions

The simulation results presented in the preceding section have demonstrated the important effect of the coupling between the control dynamics and its body dynamics on the convergence of the gait control. Based on this example, let us examine the meaning of the "well-balanced design" of the control dynamics and the body dynamics.

6.1   From the Viewpoint of Learning: Optimization of the Evaluation Function for the Learning Exploiting the Body Dynamics

We identify the problem of the convergence of the gait control with the "optimization of the *evaluation function* for the control/learning process of the

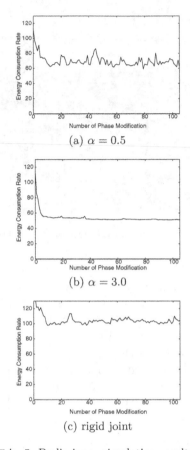

(a) $\alpha = 0.5$

(b) $\alpha = 3.0$

(c) rigid joint

Fig.5. Preliminary simulation results.

robots". Here, the "evaluation function" means a quantitative measure of the efficiency of the control method. In our case study mentioned above, this evaluation function corresponds to the energy consumption rate. As any control method is aimed to optimize a scalar evaluation function, a control can be regarded as a minimization/maximization procedure of this scalar evaluation function by adjusting the variable parameters of the robots. The essential difficulty in this optimization problem is the fact that the evaluation function usually has a *complex* landscape with many meta-stable minima. Therefore, a useful control method is an algorithm that can find the most stable minimum among these many meta-stable minima within a finite time scale.

Such a problem has been well-known in the field of statistical mechanics as a relaxation dynamics of a structurally disordered system to its equilibrium state. A typical example is the *protein folding problem*. A protein molecule is a flexible string-like material composed of an arbitrary sequence of many amino-acids. When a protein molecule is dissolved into water, the protein molecule always takes its unique folding structure. This structure corresponds to the most stable minimum of the free energy, which is expected to be a multi-dimensional complex surface with many meta-stable minima as a result of the flexibility and the randomness in the sequence of the amino-acids[4]. The "protein folding problem" is aimed to understand the mechanism of how the protein molecule finds the minimum of such a complex free energy landscape within a finite time scale[5].

Recent research[12] has revealed that the free energy of actual protein does not show a complex landscape but a *funnel-like structure* with the most stable state, *i.e.*, *native structure*, at the exit of this funnel (see Fig. 6). This ensures the following: the proteins do not have to employ a very efficient algorithm to search for the minimum of the complex landscape because the structure of the free energy landscape itself is so simple that even a naive algorithm, *i.e.*, equation of motion of each amino-acid, can find its minimum easily. In other words, only proteins that have a funnel-shaped free energy landscape can fold to a unique structure that shows the desired bio-chemical functions. If this is not the case, such a sequence of amino-acids is not called a protein but a *polypeptide*[6].

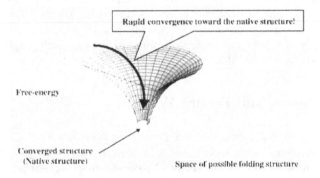

F ig . 6 . A schematic of the funnel-like landscape of the free energy observed around the converged structure.

Now, we can draw an analogy between the protein folding problem and the problem of designing the control mechanism of robots. The molecular structure of

---

[4] This free energy plays the role of the scalar evaluation function in the statistical mechanics.

[5] Anfinsen insisted that the 3-dimensional folded structure of protein cannot be derived only from the 1-dimensional sequence of amino-acids. He clarified the importance of the thermodynamic principle of the minimum free energy in the generation of the 3-dimensional folded structure from the 1-dimensional sequence of the amino-acids[11]. This concept is nowadays called "Anfinsen's dogma".

[6] Namely, only the proteins that possess bio-chemical functions indispensable for living systems have passed the hard tests for survival through the evolution process.

the protein (*i.e.* the sequence of the amino-acids) and the algorithm of searching for the minimum free energy (*i.e.* the equation of motion of the molecule) can be identified with the body dynamics and the control dynamics, respectively. This will allow the following conclusion:

> In solving the problem of controlling robots that has a complex scalar evaluation function, it is not a good idea to only search for an efficient algorithm, *i.e.*, control dynamics that minimizes the complex evaluation function. Instead, one should also adjust the mechanical structure of the robots, *i.e.*, body dynamics simultaneously so that the landscape of learning surface ends up to be a simple funnel-like structure. Such a well-balanced coupling between the body dynamics and the mechanical dynamics will guarantee the global stability and the convergence within finite time scale of the control method.

This approach is totally different from the concept of the recent "learning theories" where people accept *singularities* in the learning surface and search for a high-functional and sophisticated algorithm. If we take advantage of such degrees of freedom in the body dynamics, we can modify the landscape of the learning surface so that it does not have a singularity. This allows us to expect that even a "cheap" algorithm can satisfy the requirements for the finite-time convergence and the robustness. We should never forget the existence of the physical body of robots: the design of robots does not merely imply a designing of their "control systems".

7   Conclusion and Future Work

This paper investigated "well-balanced coupling as it should be" between control and body systems. For this purpose, a decentralized control of a multi-legged robot was employed as a case study. The preliminary experiments conducted in this paper support several conclusions and have clarified some interesting phenomena for further investigation, which can be summerized as: first, control and body dynamics significantly influence the gait convergence; second, well-balanced design in this case study can be analytically discussed in terms of the spectral radius of a matrix which specifies the property of gait convergence; third and finally, as demonstrated in the preliminary experiments, the property of gait convergence can be tuned by varying the dynamics experimentally, which suggests that there should be an appropriate coupling between the two systems.

In order to gain a deep insight into what well-balanced coupling is and should be, an intensive search of the design parameters in the control and body systems is indispensable. For this purpose, it seems to be reasonable to implement an *evolutionary computation scheme* such as a genetic algorithm to efficiently search these parameters. This is currently under investigation. In addition to this simulations, a real physical robot is currently being constructed for experimental verification. A view of this experimental robot is shown in Fig. 7.

Another important point to be stressed is closely related to the concept of *emergence*. One of the crucial aspects of intelligence is the *adaptability* under hostile and dynamically changing environments. How can such a remarkable ability be achieved under limited/finite computational resources? The only solution would be to exploit *emergence phenomena* created by the interaction dynamics among control systems, body systems, and their environment. We discussed this point by borrowing an idea from the protein folding problem. This research is a first step to shed light on this point in terms of balancing control systems with their body systems.

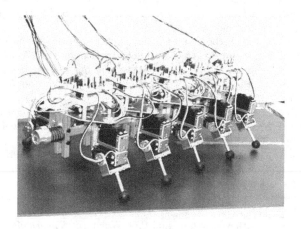

F ig.7. A view of the experimental multi-legged robot under construction.

Acknowledgements. The thorough and helpful comments of the anonymous reviewers on the preliminary versions of the article are gratefully acknowledged. The authors would like to greatly appreciate Kazuhisa Ishimaru and Koji Hayakawa for their efforts to construct the simulator and the experimental robot. The authors also would like to thank Yutaka Nakagawa at our laboratory, Josh C. Bongard at Computational Synthesis Lab. of Coenell University, and Martin C. Martin at AI lab. of Massachusetts Institute of Technology for their any helpful suggestions for the simulator. This research was supported in part by a Grant-in-Aid from the Japanese Ministry of Education, Culture, Sports, Science and Technology (No. 14750367) and a Grant-in-Aid from The OKAWA Foundation for Information and Telecommunications (No. 02-22).

References

1. R. Pfeifer and C. Scheier: *"Understanding Intelligence"*, MIT Press (1999)
2. K. Sims: *"Evolving virtual creatures"*, Computer Graphics, 28, pp.15-34 (1994)
3. K. Sims: *"Evolving 3D morphology and behavior by competition"*, Artificial Life IV Proceedings, MIT Press, pp.28-39 (1994)
4. W.P. Lee, J. Hallam, and H.H. Lund: *"A Hybrid GA/GP Approach for Co-evolving Controllers and Robot Bodies to Achieve Fitness-Specified Tasks"*, Proc. of The IEEE 3rd International Conference on Evolutionary Computation, pp.384-389 (1996)
5. C. Paul and J.C. Bongard: *"The Road Less Travelled: Morphology in the Optimization of Biped Robot Locomotion"*, Proc. of The IEEE/RSJ International Conference on Intelligent Robots and Systems (2001)
6. N. Franceschini, J.M. Pichon, and C. Blanes: *"From insect vision to robot vision"*, Philosophical Transactions of the Royal Society, London B, 337, pp.283-294 (1992)
7. L. Lichtensteiger and P. Eggenberger: *"Evolving the Morphology of a Compound Eye on a Robot"*, Proc. of The Third European Workshop on Advanced Mobile Robots, pp.127-134 (1999)
8. L. Lichtensteiger and R. Salomon: *"The Evolution of an Artificial Compound Eye by Using Adaptive Hardware"*, Proc. of The 2000 Congress on Evolutionary Computation, pp.1144-1151 (2000)
9. R. Wootton: *"How Flies Fly"*, Nature, Vol.400(8 July), pp.112-113 (1999)
10. R. Wootton: *"Design, Function and Evolution in the Wings of Holometabolous Insects"*, Zoologica Scripta, Vol.31, No.1, pp.31-40 (2002)
11. C.B. Anfinsen: *"Principles that Govern the Folding of Protein Chains"*, Science, 81, pp.223-230 (1973)
12. J.N. Onuchic, Z. Luthey-Schulten, and P.G. Wolynes: *"Theory of Protein Folding"*, Ann. Rev. Phys. Chem., 48, pp.545-600 (1997)
13. http://www.q12.org/ode/ode.html

Self-Stabilization and Behavioral Diversity of Embodied Adaptive Locomotion

Fumiya Iida and Rolf Pfeifer

Artificial Intelligence Laboratory,
Department of Informatics, University of Zurich,
Andreasstrasse 15, 8050 Zurich, Switzerland
[iida,pfeifer]@ifi.unizh.ch,
http://www.ifi.unizh.ch/ailab/

Abstract. Locomotion is of fundamental importance in understanding adaptive behavior. In this paper we present two case studies of robot locomotion that demonstrate how higher level of behavioral diversity can be achieved while observing the principle of cheap design. More precisely, it is shown that, by exploiting the dynamics of the system-environment interaction, very simple controllers can be designed which is essential to achieve rapid locomotion. Special consideration must be given to the choice of body materials. We conclude with some speculation about the importance of locomotion for understanding cognition.

1 Introduction

Normally, when dealing with locomotion, the focus is on the control aspects, as is illustrated by most of the research in the field of robotics(e.g. [1,2,3,4,5,6]). With a few notable exceptions, the physical body dynamics has not been taken into account, and has not been sufficiently exploited. As a result, most approaches still suffer from being relatively slow and lack a high degree of adaptability because of the enormous real-time computational requirements.

The idea of exploiting dynamics was introduced by the studies of Passive Dynamic Walkers [7,8,9], which demonstrated that given certain environmental conditions and a proper morphological design of the robot, biped walking is possible without, or with very little, computation and actuation. Because the Passive Dynamic Walker exploits the specific interaction with the environment to an extreme extent, its ecological niche is very narrow: it can only walk down a slope with a particular angle of inclination, and the friction coefficients must be within a small range. The exploitation of the specifics of the ecological niche always entails trade-offs: if the conditions do not hold any more, for example, if the angle of inclination is changed, the Passive Dynamic Walker will fall over.

Previously, a set of design principles of autonomous agents have been proposed [10]. In this framework, the concept of cheap design states that good designs exploit the physics of the system-environment interaction and the constraints of the ecological niche, which substantially reduces the complexity of

F. Iida et al. (Eds.): Embodied Artificial Intelligence, LNAI 3139, pp. 119–129, 2004.

the control architecture. The mechanisms underlying adaptive behavior or intelligence in general, therefore, cannot be reduced to some kind of internal representation. Rather, it is the interplay between the neural system and the "hardware" of the body that constitutes these mechanisms. So far, the interaction between body and control dynamics of locomotion has been only partially explored (e.g. [11,12,13]), and the design principles of such a mobile robot are not fully understood. Generally, cheap design, implies trade-offs which reduces the adaptability to environmental changes, because the system is relying on the environmental constraints. However, if interpreted properly as the exploitation of constraints, "cheap design" can be applied to more complex behaviors.

Based on biomechanical studies, the legged systems have been investigated, which explained the elastic components in the legs can provide the property of self-stabilization during locomotion process [14,15,16,17,18]. An interesting aspect of this approach is that the cheap design (i.e. having passive elasticity in the body) is employed not only for relaxing the control duty, but also to achieve the energy efficient and rapid locomotion.

To better understand the nature of cheap design, in this paper, we investigate the use of body dynamics with two case studies of locomotion robots, called "Stumpy" and "Puppy". We will attempt to extract the design principles for achieving behavioral diversity, which is a prerequisite for an adaptive robot. As shown in the experimental results below, a robot which properly exploits its intrinsic body dynamics and self-stabilization mechanisms, is able to display a high level of behavior diversity. We start by describing the design, the control, and the various gaits of the hopping robot "Stumpy". Then we introduce the quadruped "Puppy" and discuss the mechanisms of self-stabilization. Finally we discuss the relation of self-stabilization mechanisms and behavioral diversity.

2   Behavioral Diversity of a Hopping Machine

In this section we describe a new kind of hopping robot called Stumpy. Despite its simple structure, a salient feature of this robot is its large variety of behaviors[1]. In a set of systematic experiments, we will show how the behavioral diversity can be achieved by applying the principle of cheap design[2].

2.1   Design and Control

Stumpy uses inverted pendulum dynamics to induce biped-like locomotion gaits. Its mechanical structure consists of a wide base in the form of a rigid inverted T-shape mounted on four compliant feet (Figure 1). An important feature of the base is its springy property. An upright "T" structure is connected to this base by a rotary joint labeled "waist". The horizontal beam of the upright "T" is connected to the vertical beam by a second joint, a rotary joint that we call the

---

[1] The video clips are available at: http://www.ifi.unizh.ch/ailab/people/iida/stumpy/
[2] For further technical details, refer to the previous papers [19,20,21].

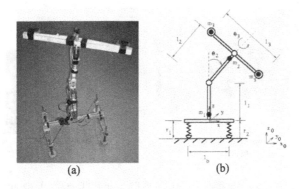

(a)   (b)

Fig. 1. The robot Stumpy. A photograph (a), and a schematic (b).

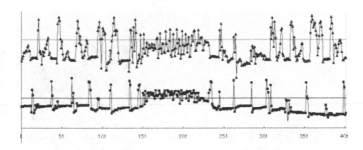

Fig. 2. Intrinsic stability of Stumpy. This graph shows the time series pressure data installed on the right and left feet (top and bottom graphs, respectively). The gait is disturbed by an external force around time step 150. A time step corresponds roughly to 1/20 seconds.

"shoulder". By using this two degrees of freedom mechanical structure, the robot is able to perform many different locomotion behaviors including hopping and walking in a straight or curved line. Note that Stumpy does not have sensors to recognize its global states and so it does not know what behavior it is currently involved in. There is only local feedback that enables it to perform synchronized sinusoidal oscillations of the two joints.

2.2   Intrinsic Stability and Behavioral Diversity

In addition to the static stability which is achieved by a wide base and four feet, the dynamical stability is one of the major features in the behavior of Stumpy. Figure 2 shows the time-series pressure data measured at the feet of Stumpy. At around time step 150 an external force disturbance is exerted. The rhythmic pattern of the ground contact is generally retrieved after a certain period of chaotic behavior, in the figure after roughly 100 time steps.

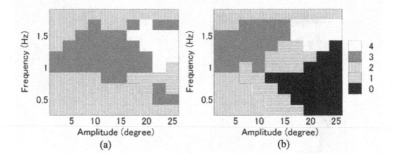

Fig. 3. Gait distribution diagrams. Gait distribution in Terrain 0 (a) and Terrain 1 (b). The shadings indicate the different gates: "4": Hopping, "3": Walking, "2": Shuffling, "1": Unstable, and "0":Fall.

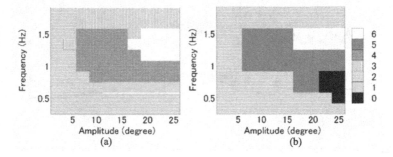

Fig. 4. Gait distribution diagrams of the lateral bounding experiments. Gait distribution on Terrain 0 (a) and Terrain 1 (b). The texture patterns indicate the different gaits: "6": Hopping to Right, "5": Hopping and Stay, "4": Hopping to Left, "3": Shuffling to Right, "2": Shuffling and Stay, "1": Shuffling to Left, and "0":Fall.

To explore this characteristic of self-stabilization further, we have conducted a systematic investigation in terms of the oscillation of the waist motor and the influence of environment. By simply changing two control parameters of the waist and the shoulder oscillation, frequency and amplitude, Stumpy exhibits a rich diversity of locomotion gaits. Figure 3 illustrates the variations of gaits when the amplitude and the frequency of the waist motor oscillation are varied in two different ground conditions (different coefficients of friction) labeled Terrain 1 and Terrain 2. The gaits are categorized in terms of the time-series ground contact data, which indicate whether both feet are off the ground at a certain period in a gait cycle (Hopping), one foot is always on the ground (Walking), or both feet are always on the ground (Shuffling) (The "Unstable" gait means that there is no periodic pattern observed in the data). In the first experiment, we set the center of oscillation to be in the center with respect to the lower body, and we recorded the foot pressure data during 10 seconds of operation for

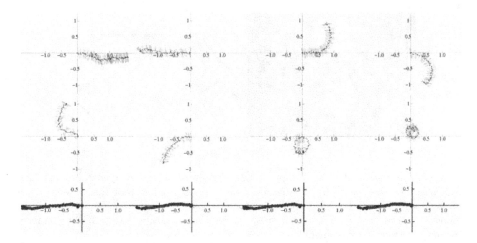

Fig. 5. Locomotion behaviors of Stumpy observed from top (The unit of these graphs is meters). Stumpy can control its movement direction, turning rate (Top and Middle panels), and lateral bounding (Bottom panels) by changing only a few control parameters. Black dots denote the trajectory of the body center, and the line illustrates the orientation of bottom base.

every parameter setting of frequency and amplitude. As a result, we observed four different gaits in this experiment. In a similar way, we also analyzed another kind of behavior called "lateral bounding". By setting the center of oscillation laterally to one side, Stumpy shows a locomotion behavior in the lateral direction. Compared to the previous experiment, the robot showed two basic gaits, i.e. Hopping and Shuffling, and the directionality of the movement depends on frequency and amplitude. Figure 4 shows the behaviors observed with respect to amplitude and frequency during the lateral bounding experiment.

So far, we used only the waist motor, but the behavioral diversity of Stumpy can be enhanced even further by adding another degree of freedom in the shoulder joint. By coupling the lateral and horizontal momentum induced by the rotary oscillation of the two motors, a hopping behavior can be achieved. While the waist oscillation generates a periodic hopping gait, the shoulder motor can control the horizontal forward/backward movement depending on the synchronization of these two oscillations. When the phase of two oscillations is reversed, the forward locomotion switches to backward. The turning rate can be controlled by biasing the speed of the horizontal oscillation: For example, faster rotation in the clock-wise direction leads to a turning movement (Figure 5).

The novelty of this kind of robot locomotion lies in its unique morphology. Because of the dynamic stability achieved by the wide springy base and the proper body design, many different patterns of physical interactions between the body, friction, actuation and control can be generated.

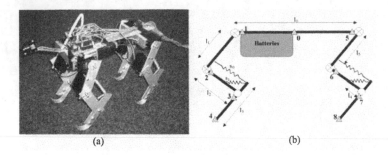

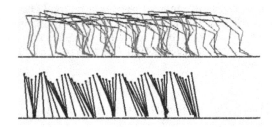

Fig. 6. The qudruped robot Puppy. A photograph of the quadruped robot (a), and schematic of the robot (b). The circles denote passive joints and the circles with a cross inside denote the joints controlled by servomotors. The triangles with numbers show the positions of LEDs which are used for visual tracking of the body geometry during locomotion experiments.

Fig. 7. Behavior analysis of a running experiment. The upper graph shows the behavior of the whole body based on the visual tracking of LEDs attached to the leg joints and the ground contacts. The lower graph shows the trajectory of a virtual linear hind leg.

3   Self-Stabilization of Quadruped Running

Cheap design is crucial for the rapid legged locomotion in order to increase the energy efficiency and reduce the computational cost. In this section, we describe a four-leg robot which exploits the elasticity of its components for running[3]. The experimental results show that the running behavior is achieved by a self-stabilization mechanism, which can be used also for the control of forward velocity[4].

3.1   Design and Control

Figure 6 shows the mechanical design of the running quadruped robot, Puppy, which is inspired by biomechanical studies. Each leg consists of two standard servomotors and one elastic passive joint in series, and the designs of all four

---

[3] The video clips are available at: http://www.ifi.unizh.ch/ailab/people/iida/stumpy/
[4] Refer to [22] for more technical details.

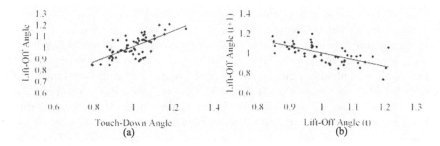

Fig. 8. Touch-down and lift-off angles of the qudruped Puppy. (a): The relation between touch-down and lift-off angles. The touch-down angles and lift-off angles are normalized by the corresponding mean touch-down and lift-off angles. (b): Relation between the normalized lift-off angles of successive leg steps.

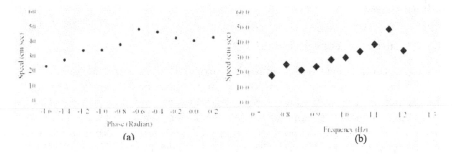

Fig. 9. Speed diagrams of Puppy for different parameter values. Average speed vs, phase (a) and average speed vs. frequency (b) parameters.

legs are identical. This robot carries eight motors, batteries, and a micro-motor controller. To demonstrate a running gait, we apply a synchronized oscillation based control scheme to four motors in the hip and shoulder, where each motor oscillates through a simple sinusoidal position control. No sensory feedback is used for this controller except for the internal local feedback for the servomotors.

3.2   Behavior Analysis of Self-Stabilization

The behavior of the robot was extracted by a standard visual motion analysis, where the trajectories of the joints were visually tracked. Figure 7 shows a typical locomotion behavior extracted from a side view. The legs exhibit simple oscillations, but through the interplay of the elastic body structure, the mass distribution, the gravity and the ground friction, a natural quadrupedal running gait occurs, which includes periods in which all four legs are off the ground. In other words, there is a clear distinction between a stance and a flight phase. We found that this kind of running behavior significantly relies on the underlying self-stabilization mechanism. Although the control of the robot is extremely simple — the controller does not recognize the stance/flight phase, acceleration, or inclination — the robot maintains a stable periodic gait. This is due to the fact that it properly exploits its intrinsic body dynamics. The self-stabilization mechanism uses a unique characteristic of legs, which can be observed by the touch-down and lift-off angles of the virtual linear leg during the stance phase. Linear leg analysis means that the virtual line between the hip and the ground contact is estimated (Figure 7b). As shown in Figure 8b, the relation between successive lift-off angles is roughly linear. This means that, when a lift-off angle is lower, the subsequent lift-off angle is larger, and vice versa, which results in a stable touch-down angle over multiple leg steps. The underlying mechanism is implicitly contained in the entire body dynamics which has the effect that there is a linear relation between the touch-down and lift-off angles of the legs (Figure 8), which implies that a lower touch-down angle results in a larger lift-off angle, on average. The data shown in Figure 8 were collected from a series of 10 runs of 6 leg steps.

3.3   Control of Forward Velocity

Owing to the intrinsic self-stabilization property, the control of forward velocity can be easily realized by varying a single phase or frequency parameter of the oscillation. Figure 9 shows the average forward velocity with respect to the phase and frequency parameters of leg oscillation, which is extracted from the visual analysis explained above. It shows that, by simply varying the phase and frequency parameters, the velocity can be changed in the range from 20 to 50 cm/sec, approximately. It is interesting to note that this control strategy of the forward velocity by means of the oscillation phase and frequency is one of the simplest possible control parameters because it can be a simple time delay in the neural substrate.

4   Discussion and Conclusion

Exploiting self-stabilization mechanisms seems to be a common strategy for legged animals in nature and some of them have been explored in biology and robotics (e.g. [14,15,16,18,23,24]). However, the case studies shown in the previous sections provide further interesting aspects for a comprehensive understanding of embodied adaptive locomotion.

Self-Stabilization and Behavioral Diversity

The simplicity of the control of Stumpy and Puppy is mostly due to the self-stabilization mechanisms. Without sensory feedback, the locomotion processes are maintained by properly exploiting the interaction of body material and dynamics (e.g. aluminium, springs, and mass distribution), environmental (e.g. friction and rough terrain), and control (e.g. amplitude and frequency). Moreover, as illustrated in these case studies, the self-stabilization mechanisms not only simplify the control, but also significantly influence the locomotion function itself. In other words, sophisticated design and control is not always required in order to achieve rich behavioral diversity. The lateral bounding behavior of Stumpy provides a good example of how behavioral diversity can be achieved by exploiting the body dynamics. By simply changing the oscillation frequency, Stumpy exhibits very different behaviors in terms of gait and direction. As another example, the control of Puppy's forward velocity is no longer possible by just varying the rotation speed of the motors, but the control parameters, phase and frequency, have to be varied in order to influence the body dynamics.

An interesting feature of the proposed approach is the fact that there are a few different control parameters instead of only one parameter, which can be used for the same purposes. For example, as shown in Figure 3, the locomotion gait of Stumpy can be controlled by both amplitude and frequency, and the same holds for the direction and the gait of lateral bounding locomotion. The forward velocity of Puppy also illustrates this point clearly, i.e. both parameters, phase and frequency, are able to control the velocity. Note that these control parameters are not controlling the locomotion function directly, but indirectly by changing the dynamics.

In the locomotion experiments shown in this paper, Stumpy and Puppy were operated mostly at the resonance frequencies of the systems in order to exploit the body dynamics. However, it is clear that the other kinds of physical interactions which influence the body dynamics should be considered as well. Not only simple linear springs and rigid materials, but properties such as damping and mass distribution need to be explored to better understand how behavioral diversity can be achieved. In addition, it should also be mentioned that the behavioral diversity could be potentially enhanced further by operating the system at non-resonance frequencies for more torque demanding stationary tasks. These two strategies of behavior control need to be explored further as well.

Toward Embodied Adaptive Locomotion

Although we have focused on the functional aspects of locomotion in this paper, this approach provides additional insight into embodied adaptive behavior or intelligence in general. The control of behavior is quite often the major research interest of adaptive locomotion, but the use of body dynamics is also a fundamental mechanism to properly understand behavioral diversity. As illustrated in the case studies of this paper, the functions of the system are no longer separable from the constraints derived from embodiment, if the behavior of the robots highly depends on its body dynamics: there is no longer a clear separation of hardware and control. In this sense, locomotion behavior is also essential for high-level cognition, as it enables the agent to construct a "body image" that on the one hand can be used to guide behavior in the real world and on the other as a basis for metaphors on top of which something like cognition can be bootstrapped.

Acknowledgements. We would like to thank Raja Dravid and Chandana Paul for the productive discussion and collaboration. This work is supported by the Swiss National Foundation, grant No. 20-68198.02.

References

1. Brooks, R. A.: A robot that walks: emergent behaviors from a carefully evolved network. *Neural Computation 1(2)* (1989) 253-262
2. Vukobratovic, M., and Stepanenko, J.: On the stability of anthropomorphic systems. *Mathematical Biosciences 15* (1972) 1-37
3. Yamaguchi, J., Soga, E., Inoue, S. and Takanishi, A.: Development of a bipedal humanoid robot - control method of whole body cooperative dynamic biped walking. *In Proc. IEEE Int. Conference on Robotics and Automation* (1999) 368-374
4. Hirose, M. Haikawa, Y., Takenaka, T., and Hirai, K.: Development of humanoid robot ASIMO. *Proc. Int. Conference on Intelligent Robots and Systems* (2001)
5. Loeffler, K., Gienger, M., Pfeiffer, F.: Sensor and control design of a dynamically stable biped robot. *ICRA 2003* (2003) 484-490
6. Arikawa, K., Hirose, S. Development of quadruped walking robot TITAN-VIII, *Proceedings of International Conference on Intelligent Robots and Systems (IROS96)* (1996) 208-214,
7. Collins, S. H., Wisse, M., and Ruina, A.: A three-dimentional passive-dynamic walking robot with two legs and knees. *International Journal of Robotics Research* 20 (2001) 607-615
8. McGeer, T.: Passive dynamic walking. *International Journal of Robotics Research* 9 (1990) 62-82
9. Wisse, M. and van Frankenhuyzen, J.: Design and construction of MIKE: A 2D autonomous biped based on passive dynamic walking. *Proceedings of International Symposium of Adaptive Motion and Animals and Machines (AMAM03)* (2003)
10. Pfeifer, R. and Scheier, C.: Understanding Intelligence. *The MIT Press*, (1999)
11. Taga, G., Yamaguchi, Y., and Shimizu, H.: Self-organized control of bipedal locomotion by neural oscillators in unpredictable environment. *Biological Cybernetics* 65 (1991) 147-159
12. Fukuoka, Y., Kimura, H., and Cohen, A. H.: Adaptive dynamic walking of a quadruped robot on irregular terrain based on biological concepts, *Int. Journal of Robotics Research*, Vol.22, No.3-4, (2003) 187-202
13. Ishiguro, A., Ishimaru, K., Hayakawa, K., and Kawakatsu, T.: Toward a "well-balanced" design: A robotic case study -How should control and body dynamics be coupled?-, in Proc. of The 2nd International Symposium on Adaptive Motion of Animal and Machines. (2003)
14. Kubow, T. M., Full, R. J.: The role of the mechanical system in control: a hypothesis of self-stabilization in hexapedal runners. *Phil. Trans. R. Soc. Lond. B* 354 (1999) 849-861
15. Raibert, H. M.: Legged robots that balance. *The MIT Press* (1986)
16. Buehler, M.: Dynamic locomotion with one, four and six-legged robots. *Journal of the Robotics Society of Japan* 20(3) (2002) 15-20
17. Alexander, R. McN.: Three uses for springs in legged locomotion, *The International Journal of Robotic Research*, 9, No. 2 (1990) 53-61
18. Seyfarth, A., Geyer, H., Guether, M., and Blickhan, R.: A movement criterion for running. *J. Biomech. 35(5)* (2002) 649-655

19. Iida, F.: Exploiting friction for a hopping robot. *Proc. of Adaptive Motion of Animals and Machines* (2003)
20. Iida, F., Dravid, R., Paul. C.: Design and control of a pendulum driven hopping robot. *Proceedings of International Conference on Intelligent Robots and Systems 2002 (IROS 02)* (2002) 2141-2146
21. Paul, C., Dravid, R., Iida, F.: Control of lateral bounding for a pendulum driven hopping robot. *Proc. of 5th International Conference on Climbing and Waling Robots (CLAWAR 2002)* (2002) 333-340
22. Iida, F., and Pfeifer, R.: "Ceap" rapid locomotion of a quadruped robot: Self-stabilization of bounding gait. *Proc. of Intelligent Autonomous Systems 8*, Groen, F. et al. (Eds.), IOS Press (2003) 642-649
23. Cruse, H., Bartling, C. H., Brunn, D. E., Dean, J., Dreifert, M., Kindermann, T., and Schmitz, J.: Walking: A complex behavior controlled by simple systems. *Adaptive Behavior 3(4)* (1995) 385-418
24. Herr, H. M., McMahon, T. A.: A trotting horse model, *The International Journal of Robotics Research*, 19, No. 6 (2000) 566-581

Removing Some 'A' from AI:
Embodied Cultured Networks

Douglas J. Bakkum[1], Alexander C. Shkolnik[2], Guy Ben-Ary[3], Phil Gamblen[3],
Thomas B. DeMarse[4], and Steve M. Potter[1]

[1] Georgia Institute of Technology
bakkum@neuro.gatech.edu
steve.potter@bme.gatech.edu
[2] Massachusetts Institute of Technology
shkolnik@mit.edu
[3] University of Western Australia
guyba@cyllene.uwa.edu.au
pgamblen@hotmail.com
[4] University of Florida
tdemarse@bme.ufl.edu

Abstract. We embodied networks of cultured biological neurons in simulation
and in robotics. This is a new research paradigm to study learning, memory, and
information processing in real time: the Neurally-Controlled Animat. Neural
activity was subject to detailed electrical and optical observation using multi-
electrode arrays and microscopy in order to access the neural correlates of animat behavior. Neurobiology has given inspiration to AI since the advent of the
perceptron and consequent artificial neural networks, developed using local
properties of individual neurons. We wish to continue this trend by studying the
network processing of ensembles of living neurons that lead to higher-level
cognition and intelligent behavior.

1 Introduction

We present a new paradigm for studying the importance of interactions between an
organism and its environment using a combination of biology and technology: embodying cultured living neurons via robotics. From this platform, explanations of the
emergent neural network properties leading to cognition are sought through detailed
optical and electrical observation of neural activity. A better understanding of the processes leading to biological cognition can, in turn, facilitate progress in understanding
neural pathologies, designing neural prosthetics, and creating fundamentally different
types of artificial intelligence. The Potter group is one of seven in the Laboratory for
Neuroengineering (Neurolab[1]) at the Georgia Institute of Technology, all working at
the interface between neural tissue and engineered systems. We envision a future in
which mechanisms employed by brains to achieve intelligent behavior are also used
in artificial systems; we overview three preliminary examples of the Neurally-

---

[1] http://www.ece.gatech.edu/research/neuro/

F. Iida et al. (Eds.): Embodied Artificial Intelligence, LNAI 3139, pp. 130–145, 2004.

Controlled Animats approach below. By using biology directly, we hope to remove some of the 'A' from AI.

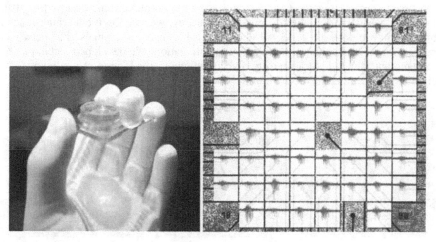

Fig. 1. Connecting neurons to multi-electrode arrays. Left: Cells are plated inside a glass multi-electrode array culture dish such as this. Right: recorded voltage traces in the lighter boxes overlay a microscope image of the neuronal network growing on a 60-electrode array (electrode diameter, 30 μm). The thick lines are the electrode leads. The voltage spikes are neural signals.

No one would argue that environmental interaction, or embodiment, is unimportant in the wiring of the brain; no one is born with the innate ability to ride a bicycle or solve algebraic equations. Practice is needed. An individual's unique environmental interactions lead to a continuous 'experience-dependent' wiring of the brain [1]. This makes evolutionary sense as it is helpful to learn new abilities throughout life: if there are some advantageous features of an organism that can be attained through learning, then the ability to learn such features can be established through evolution (the Baldwin effect) [2]. Thus, the ability to learn is innate (learning usually being defined as the acquisition of novel behavior through experience [3]). We suggest that environmental interaction is needed to expose the underlying mechanisms for learning and intelligent behavior. Many researchers use in vitro models (brain slices or dissociated neural cell cultures) to study the basic mechanisms of neural plasticity underlying learning. We argue that because these systems are not embodied or situated, their applicability to learning in vivo is severely limited. We are developing systems to re-embody in vitro networks, and allow them to interact with an environment, so that we can watch the processes contributing to learning at the cellular level *while they happen*.

We study networks of tens of thousands of brain cells in vitro (neurons and glia) on a scale of a few square millimeters. The cells in cortical tissue are separated using enzymes, and then cultured on a Petri dish with 60 electrodes embedded in the substrate, a multi-electrode array (MEA; from MultiChannel Systems) (Fig. 1) [4], [5]. The neurons in these cultures spontaneously branch out (Fig. 2). Even left to themselves without external input other than nutrients (cell culture media), they re-establish connections with their neighbors and begin communicating electrically and chemically within days, demonstrating an inherent goal to network; electrical and morphological observations suggest these cultures mature in about four weeks [6], [7], [8]. The neurons and supporting glia form a monolayer culture over the clear MEA substrate, amenable to optical imaging with conventional and two-photon microscopy [9], [10], [11]. With sub-micron resolution optical microscopy, we can observe learning-related changes in vitro with greater detail than is possible in living animals. The networks are also accessible to chemical or physical manipulation. We developed techniques to maintain neural cultures for up to two years, allowing for long-term continuous observation. For detailed methods, refer to [5].

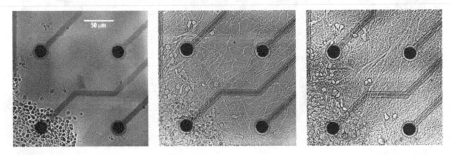

Fig. 2. Microscope images of neurites (axons and dendrites) growing across a gap. The images were taken on three consecutive days beginning the second day after plating the cells. The *black circles* are the electrodes.

A multi-electrode array records extracellular neural signals fast enough to detect the firing of nearby neurons as voltage spikes (Fig. 1, right). Neurons detected by an electrode can be identified using spike-sorting algorithms [12]. Thus, the activity of multiple neurons can be observed in parallel and network phenomena can be studied. In addition to the expression of spontaneous activity, supplying electrical stimulation through the multiple electrodes induces neural activity; we have built custom circuitry to continuously stimulate the 60 electrodes [13]. The MEA forms a long-term non-destructive two-way interface to cultured neural tissue. The recorded signals can be used as motor commands, while the stimuli represent sensory inputs, in our embodied system. These techniques allow high resolution, long term, and continuous studies on the role of embodiment throughout the life of a cultured neural network.

Wilson [14] coined the term 'animat' (a computer simulated or robotic animal behaving in an environment) in his studies of intelligence in the interactions of artificial animals. Our interfacing of cultures to a simulated environment (described below) was the first Neurally-Controlled Animat (Fig. 3) [15], [16], [17]. For cultures interfaced to physical robots, we introduce the term 'hybrot' for hybrid biological robot. Mussa-Ivaldi's group created the first closed-loop hybrot by controlling a Khepera robot with a brain stem slice from a sea lamprey [18]. In a related approach, our Neurolab colleague Robert Butera studies detailed neural dynamics by coupling simulated neurons to real neurons using an artificial conductance circuit [19], [20]. Stephen DeWeerth's group in the Neurolab develops and studies, among others things, silicon model neurons interfaced with living mollusk and leech neurons [21].

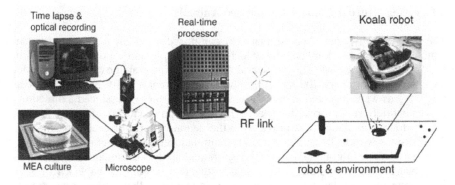

Fig. 3. Hybrot (Hybrid living+robotic) setup. Optical and electrical data from neurons on an MEA are analyzed and used to control various robotic devices, while time-lapse imaging is carried out to make movies of neuronal plasticity.

Using simulated environments is a good first step and provides easier control and repeatability compared to robotics. However, a 'real' environment's great complexity provides two advantages. First, many seemingly complex behaviors of animals are emergent: simple behavioral rules applied in a complex environment produce complex and productive behavior [22], [23], [24]. Second, a complex environment produces a robust brain to take advantage of it: among other examples, this is evident in tool use [25] and in exploiting properties such as the biomechanics of muscle tissue in repositioning an arm without excessive vibrations. It is difficult to simulate a complex environment with realistic physics. If physics plays an important role in the complex behavior of intelligent systems, then by using robots in the real world, the researcher gets the physics "for free." We believe that this merging of artificial intelligence concepts (including robotics) into neurobiological experiments can inform future AI approaches, making AI a bit less artificial.

2 Examples: Three Embodied Neural Systems

Creating a neurally controlled robot that handles a specific task begins with a hypothesis of how information is encoded in the brain. Much remains to be determined, but numerous schemes have been proposed, most based on the quantity and/or relative timing of the firing of neural signals. A neural network may be considered as a type of processing unit with an input (synaptic or electrical stimulation patterns), and an output (neural firing patterns), which can perform interesting mappings to produce behavior. Below are overviews of three such systems. These examples could have been conducted with artificial neural networks. We use biological neural networks not as substitutes to artificial neural networks, but to tease out the intricacies of *biological* processing to inform future development of *artificial* processing. In particular, we analyzed how the properties of neurons lead to real-time control and adaptation to novel environments.

2.1 Living Neurons Control a Simulated Animal

The first Neurally-Controlled Animat [16] comprised a system for detecting spatio-temporal patterns of neural activity, which directed exploratory movement of a simulated animal in real time (Fig. 4). Neural firings were integrated over time to produce an activity vector every 200 ms, representing the current activity pattern, and recurring patterns were clustered in activity space. Each cluster was assigned a direction of movement (left, right, forward, backward). Proprioceptive and exteroceptive feedback via electrical stimulation was provided to the neural culture for each movement and for collisions with walls and barriers. The stimulation induced neural activity that, in turn, was detected through the activity vectors and used as commands for subsequent movements. We created the software and hardware necessary to enable a 15-ms sensory-motor feedback latency, since we feel it is important that a tight connection between the neural system and its environment is likely to be crucial to adaptive control and learning.

Within this real-time feedback loop, both spontaneous and stimulated neural activity patterns were observed. These patterns emerged over the course of the experiment, sometimes assembling into a recurrent sequence of patterns over several seconds, or the development of new patterns, as the system evolved. The overall effect of the feedback loop on neural activity was observed from the path of the animat's movement throughout its environment (Fig. 4). As the neural network moved its artificial body, it received feedback and in turn produced more movement. The behavioral output was a direct result of both spontaneous activity within the network as well as activity produced by feedback due to the networks interaction with its virtual environment. Hence the path of the animat was indicative of current activity as well as the effects of feedback. Analyzing the change in behavior of the neurally-controlled animat provided a simple behavioral tool to study shifts in the states of neural activity. However, this first Neurally-Controlled Animat did not demonstrate noticeable goal-directed behavior, which the next example addresses explicitly.

2.2 Living Neurons Control a Mobile Robot

One of the simplest forms of 'intelligent' behavior is that of approach and avoidance. The goal of the second system was to create a neural interface between neuron and robot that would approach a target object but not collide with it, maintaining a desired distance from the target. If a given neural reaction is repeatable with low variance, then the response may be used to control a robot to handle a specific task. Using one of these response properties, we created a system that could achieve the goal [26].

Networks stimulated with pairs of electrical stimuli applied at different electrodes reliably produce a nonlinear response, as a function of inter-stimulus interval (ISI). Figure 5 shows averaged firing rate over all 60 electrodes following two stimulations separated by a time interval. At short ISI's, the response of the network following stimulation was enhanced; at longer intervals, the response was depressed. Furthermore, the variance of the data for each ISI was relatively small, indicating the effect is robust and thus qualifies as a good candidate for an input/output mapping to perform computation.

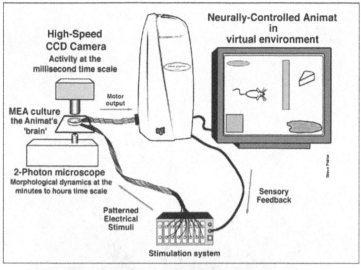

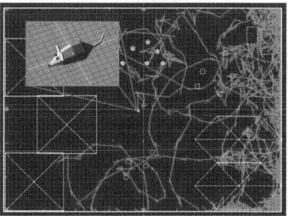

Fig. 4. Animat setup and activity. Above: neural signals are used to control the movement of an animat, whose 'brain' is exposed to microscopic imaging; feedback from the environment determines subsequent electrical stimulation of the living neuronal network in an MEA. Below: One hour of the animat's path (*curved lines*), as it moves about within its environment under neural control, with feedback. The white boxes represent various environmental obstacles.

By mapping the neural response to a given ISI as a transformation of distance to an object, we created a robot that reacts to environmental stimuli (in this case sensory information about distance from an object) by approaching and avoiding that target. To construct our "approach and follow" hybrot, sensory information (the location of a reference object with respect to the robot) was encoded in an ISI stimulation as follows: the closer the robot is to the object, the smaller the ISI. The response of the neurons to a stimulation pair, measured as an averaged firing rate across all electrodes for

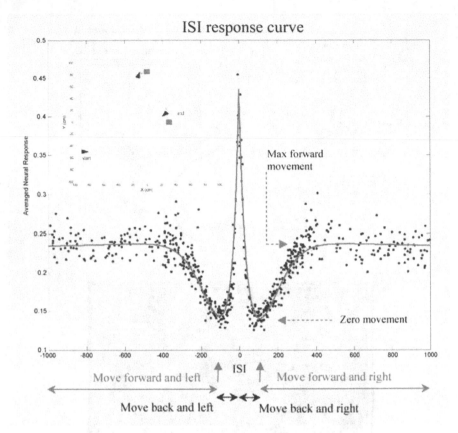

Fig. 5. Living neurons control a mobile robot. Neural firings in response to paired electrical stimulations at various inter-stimulus intervals (ISI) are plotted. In the experiments, the ISI was proportional to the distance between the neurally controlled approaching animat and its target object. It was considered positive if the target was located to the right of the animat and negative if left of the robot. The neural response determined the magnitude of subsequent animat movement; the direction of movement was determined from which quadrant the ISI fell into (see the arrows and movement key, bottom). Inset: the neurally controlled animat's trajectory (Koala robot, represented by the triangle). The target object (Khepera robot, represented by the square) was held stationary until the robot approached, and then it was moved continuously (down and to the right in the figure).

100 ms after the second stimulus, was used to control the robot's movements: a larger neural response corresponded to a longer movement (either forward or backward) of the robot.

When the robot was far away from the reference object, the ISI of the stimulation pair was long, and the neural response was large, moving the robot towards the object (Fig. 5, right). As the robot moved closer to the object, the stimulation interval decreased until it reached 150 ms. At this point, the neural response was minimal, and no movement was commanded. In other words, the robot reached its desired location with respect to the reference object. If the robot was closer to the object, the neural reaction was larger (a very short ISI), this time driving the robot away from the object. We divided the input ISI into 4 quadrants (Fig. 5, left). Each of the 4 quadrants corresponded to a directional movement: forward/right, forward/left, backward/right, and backward/left. Then, a positive ISI caused movement in a direction opposite that for a negative ISI. Given the neural response to an ISI stimulation, we decoded which quadrant the response belonged to with good accuracy (>95%).

We used the Koala and Khepera robots (manufactured by K-Team) to embody the cultured network, and to provide an environment with a moving object. The Koala robot was used as the neurally controlled robot, while the Khepera served as the reference object, moving at random under computer control. Under neural control, the Koala successfully approached the Khepera and maintained a distance from it, moving forward if the Khepera moved away, or backing up if the Khepera approached

In addition to demonstrating the computational capacity inherent in cultured neurons, this hybrot can be used to study learning in cultured neural networks. In this case, learning would be manifested through changes in the neural activity and changes at the behavioral level of the robot. Preliminary studies indicate that quantifiable behavioral traits, such as the speed with which the hybrot approaches the object, may be manipulated through mechanisms of neural plasticity.

2.3 Living Neurons Control a Drawing Arm

Meart (Multi-Electrode Array art) was a hybrot born from collaboration with the SymbioticA Research Group[2]. The 'brain' of dissociated rat neurons in culture was grown on an MEA in our lab in Atlanta while the geographically detached 'body' resided in Perth. The body consisted of pneumatically actuated robotic arms moving pens on a piece of paper (Fig. 6). A camera located above the workspace captured the progress of drawings created by the neurally-controlled movement of the arms. The visual data then instructed stimulation frequencies for the 60 electrodes on the MEA. The brain and body interacted through the internet (TCP/IP) in real time providing closed loop communication for a neurally controlled 'semi-living artist'. We see this as a medium from which to address various scientific, philosophical, and artistic questions.

Meart has brought neurobiology research to two artistic events: *Biennale of Electronic Arts Perth* and most recently at *Artbots: the Robot Talent Show* in New York. The robotic arm and video sensors were shipped to New York while the living neurons sent and received signals from Atlanta. An overview of how Meart worked may best be described by the artistic conception behind the Artbots presentation: portrait drawing. First, a blank piece of paper was placed beneath the arm's end-effector and a digital photograph was taken of an audience member. Then, communication between the arm and the neurons was begun. The neural stimulation via the MEA was determined by a comparison of the actual drawing, found using a video camera taking images of the drawing paper, to the target image of a person's photograph. Both the actual image and the target image were reduced to 60 pixels, corresponding to the

---

[2] SymbioticA: the Art and Science Collaborative Research Laboratory (http://www.fishand chips.uwa.edu.au/), based in the School of Anatomy and Human Biology at the University of Western Australia in Perth.

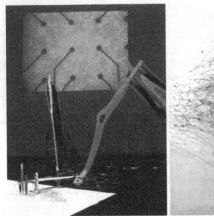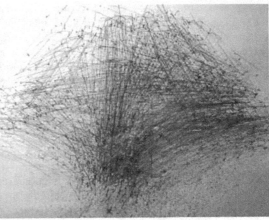

Fig. 6. Meart–The Semi-Living Artist. Left: Meart's arms used markers to draw on a piece of paper, under live neural control. In the background was a projection of the MEA and cultured net, Meart's 'brain'. Right: one drawing created by Meart in an exhibition.

MEA electrodes, and the gray scale intensity of each pixel was found. Similar to how an artist continually compares her work to her subject, the gray scale percentages for corresponding pixels on the two images were continuously compared, in this case subtracted to produce a matrix of error values. The 60 error values determined in real-time the stimulation frequency per electrode using a custom stimulation circuit built by Thomas DeMarse. Arm movement was determined by the recorded neural activity, using averaged firing rates of the induced and spontaneous activity per stimulation. Stimulation affected this neural activity, and so the communication formed a loop, with a loop time of approximately one second.

In the prior example, the sensory-motor mappings used a stable neural property to reliably control the robot. With Meart, the sensory-motor mappings are less well defined, in the hope of demonstrating a micro-scale version of the brain's creative processes. The behavioral response of the robot sheds light on the properties of the neural network and directs further encoding refinements. Thus, Meart is a 'work in progress' with the sensory-motor encoding continuously being improved to demonstrate learning processes. An example drawing is shown in Figure 6. The drawings changed throughout the life of cultures (and were different for different cultures) demonstrating neural plasticity, however, the mechanisms are still under investigation.

3   Discussion

3.1   Embodying Cultures: Theory

A Blank Slate. Since the cultured neurons were first separated and allowed to settle onto the MEA at random, they start from a 'blank slate'. Neural structure is lost and the function of neural activity is no longer obvious, yet neural network processing remains, evidenced by the complex activity patterns we have observed. For traditional in vitro neural models, function is cloudy since activity no longer relates to or causes behavioral states or actions. One cannot say 'this neuron is involved in color perception' or 'this neural structure helps to coordinate balance' as could be said for in vivo experiments. Artificially embodying and situating cultured neurons redefines their behavioral function concretely.

The structure of neuronal networks is likely to be important in neuronal processing, and changes in structure are likely to underlie learning and memory [27]. Our cultured neurons form two-dimensional monolayers; functional importance may lie in the affordances given by the three-dimensional layered nature of the cortex. We and others in the Neurolab are pursuing the construction of 3D MEAs to support three-dimensional cultures, as part of an NIH Bioengineering Research Partnership [28], [29]. However, even cultured cortical monolayers (without 3D structure nor subcortical regions) have demonstrated an ability to adapt following stimulation via potentiation and/or depression [30], [31], [32], [33]. We are exploring using these plasticity mechanisms as a means to shape the network during development, within the Neurally-Controlled Animat paradigm, so it is no longer a blank slate.

Associations. The biological brain makes associations between different phenomena observed through sensation, whether between various external stimuli or between the actions of a body and their consequences, and then commands movement accordingly. Our methods have been developed to study these processes in real time with enough resolution to capture the dynamics of these interactions. These processes can be expressed using dynamical systems theory (DST), a mathematical framework to describe systems that change in time. For example, the formation of certain functional structures (ocular dominance columns) in the visual cortex has been described using Alan Turing's reaction-diffusion equations [34]. Kuniyoshi and his group explore DST to connect sensory-motor control to the cognitive level [35]. As applied to cognition [34], DST describes the mind with a set of complex, recursive filters. This opposes the classical cognitive concept of neural processing being analogous to a digital computer, containing distinct storage and processing of symbols [36], [37]. DST contends that multiple feedback loops and transmission delays, both of which are widespread in the brain, provide a time dimension to allow higher-level cognition to emerge without the need for symbolic processing [38]. DST is a framework compatible with embodied perspectives. The dynamical systems perspective has too often been neglected in neurobiology and cognitive sciences.

In contrast to an intact brain in an animal, cultures of neurons are isolated because they do not contain the afferent sensory inputs or efferent motor outputs a body would provide and therefore no longer have a world with which to reference their activity. Under these conditions, what associations can the network make, and what would those associations mean? Moreover, what symbols are operated on? Because of this, any associations that are made must consequently be self-referential or circular and neural activity may be misleading. The network as a set of complex, recursive filters has no external signals to filter, possibly leading to the abnormal barrage activity described below. To address this major shortcoming of in vitro systems, our neural cultures are embodied with sensory feedback systems, motor systems, and situated in an environment, providing a new frame of reference. New findings about the dynamics of living neural networks might be used to design more biological, less artificial AI.

Intelligence and Meaning. By embodying cultured neurons, the 'meaning' of neural activity emerges, since this activity affects subsequent stimulation. Now the network has a body behaving and producing experiences, allowing for the study of concepts such as intelligence. We will take a behavioral definition of intelligence as our start: Rodney Brooks describes intelligence in terms of how successfully an agent interacts with its world to achieve goal directed behavior [39]. William James states, "Intelligent beings find a way to reach their goal, even if circuitous," [40]. Neurons have inherent local goals (to transmit signals, integrate synaptic input, optimize synaptic strengths, and much more) that provide the foundation to intelligently achieve meaningful behavioral goals. No doubt the basis for intelligence is inherent at birth, but an interaction with a sufficiently complex environment (learning) is needed to develop it.

In our cultured networks, the local goals of neural interaction are subject to detailed optical and electrical observation, while the execution of higher-level behavioral goals are observed through the activities of the robotic body. (Note that the behavioral goals are artificially constrained by the stimulation and recording transformations chosen.) We hope this combination will lead to a clearer definition and a better understanding of the neurological basis of intelligence, in addition to explanations of other psychological terms: learning, memory, creativity, etc. Neurobiology has given inspiration to AI since the advent of the perceptron and consequent artificial neural networks, which are based on the local properties (goals) of individual neurons. We wish to continue this trend by finding the principles of network processing by multiple neurons that lead to higher-level goals.

Network-wide Bursting. The activity of cultured neurons tends towards the formation of dish-wide global bursts (barrages) [8]: sweeps of fast, multiple neural firings throughout the network lasting between hundreds of milliseconds to seconds in duration. These barrages have been observed often in cultured neurons [41] but also in cortical slices [42] and in computer models [43]. Barrages of activity are reported in the cortex in vivo during early development, during epileptic seizures, while asleep, and when under anesthesia. These in vivo examples of barrages occur over finite periods of time. In contrast, barrages in vitro are continuous over the life of the culture. We consider the possibility that at some stage, dish-wide barrages of spiking activity are abnormal, a consequence of 'sensory deprivation' (manuscript in preparation), or a sign of arrested development [44].

For both a model system [43] and for cultured mouse spinal neurons [45], if more than 30% of the neurons are endogenously active, the neurons fire at a low steady rate of 1 to 5 Hz per neuron, while a reduction in the fraction of endogenously active cells leads to barrage activity. Endogenous activity is functionally similar to activity induced by afferent input, suggesting embodiment would lead to low steady firing rates. The hypothesis is then that the barrage activity may be due to the lack of an external environment with which to interact. We are developing animat mappings in which continuous sensory input quiets barrages, bringing the networks to a less 'sensory-deprived' state that allows more complex, localized activity patterns.

3.2 The Importance of Embodiment

The World and the Brain. Environmental deprivation leads to abnormal brain structure and function, and environmental exposure shapes neural development. Similarly, patterned stimulation supplied to cultured neurons may lead to more robust network structure and functioning than with trivial or no stimulation. The most dramatic examples of the importance of embodiment come from studies during development, when the brain is most malleable. Cognitive tests were performed on institutionalized children in Romania, children typically deprived of proper environmental and social interaction early in life [46], [47]. Compared to peers, the children showed severe developmental impairment that improved, however, after transplantation to a stable family. Those adopted prior to 6 months of age achieved nearly complete cognitive catch-up to similarly aged children, while those adopted after 6 months of age had significant but incomplete catch-up. Likewise, laboratory rats raised in environments with mazes and varied visual stimuli had 30% greater cortical synaptic density than those raised in minimalist environments, and performed better in various cognitive experiments [48], [49]. Synaptic morphology in adults [1] and adult neurogenesis is dependent on external cues [50] demonstrating that environmental interaction is important throughout life.

A disembodied neural culture, whose activity never influences future stimulation, will not develop meaningful associations to an input. In the brain, if a sensation is not useful in influencing future behavior (no association is made between the two) the percept of the sensation fades. The environment triggers an enormous number of sensory signals, and the brain develops to filter out the excess while perceiving the behaviorally relevant. All one-month-old infants can distinguish between the English L and R sounds. Five months later, Japanese infants lose the ability while American infants maintain it, because the distinction is not needed to understand the Japanese language [51]. Japanese adults consequently have great difficulty distinguishing these sounds, but perception of the distinction can be learned through targeted instruction. These studies further demonstrate how brain (re)wiring depends on environmental context and occurs throughout life: the brain focuses on perceiving the portions of the environment relevant to produce a meaningful interaction.

The Body and the Brain. The choice of how to instantiate an animat or hybrot is important to processing in cultured neural networks. For example, the body, with its various sensory apparatus and motor output, is what detects and interacts with the environment. In addition to how different environments cause differences in the brain, differences in the body will have analogous effects on the brain. Changes in the frequency or type of sensory input via practice or surgical manipulation of the body causes gross shifts in the functional organization of corresponding cortical areas (the somatotopic maps) [52]. Amputation causes a sudden change to a body, and amputees later report having at times a sensation or impression that the limb is still attached. The impression lasts for days or weeks in most cases (years or decades in other cases) and then gradually fades from consciousness [53]. These false 'phantom limb' sensations arise because the brain has wired itself for a given body that has now changed. This discrepancy further suggests the body and its interaction with the environment influence brain wiring and cognitive function. Neurally-Controlled Animats allow an unlimited variety of bodies to be studied; their structure and operating parameters can be easily varied to test effects on brain-body interactions.

4 Summary: Integrating Brain, Body, and Environment

The above paragraphs were worded as if the entities brain, body, and environment are independent. Finding physical boundaries between the three is easy, but since the brain is so enmeshed in the states of the body (influencing mood, attention, and more), which in turn are so enmeshed in the body's interaction with its environment, finding functional boundaries between the three is difficult, if possible at all [25], [54], [55]. Damasio contends that the mind depends on the complex interplay of the brain and the body, and consequently emotions and rationality cannot be segregated [56].

We have integrated our hybrots' brain (cultured network), body (robot or simulated animat), and environment (simulation, lab, or gallery) into a functional whole, even while the parts are sometimes 12,000 miles apart. Our experiments with these Neurally-Controlled Animats so far are rudimentary: we are still setting up the microscopic imaging systems to allow us to make correlations between changes in behavior and changes in neuron or network structure; we have not yet developed sensory-motor mappings that reliably result in learning. But in the process of creating this new research paradigm of embodied, situated cultured networks, we have already sparked a philosophical debate about the epistemological status of such semi-living systems,[3] and have raised a number of issues about the validity of traditional (disembodied) in vitro neural research.   We hope that others will make use of the tools we have developed such as our MeaBench software,[4] sealed-dish culture system [5], and multi-site stimulation tools [57], to pursue a wide variety of questions about how neural systems function. We expect that these inquiries will lead to fundamentally different, more capable, and less artificial forms of AI.

Acknowledgments. We thank the NIH (NINDS, NIBIB), the Whitaker Foundation, the NSF Center for Behavioral Neuroscience, and Arts Western Australia for funding. We thank Daniel Wagenaar, Radhika Madhavan, John Brumfield, Zenas Chao, Eno Ekong, Gustavo Prado, Bryan Williams, Peter Passaro, and Ian Sweetman for their many contributions to this work.

References

1.   Weiler, I. J., Hawrylak, N. & Greenough, W. T.: Morphogenesis in Memory Formation - Synaptic and Cellular Mechanisms. Behavioural Brain Research 66 (1995) 1-6
2.   Dennett, D. C.: Consciousness Explained. Little, Brown and Co., Boston, (1991)
3.   Morris, C. G.: Psychology: An Introduction. Appleton-Century-Crofts, New York, (1973)
4.   Potter, S. M.: Distributed processing in cultured neuronal networks. In: Nicolelis, M. A. L. (eds.): Progress In Brain Research: Advances in Neural Population Coding. Elsevier, Amsterdam, (2001) 49-62

---

[3]   Manson, N (2004) "Brains, vats, and neurally-controlled animats," in *Studies in the History and Philosophy of Biology and the Biomedical Sciences*, special issue on "The Brain in a Vat."

[4]   http://www.its.caltech.edu/~pinelab/wagenaar/meabench.html

5. Potter, S. M. & DeMarse, T. B.: A new approach to neural cell culture for long-term studies. J. Neurosci. Methods 110 (2001) 17-24

6. Watanabe, S., Jimbo, Y., Kamioka, H., Kirino, Y. & Kawana, A.: Development of low magnesium-induced spontaneous synchronized bursting and GABAergic modulation in cultured rat neocortical neurons. Neuroscience Letters 210 (1996) 41-44

7. Gross, G. W., Rhoades, B. K. & Kowalski, J. K.: Dynamics of burst patterns generated by monolayer networks in culture. In: Bothe, H. W., Samii, M. & Eckmiller, R. (eds.): Neurobionics: An Interdisciplinary Approach to Substitute Impaired Functions of the Human Nervous System. North-Holland, Amsterdam, (1993) 89-121

8. Kamioka, H., Maeda, E., Jimbo, Y., Robinson, H. P. C. & Kawana, A.: Spontaneous periodic synchronized bursting during formation of mature patterns of connections in cortical cultures. Neuroscience Letters 206 (1996) 109-112

9. Potter, S. M.: Two-Photon Microscopy for 4D Imaging of Living Neurons. In: Yuste, R., Lanni, F. & Konnerth, A. (eds.): Imaging Neurons: A Laboratory Manual. CSHL Press, Cold Spring Harbor, (2000) 20.1-20.16

10. Potter, S. M.: Vital imaging: Two photons are better than one. Current Biology 6 (1996) 1595-1598

11. Potter, S. M., Lukina, N., Longmuir, K. J. & Wu, Y.: Multi-site two-photon imaging of neurons on multi-electrode arrays. SPIE Proceedings 4262 (2001) 104-110

12. Wheeler, B. C.: Automatic discrimination of single units. In: Nicolelis, M. A. L. (eds.): Methods for Neural Ensemble Recordings. CRC Press, Boca Raton, (1999) 61-77

13. Wagenaar, D. A. & Potter, S. M.: A versatile all-channel stimulator for electrode arrays, with real-time control. Journal of Neural Engineering 1 (2004) 1-7

14. Meyer, J. A. & Wilson, S. W.: From Animals to Animats: Proceedings of the First International Conference on Simulation of Adaptive Behavior. MIT Press, Cambridge, (1991)

15. Potter, S. M., Fraser, S. E. & Pine, J.: Animat in a Petri Dish: Cultured Neural Networks for Studying Neural Computation. Proc. 4th Joint Symposium on Neural Computation, UCSD (1997) 167-174

16. DeMarse, T. B., Wagenaar, D. A., Blau, A. W. & Potter, S. M.: The Neurally Controlled Animat: Biological Brains Acting with Simulated Bodies. Autonomous Robots 11 (2001) 305-310

17. DeMarse, T. B., Wagenaar, D. A. & Potter, S. M.: The neurally-controlled artificial animal: A neural-computer interface between cultured neural networks and a robotic body. Society for Neuroscience Abstracts 28 (2002) 347.1

18. Reger, B. D., Fleming, K. M., Sanguineti, V., Alford, S. & Mussa-Ivaldi, F. A. Connecting brains to robots: The development of a hybrid system for the study of learning in neural tissues. In: Proc. of the VIIth Intl. Conf. on Artificial Life. (2000) 263-272

19. Sharp, A. A., Abbott, L. F. & al, e.: The Dynamic Clamp: Computer-generated conductances in real neurons. Pre-print (1992)

20. Butera, R. J., Wilson, C. G., DelNegro, C. A. & Smith, J. C.: A methodology for achieving high-speed rates for artificial conductance injection in electrically excitable biological cells. Ieee Transactions on Biomedical Engineering 48 (2001) 1460-1470

21. Simoni, M., Cymbaluyk, G., Sorensen, M., Calabrese, R. & DeWeerth, S.: Development of Hybrid Systems: Interfacing a Silicon Neuron to a Leech Heart Interneuron. In: Leen, T. K., Dietterich, T. G. & Tresp, V. (eds.): Advances in Neural Information Processing Systems 13, NIPS 2000. MIT Press, Boston, (2001) 173-179

22. Braitenberg, V.: Vehicles, experiments in synthetic psychology. MIT Press, Cambridge, Mass., (1984)

23. Arkin, R. C.: Behavior-Based Robotics. MIT Press, Cambridge, (1999)

24. Brooks, R. A.: Cambrian Intelligence: The Early History of the New AI. MIT Press, Cambridge, MA, (1999)

25. Clark, A.: Being There: Putting Brain, Body, and the World Together Again. MIT Press, Cambridge, (1997)

26. Shkolnik, A. C. Neurally Controlled Simulated Robot: Applying Cultured Neurons to Handle and Approach/Avoidance Task in Real Time, and a Framework for Studying Learning In Vitro. In: Potter, S. M. & Lu, J.: Dept. of Mathematics and Computer Science. Emory University, Atlanta (2003)
27. Engert, F. & Bonhoeffer, T.: Dendritic spine changes associated with hippocampal long-term synaptic plasticity. Nature 399 (1999) 66-70
28. Choi, Y. et al. High aspect ratio SU-8 structures for 3-D culturing of neurons. In: ASME International Mechanical Engineering Congress and RD&D Expo. Washington, D. C. (2003)
29. Blum, R. A. et al.: A custom multielectrode array with integrated low-noise preamplifiers. In: Akay, M. (eds.): Proceedings of the IEEE Engineering in Medicine and Biology Conference. (2003) 3396-3399
30. Jimbo, Y., Tateno, T. & Robinson, H. P. C.: Simultaneous induction of pathway-specific potentiation and depression in networks of cortical neurons. Biophysical Journal 76 (1999) 670-678
31. Tateno, T. & Jimbo, Y.: Activity-dependent enhancement in the reliability of correlated spike timings in cultured cortical neurons. Biological Cybernetics 80 (1999) 45-55
32. Marom, S. & Shahaf, G.: Development, learning and memory in large random networks of cortical neurons: Lessons beyond anatomy. Quarterly Reviews of Biophysics 35 (2002) 63-87
33. Eytan, D., Brenner, N. & Marom, S.: Selective adaptation in networks of cortical neurons. Journal of Neuroscience 23 (2003) 9349-9356
34. Turing, A. M.: The chemical basis of morphogenesis. Philosophical Transactions of the Royal Society of London B 237 (1953) 37-72
35. Yamamoto, T. & Kuniyoshi, Y. Stability and controllability in a rising motion: a global dynamics approach. In: International Conference on Intelligent Robots and Systems (IROS). Lausanne, Switzerland (2002) 2467-2472
36. Fodor, J. A.: Methodological Solipsism Considered as a Research Strategy in Cognitive-Psychology. Behavioral and Brain Sciences 3 (1980) 63-73
37. Vera, A. H. & Simon, H. A.: Situated Action - a Symbolic Interpretation. Cognitive Science 17 (1993) 7-48
38. Edelman, G. M. & Tononi, G.: A universe of consciousness: how matter becomes imagination. Basic Books, New York, (2000)
39. Brooks, R. A.: Intelligence without representation. Artificial Intelligence 47 (1991) 139-159
40. James, W.: The principles of psychology. H. Holt, New York, (1890)
41. Nakanishi, K. & Kukita, F.: Functional synapses in synchronized bursting of neocortical neurons in culture. Brain Research 795 (1998) 137-146
42. Corner, M. A. & Ramakers, G. J.: Spontaneous bioelectric activity as both dependent and independent variable in cortical maturation. Chronic tetrodotoxin versus picrotoxin effects on spike-train patterns in developing rat neocortex neurons during long-term culture. Ann N Y Acad Sci 627 (1991) 349-53
43. Latham, P. E., Richmond, B. J., Nelson, P. G. & Nirenberg, S.: Intrinsic dynamics in neuronal networks. I. Theory. Journal of Neurophysiology 83 (2000) 808-827
44. Tabak, J. & Latham, P. E.: Analysis of spontaneous bursting activity in random neural networks. Neuroreport 14 (2003) 1445-1449
45. Latham, P. E., Richmond, B. J., Nirenberg, S. & Nelson, P. G.: Intrinsic dynamics in neuronal networks. II. Experiment. Journal of Neurophysiology 83 (2000) 828-835
46. Rutter, M.: Developmental catch-up, and deficit, following adoption after severe global early privation. Journal of Child Psychology and Psychiatry and Allied Disciplines 39 (1998) 465-476
47. O'Connor, T. G., Rutter, M., Beckett, C., Keaveney, L. & Kreppner, J. M.: The effects of global severe privation on cognitive competence: Extension and longitudinal follow-up. Child Development 71 (2000) 376-390

48. Black, J. E., Isaacs, K. R., Anderson, B. J., Alcantara, A. A. & Greenough, W. T.: Learning Causes Synaptogenesis, Whereas Motor-Activity Causes Angiogenesis, in Cerebellar Cortex of Adult-Rats. Proceedings of the National Academy of Sciences of the United States of America 87 (1990) 5568-5572
49. Diamond, M.: Morphological cortical changes as a consequence of learning and experience. In: Scheibel, A. B. & Wechsler, A. F. (eds.): Neurobiology of Higher Cognitive Function. Guilford Press, New York, (1990) 370
50. Gross, C. G.: Neurogenesis in the adult brain: death of a dogma. Nature Reviews Neuroscience 1 (2000) 67-73
51. Kuhl, P. K. et al.: Cross-language analysis of phonetic units in language addressed to infants. Science 277 (1997) 684-686
52. Buonomano, D. V. & Merzenich, M. M.: Cortical plasticity: From synapses to maps. Annual Review of Neuroscience 21 (1998) 149-186
53. Ramachandran, V. S. & Hirstein, W.: The perception of phantom limbs - The D.O. Hebb lecture. Brain 121 (1998) 1603-1630
54. Varela, F. J., Thompson, E. & Rosch, E.: The embodied mind: cognitive science and human experience. MIT Press, Cambridge, Mass., (1993)
55. Pfeifer, R. & Scheier, C.: Understanding Intelligence. The MIT Press, Cambridge, Massachusetts, (1999)
56. Damasio, A. R.: Descartes' Error: Emotion, Reason, and the Human Brain. Gosset/Putnam Press, New York, (1994)
57. Wagenaar, D. A. & Potter, S. M.: Real-time multi-channel stimulus artifact suppression by local curve fitting. J. Neurosci. Methods 120 (2002) 113-120

Mutual Adaptation in a Prosthetics Application

Hiroshi Yokoi,[1] Alejandro Hernandez Arieta,[2] Ryu Katoh,[2] Wenwei Yu,[3]
Ichiro Watanabe[4], and Masaharu Maruishi[5]

[1] The University of Tokyo, Dept. Precision Eng., Hongo 7-3-1, Tokyo 113-8656, Japan
hyokoi@prince.pe.u-tokyo.ac.jp
[2] Hokkaido University, Complex System Eng., N13-W8, Sapporo 060-8628, Japan
{alex, katoh}@complex.eng.hokudai.ac.jp
[3] Chiba University, Eng. Dept. Medical Systems, Inage Ku Yayoi , Chiba, 263-0043, Japan
yuwill@faculty.chiba-u.ac.jp
[4] Hokkaido University, Medical Hospital, N 15, W 7, Kita-ku, Sapporo, 060-8638, Japan
wichiro-hok@umin.ac.jp
[5] Hiroshima Rehabilitation Center,
maruishi@rehab-hiroshima.gr.jp

Abstract. Prosthetic care for handicapped persons requires new and reliable robotics technology. In this paper, developmental approaches for prosthetic applications are described. In addition, the challenges associated with the adaptation and control of materials for human hand prosthetics are presented. The new technology of robotics for prosthetics provides many possibilities for the detection of human intention. This is particularly true with the use of electromyogram (EMG) and mechanical actuation with multiple degrees of freedom. The EMG signal is a nonlinear wave, and has time dependency and big individual differences. The EMG signal is a nonlinear wave that has time dependency and significant differences from one individual to another. A method for how an individual adapts to the processing of EMG signals is being studied to determine and classify a human's intention to move. A prosthetic hand with 11 degrees of freedom (DOF) was developed for this study. In order to make it light-weight, an adaptive joint mechanism was applied. The application results demonstrate the challenges for human adaptation. The f-MRI data show a process of replacement from a phantom limb image to a prosthetic hand image.

1 Introduction

Robotic applications in prosthetics are useful for supporting individuals who have been disabled, particularly those who have lost limbs, sustained muscular debilitation, or suffer nerve diseases. Advanced technology for robotics allows for controls that can discriminate between EMG signals by using a learning paradigm that can be adapted to many human activities. EMG patterns are unique to each individual. Therefore, an adaptable prosthetic controller discerns subtleties in EMG signals and, in a sense, builds a relationship through incremental learning to determine human intention. On the other hand, an individual will, in turn, adapt to prosthetic equipment. In the case of a lost forearm, an individual experiences a phenomenon called "phantom limb

F. Iida et al. (Eds.): Embodied Artificial Intelligence, LNAI 3139, pp. 146–159, 2004.

image," which remains for sometime.A phamtom limb image is the result of signals perceived to be from the forearm that originate in the motion area of the cerebrum cortex. A phantom image is a part of an individual's body image, and it is a biological example of the embodied AI representation of the end effectors. The human adaptation can be observed as an effect of the prosthetic hand application by the changes in the f-MRI images. Such a double adaptation system produces a mutual adaptation, and gives important phenomena of the embodiment issue. This paper shows the adaptable prosthetic system, and discusses the challenges for the mutual adaptation.

The EMG is a bionic signal with the potential for controlling mechanical products. The signal may be detected on the surface of a human body, and it reflects the motion of muscles. The potentials and difficulties associated with EMG signal patterns have been the subjects of many reports. [1, 2, 3] A small control package and light-weight joint mechanism are demonstrated and described in this paper.

A powered prosthetic hand with many degrees of freedom (DOF) imitates a natural hand and enhances its functionality. [4, 5, 6] Some products have already been developed for applications in the medical and welfare disciplines. [7] These previous applications are significant in the areas of medicine and welfare as well as robotics and mechanical engineering because they may lead to the development of a humanoid hand. Between industrial robotic hands and externally powered prosthetic hands, such as those that are EMG-controlled, there are large differences in the specifications related to size, weight, appearance, speed, power, and precision. For these requirements, a tendon-driven mechanism was investigated. [8, 9] A proposal is presented in this paper for the development of a prosthetic hand with 11 DOF and an adapative joint mechanism based on a tendon-driven mechanism.   A photograph of the proposed system is shown in Fig. 1.

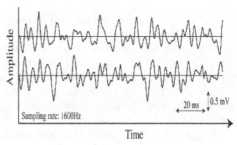

Fig. 1. EMG prosthetic hand.

Fig. 2. An EMG signal pattern detected from the surface of a forearm undergoing supination.

The proposed controller for the EMG prosthetic system is based on an on-line learning method that was developed by Nishikawa.[10] A machine that has a large DOF is difficult to control, and the EMG signal pattern is not generally stable. An example of an EMG signal is shown in Fig. 2. A good interface is necessary in order to use such a machine. The on-line learning method realizes adaptive functionality for the controller of the mechanical device. The EMG prosthetic system uses an EMG signal in order to transmit human intentions to control the mechanical device; however, the pattern of the EMG signal sometimes changes even if the motion of the hand is the same. [1, 10] This pattern is affected by the environmental temperature, time dependence, noises, and individual differences. For these unstable difficulties, the on-line learning method is applied in order to determine the target EMG signal pattern that is suitable for the motion of a hand or body.

Chapter 2 shows the background and requirements for the prosthetic hand. Chapter 3 describes the proposed adaptive joint mechanism and experimental results. Chapter 4 describes the controller of the EMG prosthetic system and its performance.

2 Prosthetic Hand

The human forearm consists of 5 fingers, a palm, and a wrist joint. Each finger has 3 joints and 4 DOF. The palm has many joints, but the motive freedom is integrated into one DOF. The wrist joint has 3 DOF. Therefore, in order to realize the functional level of a human hand, the ideal prosthetic hand should have up to 24 DOF.

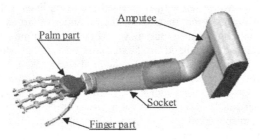

Fig. 3. Powered prosthetic hand for forearm.

However, a prosthetic hand also has physical restrictions, such as weight, size, and power. Different sizes should be produced, from small to large, to fit children and adults. The weight needs to be similar to that of a real hand as well. It is necessary to reduce the internal pressure in the socket as well as the load on the upper arm as it supports the prosthetic hand. This demonstrates that the designed prosthetic hand must be light. Therefore, conventional myoelectric prosthetic hands on the market[7] only have 2~4 functions, including grasping, hand opening, and wrist rotations (supination / pronation).

The grasping power should be strong enough to grip a glass of water, a condition that requires more than 3.5 kgcm. The objective values of the speed and torque are also critical for a prosthetic hand. Certainly, researchers have succeeded in producing a prosthetic hand with higher speed and power that can be more precisely controlled than a human hand (without including the planning of hand motion). These specifications have been achieved with no restrictions for the most part. On the other hand, prosthetic hands currently on the market can only achieve power and speed equal to those of a frail individual (up to 100N and 1.3 Hz when grasping).[7] This problem originates in the actuator's power-weight ratio. We believe that, for the purposes of normal activities, it is not necessary for a prosthetic hand to offer high speed and high power simultaneously. Considering the current potential of the actuator used for a prosthetic hand, it would be adequate to supply high speed and high torque separately through the use of a kind of torque converter. On the other hand, precision of control is not necessary for the prosthetic controller. This is because neither the driver nor the amputee can control the prosthetic hand as precisely as the central nerve system can control a human hand. One of the reasons is the low ability of the driver. Even if it could control 10~20 motion patterns or a few joints of forces, it would be inferior to the real motion patterns of a human hand. Another reason is that most drivers give the amputee less feedback information; actually, they only provide visual information. It is difficult to execute complex tasks without an adequate amount of feedback information.

In the course of our study, therefore, an electrical prosthetic hand based on a tendon-driven system was developed. We chose to give up the apparent advantage of precice control in order to build a light-weight hand and complexly controlled system. This system arranges actuators on the outside of the hand and employs wires and tubes as transmitters because the greater part of the load of a current prosthetic hand is an actuator (motor) arranged into the hand. This actuator reduces the load on the amputee by shifting the center of balance from the hand to the forearm or another part of the body. This paper presents the design of an adjustable power-transmitting mechanism that controls the torque-velocity ratio and improves the power and speed of a prosthetic hand. Moreover, through the enhancement of the grasping power, a proximal joint-assisting mechanism in which distal actuators also provide force to the proximal joint becomes available. These mechanisms make the system complex, time-delay, and non-linear. In the case that the desired trajectory is given, feedback control, represented by the Bang-Bang control, and canonical PID control have been suggested as the motor control model to realize a target-reaching behavior and are in turn employed to control the manipulators. However, when the control object contains components that would cause nonlinear and/or time-delayed responses, these kinds of controls would cause a large overshoot or oscillation phenomenon.

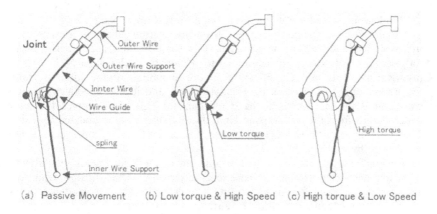

(a) Passive Movement   (b) Low torque & High Speed   (c) High torque & Low Speed

Fig. 4. Three functions of the adaptive joint mechanism.

3 Adaptive Joint Mechanism

A schematic diagram of an adaptive joint mechanism is shown in Fig. 5.[8] A spring connects a frame and a guide roll of wire. The guide roll can shift proportionally to the load. In the case of a light load, as shown in Fig. 4(b), the spring approaches the fulcrum, making its angular velocity high and its torque low. On the other hand, in the case of a heavy load, as shown in Fig. 4(c), the guide roll leaves the fulcrum, making its angular velocity low and its torque high. Accordingly, the spring-connected guide roll provides an adjustable power-transmitting function. From another viewpoint, the adaptive joint provides a "passive adaptive grasp. [9] " Dechev et al. indicate that a prosthetic hand with a hard pincher requires a high and wasteful pinch force to secure objects, hence, hands that are more flexible are needed. This mechanism adjusts a torque-angular-velocity ratio of 1:6 in the restrictions inherent in the size of a prosthetic hand. Furthermore, this mechanism has another function in addition to the adjustable function of velocity and torque. As shown in Fig. 4, the passive motion function can be obtained by precisely placing a power wire at the center of the joint rotation.

4 Main Controller for the EMG Prosthetic Hand

The requirements for the controller of EMG prosthetic hand are summarized as follows:

- The internal state and system parameter of controller should be changeable.
- The motion functions of an EMG prosthetic hand should be improved.
- The amputee should receive visual feedback of the EMG signal pattern.
- The learning mechanism should function even when the evaluation is weak.
- The learning speed should be fast enough for real-time change.

The proposed controller based on an on-line learning method consists of three units, as shown in Fig. 5. The units are Analysis, Classification, and Supervision.

Analysis: This unit extracts the feature vector V from the EMG signal S.

Classification: This unit classifies the predicted motion of the forearm from the feature vector V produced in the Analysis Unit, and it also generates the control command θ for the prosthetic hand mechanism. This unit receives the learning pattern set Ψ from the Supervision Unit, and its inner state is updated by using Ψ. The mapping functions of the feature vector and the control command are determined by Ψ.

Supervision: This unit manages the learning pattern set Ψ for tracking the alterations of an amputee's characteristics. These sets are sent to the Classification Unit.

This controller employs four management methods, which are explained as follows:

I. Best selection of sensor position by entropy evaluation

This method decides the best sensor position by calculating the conditional entropy E of learning data set Ψ, which indicates the ambiguity between an amputee's instruction θ_{teach} and its feature vector V.[11]

II. Manual addition of a learning pattern by a teaching signal

The learning pattern set Ψ is produced by the amputee's instruction θ_{teach} and the feature vectors V.

III. Automatic elimination of a learning pattern by the frequency of usage

Automatic elimination of redundant or harmful learning data based on the scored frequency of usage obtained by classification result θ and feature vector V.

IV. Automatic addition of a learning pattern by a classification result

Assuming that the motion continuous when discontinius classification results θ_d is detected; discontinius feature vector V_d is added automatically in the learning data set Ψ.

Therefore, this controller will be used to map the individual characteristics of an amputee, with specific attention given to the EMG signals and the motor command from the prosthetic hand.

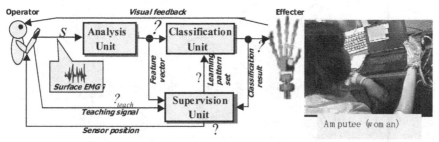

Fig. 5. The procedure of the on-line learning mechanism.

4.1   Experiment

An overview of the experimental system is shown in Fig. 6(a). The detected EMG signal from a dry-type sensor is sent to the amp, and it is amplified 10,000 times. The amplified signal is digitized into 12-bit data with a 1600 sampling rate by an AD transfer board, and it is sent to the controller as the digitized EMG signal. Dry-type electrodes are used and placed in the vicinity of the elbow, which is involved with the origin of the forearm muscles and from which the intent to move can be detected by EMG. Moreover, such applications can be used in the severest types of forearm amputation. The positions of the surface electrodes on the radius, channels one/three, and on the ulna, channels two/four, are shown in Fig. 6(b). In addition, the reference identifiers are placed on the wrist near the hand.

The classifier is implemented by software in a PC. To send the teaching signal to the supervision unit, a keyboard is used as its interface. A computer graphic (CG) representation of a human hand is substituted for the prosthetic hand because the prosthetic hand has not been developed. The subject watches the monitor and pushes the key corresponding to the teaching motion when he determines that the prosthetic hand is not moving as it should. The supervision unit generates three training patterns for each teaching signal. In order to test the performance of a classification ability of this classifier, an ability tests in which the subject controls the prosthetic hand are carried out according to instructions presented on the monitor (Fig. 6(a)). The subject executes one motion for three seconds in the ability test. Classification errors are calculated from a comparison of the instructions and the control commands in the ability test. The ability test begins when the subject determines that he is in control of the CG. The ten forearm motions that the controller classifies contain four patterns of wrist motion and six patterns of hand motion.

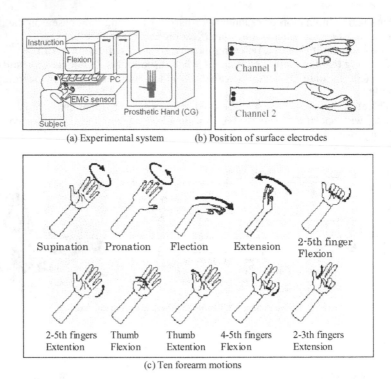

(a) Experimental system   (b) Position of surface electrodes

Supination   Pronation   Flection   Extension   2-5th finger Flexion

2-5th fingers Extention   Thumb Flexion   Thumb Extention   4-5th fingers Flexion   2-3th fingers Extension

(c) Ten forearm motions

Fig. 6. An experimental setup is demonstrated in this chapter. The proposed controller is implemented by PC software. A subject pushes the keyboard to teach and watches computer graphics (CG) on the monitor instead of the prosthetic hand. In the ability test, a subject executes instructed motions.

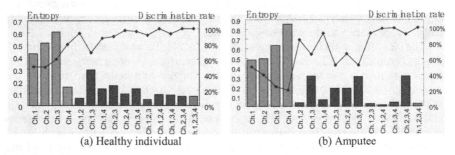

(a) Healthy individual   (b) Amputee

Fig. 7. A variation in entropy due to changes in the position of a surface electrode for three subjects. The ten motions contain four patterns of wrist motion and six patterns of hand motion.

4.2   Experimental Results 1: Selecting the Position of the Surface Electrode by Entropy Evaluation in a Supervision Unit

The correlation between entropy evaluations calculated from all possible combinations (2^4 -1) of four sensors on three subjects and the classification rates is shown in Fig. 7. These results showed a strong correlation between the entropy and the classification rate (a correlation factor of -0.89). The classification rate was 90% or more when the entropy was 0.14 or less.

Moreover, there are combinations of surface electrode positions allowing low entropy values to be obtained even when there are fewer measurement points. This means that the selection of surface electrode positions is important for information extraction.

5   EMG Prosthetic Hand

The proposed EMG prosthetic hand has 5 fingers and a wrist. Each finger has 3 joints; however, the DIP joint and PIP joint are actuated by a common tendon wire. The MP joint is actuated by one motor. Thus, each finger has 2 DOF as an active motion control. The only thumb cmc joint is directly connected by a servo motor in order to realize the abduction of motion. The wrist is supported by two motors that actuate pronation/supination and extention/flexion. Therefore, the proposed EMG prosthetic hand has 12 DOF as an active motion control. Each joint that is actuated by a tendon wire has an adaptive joint mechanism, such as the one shown in Chapter 3. The entire weight of the RC servo motors is 280g. The weight of the aluminum body of the hand is 204g. The weight of the small controller and battery is 100g. The weight of the socket and cables is 623g. The total weight is 1207g, which is almost identical to the weight of an adult woman's forehand.

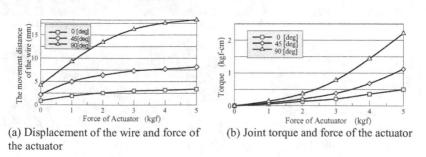

(a) Displacement of the wire and force of   (b) Joint torque and force of the actuator
the actuator

Fig. 8. Static of the adaptive joint mechanism

The experimental results show the statics of a proposed finger. The relationship between the displacement of wire and the force of the actuator is shown in Fig. 8(a). The angle (0 degrees, 45 degrees, and 90 degrees) is measured between a finger element and the wire direction. This result shows that a bigger angle requires a longer wire movement. The relationship between the joint torque and the force of the actuator is shown in Fig. 12(b). This result shows that a bigger angle produces more joint torque. If the maximum torque of an RC servo motor is 3.6 kgcm at 6V, the maximum torque of a finger joint is 1.1 kgcm. If there is no obstacle of finger motion, only 5 mm of wire movement would be sufficient to rotate the finger joint at 90 degrees. The maximum velocity of the rotation of a finger joint was 200 degrees per second by using an RC servo motor. The maximum frequency of the tapping motion (0 to 90 degrees) of a finger obtained 1 Hz.

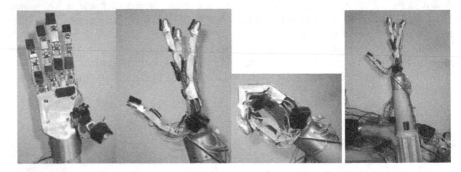

(a) Aluminum body of a finger part and a palm part   (b) Whole products

Fig. 9. Proposed EMG prosthetic hand.

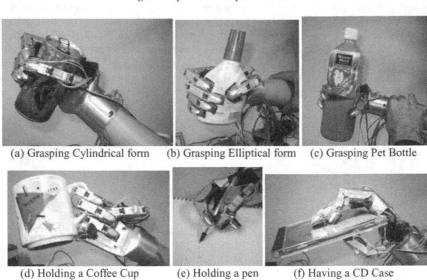

(a) Grasping Cylindrical form   (b) Grasping Elliptical form   (c) Grasping Pet Bottle

(d) Holding a Coffee Cup   (e) Holding a pen   (f) Having a CD Case

Fig. 10. The proposed prosthetic hand

A prototype of a prosthetic hand was developed with the use of this finger mechanism. A snapshot of the proposed EMG prosthetic hand is shown in Fig. 9. In Fig. 16, the following positions are shown: (a) the grasping of a cylindrical form, (b) the grasping of an elliptical case, and (c) the grasping of a pet bottle with juice. The proposed hand can hold 1000 cc juice in a pet bottle in a stable state. The objects were shaken to test the grasping torque. The maximum capacity amount of juice that could be held in the pet bottle was 400cc. The position for holding a coffee cup is shown in Fig. 10(c). The proposed hand grasps the cup with two fingers. The position of the hand holding a pen to write the letter "A" is shown in Fig. 10(d). Fingers grasping a CD case are shown in Fig. 10(e).

6 Tactile Sensory Feedback

One of the current problems with prosthetic devices is the paucity of tactile information, which results in difficult manipulation of the prosthesis and clumsy actions from the device. This lack of feedback limits reaction and adaption to the changing environmental conditions. With more tactile information, robotic hands perform better.[12, 13, 14] The application of tactile sensing on prosthetic devices has been addressed, [15, 16] giving feedback to the controller in the case of electrical devices. In order to obtain tactile feedback, the prosthetic hand has been equipped with pressure sensors based on conductive silicon rubber. [17] The sensors are placed in the fingertips and in the base of each finger (black Rubbers in Fig. 10). The signal provided by the sensors has the following characteristics: non-linear and with hysteresis, a working range between 15 gf and 400 gf, high sensitivity. The sensor response graph is shown in Fig. 11. The signal from the sensors does not need to be linear to provide enough information to regulate the grasping task.

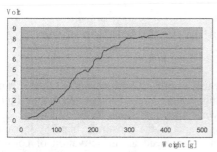

Fig. 11. Average of the sensor element.   **Fig. 12.** Biofeedback using functional electric stimulation.

Currently, for research purposes, an A/D acquisition card, with 12 bits of resolution, makes the interface with the sensors. For practical purposes, the sensor's signal-acquisition process can be transferred to a microcontroller. In this study, an H8-tiny microcontroller from Hitachi was used. The application of tactile sensors increases the performance in the grasping tasks, enabling the controller to react to the environment. However, still, there is a need to provide information to the user in order to gain subconscious control of the device.[18] Having visual feedback as the only means to confirm the performance of the task places a great burden of concentration upon the user, making the manipulation of the prosthesis difficult and tiresome. In hybrid systems, as well as in body-powered devices, the user has direct contact with the prosthesis, increasing the number of channels from where the user is acquiring information about the status of, for example, a prosthetic hand, and allowing the generation of new paths in the brain for its control. Still, the movement of these devices is limited, as are their applications. Some research on biofeedback through vibro-stimulation has been done; [19] however, habituation prevents any long-term sensitivity to vibro-stimulation. In order to solve this problem, we proposed the use of functional electrical stimulation (FES) as a biofeedback signal to the system user, as the usefulness of electrical stimulation as a means of providing feedback in humans has been demonstrated. [20] In particular, the use of the interferential current method, in order to provide a painless signal, has also been demonstrated. [21] Due to the capacitive properties of the skin, the resistance to an electrical current keeps an inverse relationship with the frequency (about 4 kHz). The higher the frequency, the lower the resistance. By applying a high frequency signal at the skin level, the reduction of the resistance of the skin eliminates the pain reaction, thus, reducing the rejection by the user. By using the interference, the envelope resulting from the intersection of the two signals applied has the same frequency as the difference between the two signals.

Our systems consist of a microcontroller, which generates the frequency signal, and a motor driver as a transducer between the commands and the generation of the electrical signal. We are using PALS neurostimulation electrodes (AXELGAARD), for their reusability, self-adherence, and compliance with the non-invasive policy of our system. The electrodes are placed as shown in Fig. 12. Currently, the commands are generated from a personal computer, but the system can be transferred completely into a microntroller array that can generate biosignals from a tactile sensory signal in an independent manner.

7 Discussion: Imaging Experiment in the Adaptation of the Prosthetic Hand

The prosthetic hand is a new tool for a forearm amputee. We investigated the learning process to control such a machine with a high DOF. The prosthetic hand is designed as an adaptable machine that can predict what a patient wants to do by EMG signal processing. On the other hand, an amputee adapts to the prosthetic hand by learning how to control it. The amputee's reaction is observed as an activation of cortical mapping by using functional magnetic resonance imaging (f-MRI). [22] The topographic image of f-MRI shows broad activation and demonstrates the effect of the application of an EMG prosthetic system. In an f-MRI room, the motion of a prosthetic hand is projected on a screen in front of the amputee. The activation of a primary sensory-motor area in each condition is demonstrated in Fig. 13. Before the use of prostheses, for amputees who experienced phantom pain in a grasping task, [23] the f-MRI demonstrated broad activation on the contralateral M1 (filling-in). On the contrary, f-MRI images of amputees without phantom pain demonstrated an activation shifting medially than the contralateral activation of healthy hand in Fig. 13. The use of the prosthesis affected the location of the activation in the M1.

The experiment shows an example of cortical plastic change during the use of an EMG prosthetic hand. Furthermore, it demonstrates the possibilities of neuro-modulating technology. [24] This change requires mutual adaptation, and it produces an unstable relationship between man and machine. In addition, it produces a local minimum of parametric optimization for the classification unit of the controller. The f-MRI image is a somatosensory status indicator of the skill level of the prosthetic system. The activation of the visual cortex shows the requirement for a feedback system. Therefore, we have new challenges for the practical application of a prosthetic system:

a) Active search is necessary for the mutual adaptation in order to build an adaptable prosthetic system.

b) Evaluation of the skill level is necessary for the active search (Proprioception).

c) Tactile sensory feedback should replace visual feedback.

Images were collected using a 1.5T MRI system (Signa, GE, USA) equipped with echo-planar imaging (EPI) capabilities and a radio frequency (RF) surface coil. Sequence parameters: gradient-echo EPI, repetition time (TR) = 5000 ms, echo time (TE) = 60 ms, field of view (FOV) = 200 mm, resolution = 64 by 64. Eighteen contiguous axial slices from the most rostral part of the brain with a slice thickness of 5 mm each were acquired. The subjects' heads were immobilized using foam pads. Forty images were obtained per slice over a four-minute period, during which subjects alternated between 30-second periods of rest and activity. Images were analyzed by statistical parametric mapping (SPM96) analysis.

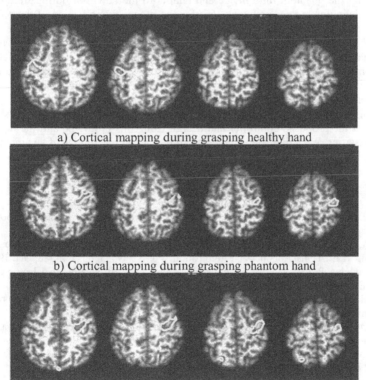

a) Cortical mapping during grasping healthy hand

b) Cortical mapping during grasping phantom hand

c) Cortical mapping during grasping EMG prosthetic hand

Fig. 13. f-MRI images of the plastic change due to amputation

8 Summary

This paper contains a proposal for an adaptive joint mechanism and a controller based on EMG signals for an adaptable prosthetic system. The experimental results demonstrate that the proposed mechanism can achieve powerful enough grasping to hold a bottle with 500 cc juice, and, in addition, to hold a CD case using two fingers. The positions of the fingers can retain a stable grip while exterting a sliding motion between the thumb and other fingers. The controller realized 10 patterns of finger motion based on EMG signal patterns. The total weight was 1.2 kg, which included the controller and batteries.

The adaptable prosthetic controller is proposed and applied to the EMG prosthetic hand. The experiment shows that the proposed controller realized 10 patterns of prosthetic hand motion based on EMG signal classification. This controller technology gives a new possibility to smooth the manipulation of the machines without a switch and lever. In the experimental results, the f-MRI image demonstrated a mutual adaptation among man and machine. The amputee feels the existence of the forearm by the phantom limb image and adapts to the new tool during training. The mutual adaptation creates new questions for the practical application of the prosthetic system presented in the discussion part of the paper.

The concept of the proposed system is applied to another application to show additional possibilities. The adaptive joint mechanism and controller are applied to the development of a power assist device [25] for the lower back on the hip joint. The prototype was developed, and it shows that an EMG-signal, which was detected from the upper leg, could control the proposed device and could assist with a 30 kg load. The total weight of the developed device was 7 kg. In future studies, we will show the validity of the proposed controller and design using this prototype and the developed EMG classifier. We will also examine and study the daily activities of an amputee.

References

1. M. Uchida, H. Ide, S. P. Ninomiya, "Control of a robot arm by myoelectric potential," Journal of Robotics and Mechatronics, Vol. 5, No. 3, pp. 259-265, 1993.
2. K. A. Farry, I. D. Walker, R. G. Baraniuk, "Myoelectric teleoperation of a complex robotic hand," IEEE Trans. Robotics and Automation, Vol. 12, No. 5, pp. 775-788, Oct. 1996.
3. B. Hudgins, P. Parker, R. N. Scott, "New strategy for multifunction myoelectric control," IEEE Trans. Biomedical Engineering, Vol. 40, No. 1, pp. 82-94, Jan. 1993.
4. H. H. Sears, J. Shaperman, "Electric wrist rotation in proportional-controlled systems," Journal of Prosthetics and Orthotics, Vol. 10, No. 4, pp. 92-98, 1998.
5. N. Dechev, W. L. Cleghorn, S. Naumann, "Multiple finger, passive adaptive grasp prosthetic hand," Mechanism and Machine Theory, Vol. 36, pp. 1157–1173, 2001.
6. M. Neal, "Coming to grips with artificial hand design," Design Engineering, pp. 26-27,29,32,34, March 1993.
7. "SensorHand technical information booklet," Otto Bock Co., Ltd., 2001, http://www.healthcare.ottobock.com/
8. Y. Ishikawa, W. Yu, H. Yokoi, Y. Kakazu, "Development of robot hands with an adjustable power transmitting mechanism," Intelligent Engineering Systems Through Neural Networks, Vol. 10, pp. 631–636, ASME Press, 2000.

9. S. Hirose, S. Ma, "Coupled tendon-driven multijoint manipulator," Proc. IEEE Intl. Conf. on Robotics & Automation, pp. 1268-1275, 1991.
10. D. Nishikawa, et al., "On-line learning based electromyogram to forearm motion classifier with motor skill evaluation," JSME Intl. Journal Series C, Vol. 43, No. 4, pp. 906–915, 2000.
11. R. Katoh, et al., "Evaluation of biosignal processing method for welfare assisting devices - Evaluation of EMG information extraction processing using entropy -," Journal of Robotics and Mechatronics, Vol. 14, No. 6, pp. 573-580, 2002.
12. S. Crinier, "Behavior-Based Control of a Robot Hand using Tactile Sensors." Master's thesis written at the Center for Autonomous Systems, Royal Inst. Tech. in Sweden, 2002.
13. G. Buttazzo, et al., Robot Tactile Perception. In C. S. G. Lee, editor, Sensor Based Robots: Algorithms and Architectures. Springer Verlag, Berlin Heidelberg, Germany, 1992.
14. A. Bicchi, "Optimal Control of Robotic Grasping." In Proc. American Control Conf., 1992.
15. M. C. Carrozza, et al., "A wearable artificial hand for prosthetics and humanoid robotic applications," IEEE-RAS Intl. Conf. on Humanoid Robots, Waseda Univ., 2001.
16. P. J. Kyberd, P. H. Chappell, "The Southampton hand: An intelligent myoelectric prosthesis." J. Rehabilitation Research and Development, Vol. 31, No. 4, pp. 326-334, 1994.
17. Conductive rubber silicon sheet. Characteristic Performance CS57-7RSC(CSA).
18. D. H. Plettenburg, "Prosthetic control: A case for extended physiological proprioception." Univ. New Brunswick, MyoElectric Controls/Powered Prosthetics Symposium 2002.
19. A. Rios Poveda, "Myoelectric prosthesis with sensorial feedback." University of New Brunswick, MyoElectric Controls/Powered Prosthetics Symposium 2002.
20. Y. Handa et al., Sensory feedback on the FES, J. Biomechanism, Vol. 12, No. 1, 1988.
21. M. Yoshida, Y. Sasaki, N Nakayama, "Sensory feedback for biomimetic prosthetic hand," BPES 2002, The 17th living body and physiology engineering symposium. (in Japanese)
22. H. Flor et al., Taub.Phantom-Limb Pain As A Perceptual Correlate Of Cortical Reorganization Following Arm Amputation. Nature, Vol. 375, pp. 482-84, June 8, 1995.
23. M. MacLachlan et al., "Psychological correlates of illusory body experiences," Journal of Rehabilitation Research and Development, Vol. 40, No. 1, pp. 59-66, 2003.
24. Y. Mano et al., "Central motor reorganization after anastomosis of the musculocutaneous and intercostal nerves following cervical root avulsion." Ann. Neurol. Vol. 38, pp. 15-20, 1995.
25. K. Naruse et al., "Development of EMG-based force sensing system in virtual reality system," Proc. of The Ninth Intl. Conf. on Advanced Robotics, pp. 185-190, 1999.

A Human-Like Robot Hand and Arm
with Fluidic Muscles:
Biologically Inspired Construction and Functionality

Ivo Boblan[1], Rudolf Bannasch[1,2], Hartmut Schwenk[1],
Frank Prietzel[2], Lars Miertsch[2], and Andreas Schulz[1]

[1] Technische Universität Berlin, FG Bionik und Evolutionstechnik, Ackerstr. 71-76,
13355 Berlin, Germany
{boblan, bannasch}@bionik.tu-berlin.de
http://www.bionik.tu-berlin.de
[2] EvoLogics, F&E Labor Bionik, Ackerstr. 71-76,
13355 Berlin, Germany

Abstract. Humanoid robots are fascinating from two points of view, firstly their construction and secondly because they lend life to inanimate objects. The combination of biology and robots leads to smoother and compliant movement which is more pleasant for us as people. Biologically inspired robots embody non-rigid movement which are made possible by special joints or actuators which give way and can both actively and passively adapt stiffness in different situations. The following paper deals with the construction of a compliant embodiment of a humanoid robot arm, including a five-finger hand with artificial fluidic muscles. The biologically inspired decentralized control architecture allows small units to be responsible for each main movement task. The first section gives a short introduction as to how bionics engineers think and tries to motivate us to build compliant machines. The second section looks at mechanical aspects, limitations and constraints and furthermore describes a human-like robot arm and hand. Section 3 presents the fluidic muscle actuator of the company FESTO[1]. The fourth section describes the decentralized control architecture and the electronic components. The last section concludes the paper while looking at further prospects.

1 Introduction

Nature has been creating sub-optimised individuals over a period of millions of years. Therefore, in a technical sense nature itself is a massive environment of optimisation. The question is, is it possible to understand and derive the methods underlying Darwinian evolution teaching and if so, can we generally manufacture products for specialized application which optimise the use of energy. Two directions are possible:

---

[1] This project was supported by Dr. Werner Fischer and Univ. Prof. Axel Thallemer, FESTO AG & Co. KG.

F. Iida et al. (Eds.): Embodied Artificial Intelligence, LNAI 3139, pp. 160–179, 2004.
© Springer-Verlag Berlin Heidelberg 2004

- To use the optimisation method of nature, the "Evolution Strategy" [1] and fulfilment of nature's evolution in vitro.
- To extract the underlying methods of optimised phenotypes directly from nature and use the underlying ideas to develop technical products.

The field of engineering science called "Bionics" is concerned with decoding 'inventions' made by living organisms and utilising them in innovative engineering techniques. Bionics is a made-up word that links biology and technology. However, nature does not simply supply blueprints which can merely be copied. Findings from functional biology have to be translated into materials and dimensions applicable in practical engineering.

In order to build humanoids we have to look at individuals in nature with the same proportions and environmental conditions and try not to scale the joints of a beetle, for example, which were not designed to carry heavy weights. Nature always develops optimally, based on the respective surroundings conditions. A parakeet in the jungle is subjected to different conditions than an eagle living in high mountainous regions. The law of survival of the fittest determines natural selection and consequently how the individual adapts to its living space. The parakeet, for example, is not optimised to cover long distances, but rather to be beautiful and to appeal the females.

At present there is no accepted theory or system to find bionic solutions, nor is there an accepted approach to systematically screen for systems. Bionic designs which currently exist owe their creation mostly to luck or scientific research over many years.

What can we learn from nature about morphology and physiology for the design of humanoid robots? If we concur with the law of survival of the fittest, then we believe that only optimised individuals can exist in nature in their respective surrounding conditions. Bionics initial task is to search for individuals in nature which have the same characteristics as the object to be developed. In our case, we are searching for a model of a humanoid robot arm and hand. We are thus looking for animals which are able to hold and/or carry several kilograms and which have human-like proportions with respect to weight and inherent compliance. When looking at the problem more closely, the intrinsic problem is how can we produce a multiple of force which are able to hold objects that are heavier than their own weight. This is a so-called power-weight ratio; this ratio is about one to one for electric motors. We have found other solutions for actuators in nature, particularly linear actuators that produce tractive force. The power-weight ratio of these actuators is multiplicatively higher than those known for technical actuators. Thus, it seems that nature has a better solution for our technical problem under the given terms and conditions.

We will not look at industrial robots here, as they carry out rigid tasks among themselves, or in contact with a technical environment. This field, called contact stability [2-5], has been widely investigated and has presented large problems for robotic manipulation tasks till date. Starting or dampening oscillation and performing a task requiring rigid contact from a free movement are related questions. The problem of contact stability arises, if one operates with rigid manipulators without spring-like or compliant properties.

We will instead focus on human-like robots and their interaction with humans and the environment. This contact or physical touching between robot and human is subject to special requirements as regards softness and compliance of motions. The goal of humanoids is not to assemble printed circuit boards that are also hard for humans, but also to master soft and energy-optimised movement in different situations of life.

If we look at the grasp movement of our own hand, we observe a transient effect and if necessary, feelings or vision-based adjustment of the hand. These special characteristics utilised when we touch demands new, innovative embodiment (morphology) and actuators (physiology).

The difference between a machine and a humanoid is its morphology. A human is living and can fulfil several different tasks which have special requirements in construction, freedom of movement and arrangement of weight. If we assume that the human body is an optimised structure, we have to study the load-bearing skeleton and the load transmission via the muscle-tendon system. Both criteria together form a unit which cannot be treated separately.

The study of the physiology of the muscle-tendon system [6-9] and its activation by the central nervous system gives us insight into the functions and activities of the human body. Current walking robots are heavy-weight, unproportional and unable to accomplish human-like performance. The motor actuators located in the joints increase the masses moved and accordingly the torque as well. The human muscle has a high power-weight ratio and transmits tractive power via a tendon across special parts of bones. There are located on the top or proximal to the centre of rotation. This leads to less torque and the ability to carry out fast movement with respect to energy need.

A current humanoid robot project in Germany is the development and construction of a two-arm robot called the "Zwei-Arm-Roboter" (ZAR3) in German. The third prototype has been constructed where a right arm with hand has been attached to a rigid spinal column.

The robot is 190 cm tall and the proportions are similar to humans of this size. Attention has been concentrated on its human size, anthropoid proportions and functionality of the actuators. The radius of action as well as the velocity of movement is anthropoid. The company FESTO has provided the linear actuators of the fluidic muscles. Tendons of Dynema filaments are used to convey the tractive force to the joints as regards tensile strength, lightweight and little bending radius.

The next section will describe the mechanical body with reference to skeleton, joints and tendons.

2 Mechanical Aspects

The whole body has been designed by AUTOCAD and the date translated to the special Computer Numerical Control (CNC) code and transferred to a 3 axes CNC milling machine. All parts, about 950 not including the purchased parts, have been manufactured from aluminium. Aluminium is lightweight, strong enough and easy to machine.

ZAR3 consist of a base which can roll, a rigid spinal column, an upper arm, a forearm and a five-finger hand (see figure 1).

Fig. 1. This shows a photograph of the current version 3 of the humanoid robot ZAR3

2.1 Base

The mobile base houses the control PC, the electronics, valves for the body actuators and the power supply for the whole robot.

The PC in the middle of figure 2 is a geriatric Pentium I with 400 MHz but fast enough to perform the following tasks immediately:

- Managing of the data bus activity and adhering to the time schedule
- Sending of defined goal angle and pressure data to each micro controller (intermediate steps are calculated locally)
- Monitoring of sensor data (angle, pressure) and error processing

A 15″ TFT panel is located in the middle of the front cover and along with a keyboard and a mouse make up the interface to the operator.

A 5/3-port directional control valve is needed to drive each muscle. The same functionality is obtained with two 3/2-port valves, which are space saving and are assembled as a valve cluster. Fast relay valves of the company FESTO with a discharge of 100 l/min and a maximum switching time of 2 ms of the type MHE2 are used. Integrated electronics are provided with each valve are shown in figure 2 as a black add-on on the white valve, this facilitates a fast switching operation at increased current

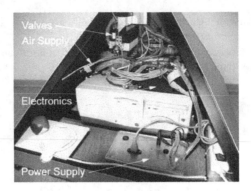

Fig. 2. The photograph above shows an inside view of the base which contains the power supply for 24 V and 5 V, the electronic devices for the shoulder and arm, the air tubes for supply and delivery directly connected to the valve cluster and the on-off valves for the shoulder and arm placed on a valve cluster.

consumption. A terminal block with two valve packs on each side of the block is used to increase packing density. The inflating valves are located on the left side the deflating valves are on the right. Only the valves for the body muscles are located in the base, thus there are 16 valves for 8 body muscles.

The air supply is directly connected to the valve cluster (see figure 2) and is partitioned into two separate air tubes, one for the body and one for the hand. This becomes necessary as there are body muscles which can be driven with a higher pressure than the small finger muscles. The outgoing air is routed to a common tube and is actually not won back. We presently use two different air supply alternatives. Both alternatives are not really suitable for mobile use. Our in-house compressed air line with 6 bar is used for stationary operation whereas we utilise standard 10 litres 200 bar compressed air bottles encased in a smart aluminium case for 'mobile' use. Current small sized and noiseless air generators cannot produce the required amount of volume flow to fill up the bigger muscles.

To increase the reliability, the power supply is physically split into one for the electronic devices with 5 V and one for the valves with 24 V. We use the switching power supply (SPS) SPS 100PX with an output of 5 V / 10 A. The 24 V output of the SPS does not supply the required current start-up peak of the electronic driven valves. A disadvantage of SPS is the break-down of the voltage by overload a special power supply has been assembled for this task and facilitates the delivery of up to 20 A by 24 V.

The third version of the ZAR comprises a right hand and the associated arm and the shoulder. The hand and arm with shoulder constitute independent units and are steered separately. This basic concept of decentralization by many small 'intelligent' units is found in nature and also has advantages in technical realization. The decentralized control architecture and the associated electronic components are explained in more detailed in section four.

2.2 Torso and Shoulder

The torso of ZAR3 only consists of the muscle assembly of the shoulder joint.

The shoulder is the most flexible joint in the human body which it achieves at the expense of stability, less guidance of motion and less arranged limit stops as, for example, the hip joint. The human shoulder joint allows for the placing and rotating of the arm in many positions in front, above, to the side and behind the body. This flexibility also makes the shoulder susceptible to instability and injury. Figure 3 shows the complexity of human shoulder joint.

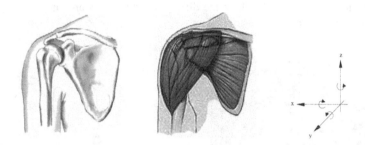

Fig. 3. This shows a human shoulder. *Left*: Skeleton only; *Middle*: Skeleton with muscles; *Right*: All movements of the shoulder joint may be understood as a combination of the motions of rotation and translation in the particular plain [10].

The human shoulder is a ball and socket joint. The ball is the head of the upper arm bone (humerus) and the socket is a part of the shoulder blade (scapula). The ball at the top end of the arm bone fits into the small socket (glenoid) of the shoulder blade to form the shoulder joint (glenohumeral joint). The socket of the glenoid is surrounded by a soft-tissue rim (labrum). A smooth, durable surface (articular cartilage) on the head of the arm bone, and a thin inner lining (synovium) of the joint facilitates the smooth motion of the shoulder joint.

A technical replica has proven to be a bold venture; this is because the construction involves a group of muscles (rotator cuff) which covers the shoulder joint (see figure 3 middle) which help keep the shoulder in the socket and enable the movement of the arm. A muscle area or the placing of muscles around the joint to imitate the human shoulder muscle-tendon system is awkward to construct and susceptible in operation.

A better way to build a complex shoulder joint is to split the multi-freedom joint into separate rotational joints each of which have one degree of freedom. These single joints are easier to construct, can be attached directly to the muscle-tendon system and are more rugged in use. Each of the three rotational joints spans a 2D vector space around an axis of the Cartesian coordinate system.

Electric motors are often used to drive the rotational joints. The motor is positioned directly on each axis which results in size increase and means that the design becomes larger than human scale. Another method would be to move the motors away from the joints and convey engine torque via driving belts. This approach is legitimate and appropriate for industrial robots which do not need to move away.

Our approach focuses on anthropoid aspects which comprise biological inspired sensors, actors, design and freedom of motion in consequence of lightweight construction and functional morphology. There are no 'natural' rotary machines in the animal world. Human construction utilises linear actuators in terms of muscles which are able to contract and are consequently then shortened in length.

For one surface of revolution, two muscles are necessary for an active conducted animation. The muscles of the x- and y-axis are arranged to revolve, rotated by the muscles of the z-axis. The actual application of the shoulder joint is shown in the photograph below (figure 4) where the different redirections are clarified in order to be able to complete a 3D radius of action.

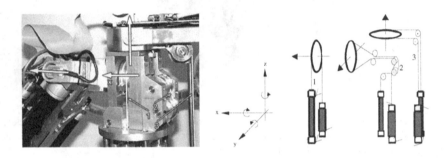

Fig. 4. *Left*: This is a photograph of the shoulder joint of ZAR3. The numbers 1,2 and 3 indicate the tendons of the x-, y- and z-axis of the joint. *Middle*: Shows the relation of the rotary directions to Cartesian space. *Right*: Clarification of the muscle-tendon systems and the redirections caused by the mechanical constraints and the acting pulley to drive the distal segments

The construction of the x-axis (see the diagram on the right-hand side of figure 4) of the shoulder joint allows to be able to directly calculate force and torque. The radius of action ranges from -30° when the arm is hanging down vertically (0°) and slightly backward to 150° when the arm is stretch up vertically (180°) and slightly forward. An extended radius would be desirable, but the actual angle measuring electronic can only provide for a radius of 180°. The diagrams show the compressed circle and the deflection pulley where the muscle tendon system drives the belonging distal limbs.

The y-axis, the tendons guidance system, is complex due to the arrangement of the muscles and tendons and a common origin of the coordinate system. The tendons of the y-axis are guided via several deflection pulleys and through the centre of the wheel of the x-axis. The freedom of motion ranging from 0° in the vertically hang down position up to 180° vertically stretching above.

The muscles of the z-axis rotate the whole revolver of the x- and y-axis muscles from 45° horizontally forward up to -45° backward, limited by mechanical limit stops to meet human restrictions.

The aim of the arrangement of the shoulder joint and the rotational revolver is to concentrate the mass of the actuators proximal to the centre of the torso. The smaller the distance between mass and centre of rotation, the smaller is the inertia. This is always a balance between displacement of mass and level of complexity. This type of construction of the shoulder joint only allows the muscle actuators for the elbow, wrist and hand to be placed on the arm. This results in smaller inertia, more speed of movement and less effort required to control the movement.

The muscle pair attached to a joint in a human body is always placed proximally. Therefore, the muscles only actuate the lower parts of the chain (distal segments) and can be powerless. The rule is the correct placing of the actuators so that they don't lift themselves. The other parts of the arm have to be consequent in dealing with this fundamental aspect.

2.3 Arm

The arm is divided into upper arm and forearm. The muscle pair for the elbow joint is placed on the upper arm. Up to the current version of humanoid ZAR, the valves for the rest of the arm (forearm and hand) have been placed on the outside of the upper arm. This design has both advantages and disadvantages. On the one hand, the distance between muscle and valve should be as short as possible to compensate for small speed loss by relay operations caused by the inertia of masses and compressibility of the air. Reducing of the air hose length also leads to a reduction of unused air in the system and the calculation effort which should primarily only depend on muscle volume. On the other hand, the unnecessary mass on the arm increases the centrifugal force and therefore the control effort.

Figure 5 (left) shows a photograph of the original adapted elbow joint, the diagram on the right outlines the extracted special moving directions.

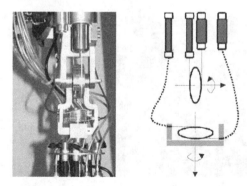

Fig. 5. *Left*: A photograph of the actual elbow joint is shown. *Right*: The redirection of the muscle-tendon systems and the displacement of the Bowden cables (dotted lines) are traced

Our first effort at producing an elbow joint tries to imitate the human elbow joint using a technical solution. This turned out be difficult as the versatile joint or the link between ulna and radius is too complex to be able to exactly copy. The analysis of the resulting degrees of freedom facilitated the assembly of the muscle-tendon-pulley-limb system shown in figure 5. The dotted lines in figure 5 (right) depict Bowden cables, which allowed the tendon to be guided without the use of pulleys. This brilliant invention from the bicycle world facilitates the configuration of the actuators in the best possible way and is dependent on mechanical contraction and human design.

The front muscle of the horizontal axis of the elbow joint is the biceps, the back muscle the triceps that move the forearm. The biceps-triceps system was constructed according to the human system. The elbow joint is technically a hinge and allows bending and straightening but does not rotate. The coordinate system is zero on this axis when the forearm hangs down. There are humans who can overstretch their elbow joint, but in order to take into account what is generally possible, the joint is mechanically fixed at the stretched position. That allows 180° up to where the upper arm and forearm contact and constitutes the mechanical limit.

The human twist behaviour of the ulna-radius system is a rotary motion of the wrist which can be simplified by a joint with pulley and vertical rotation axis. This is shown in figure 5. The range of movement is designed to be 45° in both directions.

Therefore, the forearm can be rigid and carry other equipment. In this version of ZAR, the forearm housed the finger-muscle-revolver. The term 'revolver' means the assembly of the 16 muscles around the forearm. If we consider the human model to be ideal, all the imaginable muscles of the hand are located on the forearm bones ulna and radius. This leads to a filigree assembly of the five-finger hand and reduces the amount of mass. The tractive forces of the flexors and extensors of the fingers are transferred by tendons which are embedded in connective tissue for guidance. Bowden cables are used to install the appropriate muscle-tendon systems to the finger joints (phalanx). Figure 6 shows the arrangement of Bowden cables connected to the five-finger hand.

Fig. 6. *Left*: A photograph of ZAR's wrist joint with axes 1 points to a Bowden cable which connects the muscle actuator (fixed end) with a finger root joint inside the palm. 2 points to the tendon which drives the hand lift joint and 3 to the tendon attached to the tilt joint of the wrist. *Right*: The depiction of the corresponding Cartesian coordinate system of the wrist

The challenge of this joint is to duplicate full functionality of the human wrist with a simultaneously simple and durable construction. All the Bowden cables have to be concentrated in the middle of the rotation axes. The mechanical resistance in the joint arise from the guidance of the Bowden cables to the sockets of the fingers. In particular, the tilt and lift muscle works against this rising mechanical resistance, see arrows numbers 2 and 3 in figure 6. For this reason, we have only been able to achieve a degree of movement of 20° in each direction. Two muscles (flexor, extensor, respectively) are used to tilt and lift the joint and are arranged as pairs of antagonists. In the technical sense one speaks of an ellipsoid joint which is a less flexible version of a ball-and-socket joint (shoulder).

2.3 Five-Finger Hand

The first artificial hand developed and constructed based on the archetype of the human hand was the Waseda Hand (WH-1) in 1964. Since this there have been a multitude of artificial hands which are more or less anthropomorphic, anthropoid, humanlike or humanoid. The academic question regarding humanoid hands, which are not actually humanoid in construction and function, will not be discussed here. The following small survey of artificial hand constructions is not exhaustive.

Many three and four finger hands with more-or-less humanoid proportions have been designed. The Utah/MIT dextrous hand [11, 12] has a four-finger system with 16 DOF and is powered by 32 pneumatic actuators. The actuator pack is placed remote from the robot hand and connected by antagonistic polymeric tendons. The Karlsruhe dextrous hand II [13, 14] can be considered to be a non-anthropomorphic approach. Tendons drive the four-finger autonomous gripper. Other artificial hands are the Stanford-JPL hand [15, 16], the Omni hand [17], the NTU hand [18], the DLR hand [19, 20] with a semi-anthropomorphic design, the cybernetic hand prosthesis by IST-FET [21] and the DIST hand by Genoa Robotics [22-24]. These hand projects do not fulfil the requirements for the number of fingers, joints in the fingers and humanlike movements. However, the professional design, control architecture and functionality of a couple of them is convincing.

Several artificial anthropomorphic five-fingered hands have been designed with servomotors which are built into the fingers, for example, the "Gifu hand" I-III [25-27] has 20 joints with 16 DOF and is equipped with a six-axes force sensor at each fingertip. The Gifu hand is intended to be a prosthetic application for handicapped individuals. The "Robonaut" [28], designed by NASA's Johnson Space Center and DARPA, is a dexterous five-fingered hand with 14 DOF and a human-scale arm. The forearm houses all fourteen brushless motors and all of the wiring for the hand. The prosthetic hand described in [29, 30] has 24 DOF and is controlled by EMG signals detected from the forearm of a human handicapped individual. A tendon driven adaptive joint mechanism adjusts velocity and torque functions by use of a spring type wire as an elastic guide. The "Blackfingers" hand prosthesis [31, 32] is a five-fingered hand with traditional pneumatic cylinders which function as linear actuators. The so-called bionic five-fingered hand by FZK (IAI) [33, 34] has 13 DOF and utilises flexible fluid actuators [35]. This fluid actuators approach is the attempt to design muscles similar to those of the human, but which do not have the human-like power-weight ratio. This ratio has been improved by the "Smart Award Hand" from SHADOW [36]. This artificial robotic five-fingered hand has 24 DOF and is complete driven by air muscles from the company SHADOW. The muscle pack of the hand is located on the forearm and use tendons to power transmission. This design and philosophy of a humanoid hand goes in the same direction as those of ZARx.

The hand is the most complicated component of the ZAR3. Not only the small limbs and joints of the fingers, but also the guidance of the tendons in human size proportions render the hand the most elaborated part of the project. The hand was assembled separately, tested on a vice and was finally attached to the arm.

The ZAR3 hand has 12 DOF without the wrist. Taking into account the diameter size of the smallest muscle from FESTO, we decided to only attach the flexor muscle to each finger limb and lay on the extensor as the pullback spring. This construction does not constrict the task of grasping, but only active releasing. However, this results in the forearm revolver being reduced in size and mass and, due to this, to a smaller inertia of masses and control effort. A disadvantage of this concurrence is the unnecessary additional expenses of providing tractive force via the small muscles to overcome the resilience of the springs. See section 3 as regards the dimensions of the muscles.

Biological Motivation

The hand is the human beings' door to the outside world. The loop of interaction with the environment is that the brain manipulates the information provided by the sense organs which then are executed by actuators to the extremities. The hand has to accomplish a variety of positions, operations and activities in the life of a human, to survive the rat race. The hand has been optimised to fulfil these manifolds task in the hundred million years of human life. The hand is able to sign, to grasp, to hold and carry, to interact with itself, to dig, to write, to play and a lot more. It is still however lightweight enough to run with a complete runner the 100 m in less than 10 sec. A full-grown human hand weighs approximately 500 g and has a far greater degree of freedom than 16.

Trials to copy the human hand have failed due to the concatenation of the many small bones of the palm. The combination of these bones enables the palm to form a cavity. The intention to build a human-like or biological inspired robot is to carry out the science of Bionics. This means to abstract the amount of degrees of freedom and to deviate from joint structures which are too complicated. The question has to be, what joint which is easy to construct can provide the greatest degree of functionality? Is it necessary for a robot hand to form a cavity? I do not think so. I think it is more important to be able to hold a glass and handle it. In addition, the ability of a finger to move in a circle around the root can be neglected.

All other joints of a human hand have been implemented to the greatest possible extent. Each of the four long fingers has three hinge joints. The outer first and middle joint of each finger is coupled because only very few humans can move these joints separately. Consequently, eight muscle actuators are required. All four long fingers are coupled at their roots by a spreading mechanism actuated by one muscle. The fingers fan each other at the same angle around the middle finger which constitute the fixed base. This artifice simplifies the matter and retains the relation. The different spreading of the fingers is also a challenge for humans. One can observe that the middle finger is fixed on one's own hand. The thumb has two hinge joints and a saddle joint at the root; therefore only three muscle actuators are required. Altogether, 12 muscle actuators fulfil full functionality of a real human hand. Figure 7 (below) shows the hand of ZAR3 in comparison to bones of a real hand.

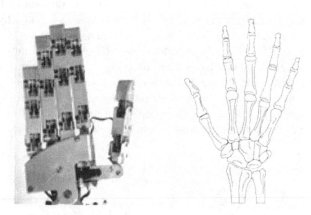

Fig. 7. *Left*: A photograph of the five-finger hand in home position. The dark spots between lighter surfaces are recesses to afford the bending of the phalanxes. *Right*: A view of the bones of a human hand is shown; the similarities are clearly visible

The size, weight, morphology and functionality are similar to the human hand and as well the radii of action. The artificial hand can grasp things and hold several poses.

3   Fluidic Muscle Actuator

The idea of an inflatable rubber tube to facilitate shortening is not new.

The McKibben muscle actuator [37] was developed in the 1950s and 1960s. The deflated rubber tube was not stiff enough to hold the shape itself, which means without an amount of air inside, the muscles kink off and have to firm up additionally.

The company SHADOW attempted another approach. This muscle actuator is also flexible, but is wrapped in a tough plastic weave to hold the cylindrical form. However, an exact deformation across the whole length and diameter and according to this a geometric measurement is not possible.

A large company called FESTO have constructed a fluidic muscle actuator over the last few years using the above-mentioned characteristics. This muscle sufficiently meets the requirements of dimensional stability, quantity of shortening and lightweight construction.

A muscle actuator works as a linear actuator and has advantages compared to a hydraulic cylinder and an electric motor with leverage. The hydraulic cylinder has significantly more weight, can start without jerking and has no disagreeable leakages. The electric motor can be placed directly at the joint without leverage which leads to an increase of mass and consequently, to greater control effort. A motor does not fit the necessary requirements for a humanoid or human-like robot. The task is to try to emulate or to pattern the functionality, physiology and morphology of the muscle-tendon-bone system of a human. This consequent approach can lead to a rather more human-like robot if we agree with the law of Darwinians survival of the fittest in natural evolution. To address the issue of why this is the case and why an electric motor does not meet these requirements will not be discussed here.

The company FESTO officially provides three different sizes of muscle actuators, namely MAS-40/20/10. A smaller version, MAS-5, is currently being prepared for realise. Only the MAS-5/10/20 is used in our robot ZAR3. The number 5 indicates the inside diameter in millimetres. All muscles have the same characteristic, that is the shortening contraction to the acting force dependent on the level of compressed air inside the muscle. This relationship is shown in the following (figure 8).

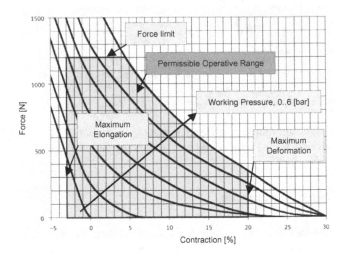

Fig. 8. This diagram shows the relationship of the possible produced tractive force in Newton to lift up something to the affordable contraction rate expressed in percentages of the basis length by a given working air pressure in bar of the fluidic muscle MAS-20

This non-linear interrelationship is commonly depicted as force F in Newton over contraction Δl in percent with supplied air p_{air} in bars as constant parameter. The greater the affected force by a constant air pressure, the smaller the shortening referred to as base length L_0 of the muscle rubber tube. Moreover, the higher the air pressure by a constant force, the greater the shortening. These relationships can roughly be described as follows

$$F \propto \frac{p_{air}}{\Delta l / L_0} \; . \tag{1}$$

The McKibben muscle has been extensively researched as regards static modelling and geometric calculations [38-40]. Static physical modelling can take over the characterization of the fluidic muscle from FESTO, however it uses the new measured data and some adapted details of the behaviour of the MAS. The dependence of the produced force of the muscle on geometric quantities such as volume, braid angle and diameter is common to models and is merely of theoretical value.

Although we have not undertaken this in our paper, in order for exact modelling the measured sets of force, air pressure and contractions concerning the time are required. In our opinion, the braid angle at a certain length of the muscle to predict the produced force in this position is not need. Based on the relationship of force, pressure and length determined by a proper invertible model, we have been able to make a model and then control the muscle actuator. Such an approach results in a non-linear interrelationship which can be dealt with in several ways.

The most acceptable approximation is achieved by an engineering approach using a spring system [38]. The actuator can be considered as an elastic element of variable stiffness where the force is a function of the pressure and the length. Stiffness $k = dF/dL$ is proportional to the pressure and stiffness per unit pressure $k \sim dk/dP = k(p)$ which results in

$$l(F, p) = L_{max} - \frac{F_{max} - F}{k(p)} \; . \tag{2}$$

The length L_{max} is the theoretically possible maximum length when F at its maximum. Due to the decreasing of muscle stiffness when air pressure is increased, the maximum values of force and length have been used. This dependency is the inverse of the behaviour of a general spring. Stiffness directly depends on the air pressure. Stiffness in respect to force can be neglected in a first approximation. The emphasis in this approach is to concentrate on the modelling of the variable stiffness.

The maximum specified air pressure for the FESTO muscle is 9 bar for the MAS-10 and 7 bar for the MAS-20. The operating range expressed in terms of force is 400 N for the MAS-10 and 1200 N for the MAS-20. MAS-5, the smallest muscle, has not yet been specified. Detailed information can be found on the website of the company FESTO (www.festo.com).

The dimensioning of the muscle type, length and the deflection pulley are the most important tasks in order to fulfil the requirements as regards radius of action, velocity of movement and, in the end, the dimension of the possible weight to be lifted. Due to being scaled to human proportions, the type and the length of the muscle is limited. The relationship between muscle length and radius of the deflection pulley has been well defined and is calculated beforehand. The smaller the pulley, the smaller the length of the muscle can be, however the muscle must be the most powerful. If C is the centre of rotation of the joint, F_{FM} the produced force of the fluidic muscle, G the weight of the actuated limb and F_L the load force, then the equation of torque can be depicted as follows:

$$\sum M_C = 0 = F_{FM} \cdot l_{FM} - G \cdot l_G - F_L \cdot l_L \,. \tag{3}$$

The values of G, l_G and l_L are fixed and cannot be changed by human proportions and known mass of aluminium and equipment. The estimate of F_L depends directly on the carrying power of the humanoid and has to be completed before designing the robot whole. The other two variables have to determine iteratively.

Shoulder
The more powerful MAS-20, 400 mm in length, has been assembled for use as the shoulder joint. A length of 250 mm is sufficient for the smaller range of the z-axis. When considering the required space and that a second arm will be added in the future, the MAS-20 seems to be the best choice as regards diameter size, particularly when all muscles are inflated.

The question is now how long should the muscle be and what should the diameter of the pulley be. A reasonable trade-off is that all the joints of the shoulder should have a diameter of 50 mm. This allows the muscles to have a short length of 400 mm but ensures that they have enough power to lift the payload in the critical weight range. The lifter muscle (flexor) of the x- and y-axis in particular has limitations as regard load. The extensor muscle guides the descent of the arm with the help of gravity. The extensor muscle's major task is to control stiffness and compliance of motion. The more this muscle pulls against the flexor, the stiffer the motion. This procedure puts the fringe range of the produced amount of force of the flexor into perspective. The most advantageous thing is that the critical area of muscle shortening has not been attained even when the arm has been extended forward to a 90° angle, that is where the extensor muscle has to generate maximum power. The muscle contraction only reaches the critical level once the arm has reached an angle of around 120°. The muscles of the z-axis of the shoulder can be designed to be smaller as the torso itself holds the mass of the arm and only the horizontal motion has to be executed.

Elbow
The elbow joint can be calculated similar to the y-axis of the shoulder. As the one of the ZAR3's tasks is to be able to lift a glass of beer, the elbow joint is also assembled using the MAS-20. The shoulder hangs and only the biceps lift up the payload, including the revolving forearm and hand. The maximum angle for lifting is controlled to 135° to allow an ulna-radius action which doesn't become mechanically stuck. The diameter of the pulley is set to 50 mm and the muscle length to 220 mm. The smaller radius of action allows a shorter muscle length to achieve this human motion. A pair of muscles called 'agonist' and 'antagonist' drives the motion of rotation of the ulna-radius system. This joint works as well as the elbow joint, the difference being in the axis of rotation. The assembly is described above in section two.

The dimensioning of the muscle type, length and diameter of the pulley follows the same principles as above. The rotation process does not have to lift or hold a mass, but is responsible for adjusting the hand's posture and to act with the payload. Due to the number of air pipes which guide via this joint, the diameter of the pulley has to be limited to 30 mm in order to achieve human-like proportions. Consequently, the shorter muscle length of 200 mm and the power of a MAS-10 seem to be sufficient for this task, also as regards the redirection of force using a Bowden cable (see figure 5).

Wrist and Hand

The MAS-5 muscle is the only way a hand as sufficiently compact to be of human scale can be achieved. The extent of the 16 fully inflated muscles and mechanical fixings is minimally thicker in diameter to that of a full-grown male. The length of the muscles varies in two steps, from 80 mm to 110 mm. The four muscles in agonist-antagonist construction of hand up/down and hand tilt left/right are longer to afford more force enabled by a larger level arm and by the use of larger 16 mm pulleys.

According to the developmental department of FESTO, the MAS-5 can pull up to 50 N. This specification has established by a vertical experiment in ideal conditions without deflection pulleys. In real-life applications, only a fraction of this tractive force of a MAS-x can be achieved and can be calculated in terms of equation (3).

4 Electronics and Control Architecture

The electronic components, the communication to the controlled PC together with the architecture to manage and control tasks which is what defines when a machine is a robot and is the counterpart to the human brain and the central nervous system. Engineers till date have not been able to reproduce this data flow and communication network in vitro. The task will be to assemble, place and manage electronic parts in the same way as to achieve results similar to that of the human. Many small activities and reactions are not controlled by the brain, but rather initiated by the spinal cord or local reflexes. The advantage of this is faster reaction time; specialized distributed units can be used as a paradigm to design decentralized control architecture. This approach applied to a technical system is tolerant of failure, enables short distances in the sensor-control-actuator loop and provides for command structure and control hierarchy.

The robot ZAR3 is divided into two units, completely separately assembled and controlled, one for the five-finger hand and one for the arm and shoulder. Both units have identical circuit devices and functional range. Each functional unit consists of two communication directions and can be addressed both separately and independent of each other. The differences lie in the amount of driven outputs, the physical subdivision of input-output channels and the user-defined software of the controller. A diagram of the structural components and communication channels are shown below (figure 9).

The body electronics for reading sensors and driving the valve-muscle actuators are located in the base and is arranged on one printed circuit board.

The hand electronics are separated into a sensor input board and an actuator output board to drive the muscle valves. The hand electronics, located on the upper side of the palm, process the data signals from each measured finger joint. The associated output board is placed near to the upper arm valve block on the shoulder.

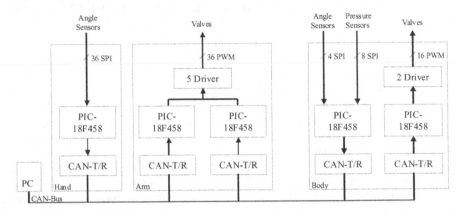

Fig. 9. The above shows a schematic plan of the connections of the hand (left) and shoulder-arm (right) electronic devices. The hand consists of two boards, one for sensor inputs and one for valve outputs, which communicate via CAN. The shoulder-arm electronic, in brief 'body', is configured as one printed board and located in the robot's base

The angle sensor uses a magnet, placed on the distal part of the joint, which rotates closely below a sensitive array. This array is implemented as integrated circuit to detects the changing magnetic field and works as a magneto-resistive sensor. This non-linear relation compensates for temperature and is linearized at the sensor spot. The communication protocol Serial Peripheral Interface (SPI) from each angle transmitter is used to transmit the digitalized angle sensor data directly to the PIC micro controller 18F458 from the company MICROSHIP. The SPI interface is used as it requires less effort to wire, has a high data rate and as it provides the possibility of connecting to the controller. The three-wire-bus consist of two data and one clock signal and works in the master-slave-mode.

The two PIC 18F458 controllers, each concerned with one signal path, communicate via the Controller Area Network (CAN) bus and shares the effort of data processing, executing of control loop and generating of Pulse Width Modulation (PWM) signals. The CAN transceiver/receiver allocates the signal level to the physical bus. Driver devices, each of which have eight outputs, realize the 24 V output level for the electronic driven valves and must provide up to 1 A inrush current per valve. To drive each valve, electronics are needed on the one hand to supply current demand and, on the other hand, to enable the height switching time of the PWM output.

The strict separation of different components and data directions enables speedier troubleshooting and is a first step towards of decentralization. The distribution of responsibilities and the break down of information handling reduced data activities on the bus and the complexity of the units. The fast response time of an unit in a control loop in case of emergency cannot be affected by a fewer crucial task of monitoring or finger play. The remote unit receives a command from the control PC or from another unit via CAN-bus and decides about which operations to be done. Without any errors, the unit will initiate the appropriated control loop to reach the demanded goal angle. This stand-alone execution can be interrupted by the control PC or by an exceeded sensor limit value. The CAN-bus only serves as asynchronous communication channel of control and information messages not for the synchronous control loop between sensor, controller and actuator. The transmission of the entire control loop data via CAN-bus leads to an exceeding of the data rate specification of CAN of 1 Mbit/sec at the latest by triggering of the second arm. However, there is a possibility to use the CAN-bus which is carried out between the palm and shoulder board for the hand control loop. The next generation of ZAR will prevent this issue.

5 Conclusion and Future Prospects

It is far more difficult to design a practicable human-like robot than it would at first seem to be. Being constrained to human-like proportions increases the manufacturing effort which is compounded by being able to find practicable analogies and solutions for geometrical and functional interrelationships in human morphology and physiology. This has to lead to a completely new process of thought. The science of Bionics aims at analysing the methods behind the processes and to translate them into a practicable technical solution; this helps to construct machines which are similar to the model in nature, particular as regards excellence in shape and function.

This manuscript introduces the humanoid robot ZAR3, the mechanical design and development process is explained and constraints and limitations pointed out. A practicable artificial fluidic muscle is briefly proposed and the fundamental correlation of length, force and pressure introduced.

Evident constraints such as the valve block on the upper or the too faint biceps muscle have arisen already during the construction and test phase. These features will be modified in the next version, the ZAR4. In addition, the mechanical effort in producing the many small parts will be decreased as well as increased simplification of the joints will be promoted. Once the second arm is completed, attention will have to be turned to the control architecture, to converge the conventional information processing in the human nervous system and neuronal processing. The participation of more units or subunits increases traffic on the signal bus and the increase in detail could be the next assignment to meet the requirements for fault tolerance, reliability and prioritisation of the data.

Acknowledgement. The company FESTO AG & Co. KG which supports the work on the various versions of the robot ZAR.

References

1. Rechenberg, I., *Evolutionsstrategie '94*. 1994: fromman-holzboog. 434.
2. Šurdiloviæ, D.T., *Synthesis of Robust Compliance Control Algorithms for Industrial Robots and Advanced Interaction Systems*, in *Mechanical Engineering Faculty*. 2002, University in Niš. p. 475.
3. Okada, M., Y. MNakamura, and S. Ban. *Design of Programmable Passive Compliance Shoulder Mechanism*. in *Proceedings of the 2001 IEEE International Conference on Robotics & Automation*. 2001. Seoul, Korea.
4. Ahmadi, M. and M. Buehler, *Stable Control of a Simulated One-Legged Running Robot with Hip and Leg Compliance*. 1996, Department of Mechanical Engineering, Centre For Intelligent Machines, McGill University: Montral, QC, Canada. p. 9.
5. Rocco, P., G. Ferretti, and G. Magnani, *Implicite Force Control for Industrial Robots in Contact with Stiff Surfaces*. Automatica, 1997. **33**(No.11): p. 2041-2047.
6. Fenn, W.O. and B.S. Marsh, *Muscular force at different speeds of shortening*. Journal of physiology, 1935. **85**(3): p. 277-298.
7. Gordon, A.M., A.F. Huxley, and F.J. Julian, *The variation in isometric tension with sarcomere length in vertebrate muscle fibres*. Journal of Physiology, 1966. **184**: p. 170-192.
8. Carlson, F.D. and D.R. Wilkie, *Muscle Physiology*. 1974: Prentice-Hall, Inc.
9. Huxley, A.F., *Muscular Contraction*. Journal of Physiology (London), 1974. **243**: p. 1-43.
10. Gohlke, F., *Biomechanik der Schulter*. Orthopäde, Springer-Verlag 2000, 2000. **29**: p. 834-844.
11. Jacobsen, S.C. and e. al., *The Utah/MIT dexterous hand:Work in progress*. Int. J. Robot. Res., 1984. **3**(4): p. 21-50.
12. Jacobsen, S.C., et al., *Design of the Utah/MIT Dextrous Hand*. Proceedings IEEE Int. Conf. on Robotics and Automation, USA, 1986: p. 1520-1532.
13. Fischer, T. and H. Woern, *Structure of a robot system: Karlsruhe dextrous hand II*. Proc. of Mediterranean Conf. on Control and Systems, 1998.
14. Osswald, D. and H. Wörn, *Mechanical System and Control System of a Dexterous Robot Hand*. Proceedings of the IEEE-RAS International Conference on Humanoid Robots, 2001: p. 8.
15. Salisbury, J.K., *Articulated Hands: Force Control and Kinematics Issues*, in *Stanford University*. 1982: Stanford.
16. Salisbury, J.K., *Design and Control of an Articulated Hand*. Int. Symposium on Design and Synthesis, Tokio, 1984.
17. Rosheim, M., *Robot Evolution*. New York: Wiley, 1994: p. 216-224.
18. Lin, L.R. and H.P. Huang, *"Integrating fuzzy control of the dexterous National Taiwan University (NTU) hand*. IEEE/ASME Trans. Mechatron, 1996. **1**: p. 216-229.
19. Butterfass, J., et al., *DLR's Multisensory Articulated Hand Part I: Hard- and Software Architecture*. IEEE International Conference on Robotics and Automation, 1998: p. 2081-2086.
20. Butterfass, J., et al., *DLR-Hand II: Next Generation of Dextrous Robot Hand*. Proc. IEEE Conf. on Robotics and Automation, Seoul, Korea, 2001: p. 109-114.
21. Dario, P., et al., *On the development of a cybernetic hand prosthesis*.
22. Bernieri, S., et al., *The DIST-Hand Robot*. IROS '97 Conf. Video Proceedings, Grenoble, France, 1997.
23. Caffaz, A. and G. Cannata, *The Design and Development of the DIST-Hand Dextrous Gripper*. Proc. IEEE Int. Conf. on Robotics and Automation, Leuven, Belgium, 1998: p. 2075-2080.
24. Caffaz, A., et al., *The DIST-Hand, an Anthropomorphic, Fully Sensorized Dexterous Gripper*. IEEE Humanoids 2000 Boston, MIT, USA, 2000.

25. Kawasaki, H. and T. Komatsu, *Mechanism design of anthropomorphic robot hand: Gifu hand I.* J. Robot. Mechatron., 1999. **11**(4): p. 269-273.
26. Kawasaki, H., T. Komatsu, and K. Uchiyama, *Dexterous Anthropomorphic Robot Hand With Distributed Tactile Sensor: Gifu Hand II.* IEEE/ASME Transactions on Mechatronics, 2002. **7**(3): p. 296-303.
27. Mouri, T., et al., *Anthropomorphic Robot Hand: Gifu Hand III.* ICCA, Muju Resort, Jeonbuk, Korea, 2002: p. 6.
28. Lovchik, C.S. and M.A. Diftler, *The robonaut hand: A dextrous robot hand for space.* Proc. IEEE Conf. on Robotics and Automation, 1999: p. 907-912.
29. Hirose, S. and S. Ma, *Coupled tendon-driven multijoint manipulator.* Proceedings of IEEE International Conference on Robotics & Automation, 1991: p. 1268-1275.
30. Ishikawa, Y., et al., *Development of Robot Hands with an Adjustable Power Transmitting Mechanism.* Intelligent Engineering Systems Through Neural Networks, ASME Press, 2000. **10**: p. 631-636.
31. Folgheraiter, M., et al., *Blackfingers a Sophisticated Hand Prothesis.* p. 4.
32. Folgheraiter, M. and G. Gini, *Blackfingers an artificial hand that copies human hand in structure, size, and function.* Proc. IEEE Humanoids 2000, MIT, 2000: p. 4.
33. Schulz, S., C. Pylatiuk, and G. Bretthauer, *A new ultralight anthropomorphic hand.* Proceedings of the 2001 IEEE International Conference on Robotics & Automation Seoul, Korea, 2001: p. 2437-2441.
34. Pylatiuk, C. and S. Schulz, *Neuentwicklung einer Prothesenhand.* Prothetik, 2002. **8**: p. 4.
35. Pylatiuk, C. and S. Schulz, *Entwicklung flexibler Fluidaktoren und ihre Anwendung in der Medizintechnik.* Med. Orth. Tech., 2000. **120**: p. 186-189.
36. *Design of a Dextrous Hand for advanced CLAWAR applications.* 2003, Shadow Robot Company: 251 Liverpool Road London ENGLAND.
37. Schulte, H.F.J., *The characteristics of the McKibben artificial muscle.* The application of external power in prosthetics and orthotics, National Academy of Science-National Research Council, Washington D.C., 1961.
38. Chou, C.-P. and B. Hannaford, *Static and Dynamic Characteristics of McKibben Pneumatic Artificial Muscles*, in *IEEE*. 1994, Department of Electrical Engineering, FT-10, University of Washington: Seattle, Washington 98195. p. 281-286.
39. Colbrunn, R.W., *Master of Science*, in *Department of Mechanical and Aerospace Engineering*. 2000, Case Western Reserve University. p. 141.
40. Tsagarakis, N. and D.G. Caldwell, *Improved Modelling and Assessment of pneumatic Muscle Actuator.* Proceedings of the 2000 IEEE International Conference on Robotics & Automation, San Francisco, CA, 2000: p. 3641-3646.

Agent-Environment Interaction in Visual Homing

Verena V. Hafner*

Artificial Intelligence Lab,
Department of Information Technology,
University of Zurich, Switzerland,
vhafner@ifi.unizh.ch

Abstract. This study illustrates how obstacle avoidance can emerge from a visual homing strategy, caused by the intrinsic geometric structure of the environment. An example is shown where an agent performs visual homing in a virtual environment with several obstacles which also serve as visual landmarks. The agent has omnidirectional vision similar to many prey animals. The applied visual homing strategy is the Average Landmark Vector (ALV) model by Lambrinos et al.[1]. When observing the homing trajectories of the agent, it can be seen that it performs obstacle avoidance without having this behaviour explicitly encoded. It will be shown that the dynamic feedback the agent gets from its environment is crucial for this kind of behaviour.

1 Introduction

Obstacle avoidance in animals evolved soon after the first living creatures started to move. Even bacteria such as *E.coli* perform attractant or repellent behaviour. By doing so, they can move towards certain objects (or chemicals in this case), avoid others, and consequently perform obstacle avoidance. In artificial agents, obstacle avoidance is considered as one of the most basic behaviours, and was already present among the earliest autonomous mobile robots, such as the turtles 'Elsie' and 'Elmer' built by Grey Walter [2] (see Figure 1). For moving, they had two propulsion wheels and one steering wheel; as sensors, one light and one touch sensor were used. The robots' control system was completely analogue, and they could perform two main actions: obstacle avoidance, by retreating on contact, and light following. These simple behaviours in combination with an interaction with the environment led to other kinds of behaviour which are perceived as very complex and even intentional by outside observers. For example, by placing a lamp on each turtle's shell, a kind of 'social behaviour' emerged from the interaction of the two turtles.

Since the pioneering work of Grey Walter, increasing numbers of robots have been built in order to understand adaptive behaviour in the real world. The new interest in *embodied artificial intelligence (EAI)* builds on this early work, and recognised the importance of interaction with the real world. In this chapter, I will present an example of agent behaviour that profits from the interaction with the environment in a way

* Now at Sony CSL Paris, France, *hafner@csl.sony.fr*

F. Iida et al. (Eds.): Embodied Artificial Intelligence, LNAI 3139, pp. 180–185, 2004.
© Springer-Verlag Berlin Heidelberg 2004

Fig. 1. Second generation turtle designed by Grey Walter. Its main two behaviours are obstacle avoidance and light seeking (picture taken from the Grey Walter Online Archive [3]).

that has not been anticipated by the designer. The example shows an agent navigating back to its home location by means of visual homing. What makes this experiment particularly interesting is the fact that this agent gets obstacle avoidance "for free" through the interaction with the environment and intrinsic geometric properties.

2   An Example: Navigation and Emergent Obstacle Avoidance

In this section, I introduce the emergence of obstacle avoidance induced by a behaviour which is usually considered as being on a higher level than pure obstacle avoidance: navigation[1]. I investigate the emergent obstacle avoidance properties of a visual homing method using snapshots called the Average Landmark Vector (ALV) model [1]. The underlying principle of snapshot visual homing is the following: Omni-directional one-dimensional snapshots along the horizon are taken at two different positions. Those are usually the home position and the current position of the agent. The snapshots are aligned toward a common global orientation. The visual homing model allows to infer the vector of displacement for the two positions of the snapshots. Note that this homing strategy is local and therefore a large subset of the landmarks has to be visible from both snapshot positions.

The ALV model can explain certain aspects of the navigation behaviour observed in insects. It calculates the homing vector h by subtracting the AL vector at the target position from the AL vector at the current position:

$$h = a_c - a_t,$$

---

[1] This experiment has been described first in the context of emergence together with the example of a holonomic robot in [4].

where $a_t = \sum_{i=1}^{n} l_i^t$ and $a_c = \sum_{i=1}^{n} l_i^c$ are the sums of the landmark vectors at the target and the current position respectively. The landmark vectors l_i^c and l_i^t have unit length since distance information is not used. For simplicity, the AL vectors are expressed as the sum (not the average) of the landmark vectors. The ALV model reduces the image to a one-dimensional binary array, where each landmark is represented by one pixel at the position pointing toward the centre of the landmark (or alternatively two pixels pointing toward the left and right edge of the landmark). The ALV model has been successfully implemented on a mobile robot built completely in analogue hardware [5] using two capacitors in order to store the AL vector at the home position. In a natural environment, it is difficult to separate the landmarks from the background. An alternative version of the ALV model with continuous image processing has been introduced [6]. It works on normalised, low-pass-filtered greyscale images, where a vector pointing towards the centre of mass is used rather than an AL vector.

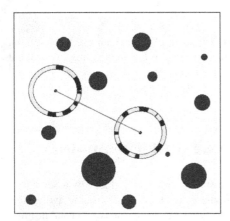 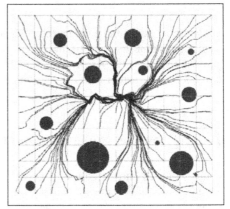

Fig. 2. Left: Schematics of an agent within a virtual environment at two different positions. The omni-directional one-dimensional visual field of the agent is represented by a ring, containing the projections of the landmarks. These rings do not represent the size of the agent. Right: Trajectories from different starting positions on a grid toward a goal position at the centre where a snapshot had been taken. The homing strategy used here is the average landmark vector (ALV) model, using the edges of the cylinders as landmarks.

In Figure 2 left, the schematics of an agent within a virtual environment with sparsely distributed cylindrical obstacles at two different positions are displayed. The omni-directional one-dimensional visual field of the agent is represented by a ring, containing the projections of the landmarks. The robot trajectories during visual homing using the ALV model from different starting positions on a grid toward the goal position can be seen in Figure 2 right. The home snapshot has been taken at a position near the centre of the virtual world. The homing algorithm takes a new snapshot at the current position at each time step, calculates the homing vector, and moves a small step in this direction. As can be seen in Figure 2 right, the homing trajectories move around the obstacles. This behaviour is not explicitly encoded in the ALV model.

It has been shown by Hafner[6] and Hafner and Möller[7] that using the two snapshots as inputs and the homing vector as desired target to a feed-forward neural network, the ALV model can be learned in a self-supervised manner. An interesting aspect shows up if we consider the learned visual homing model, which resembles the original ALV model very closely. The neural network is trained with a set S of snapshot pairs $(s_i, s_j)^P$ and vectors v_h^P, which directly point from position i^P to position j^P, regardless of whether there are any obstacles in between the two snapshot positions or not. The resulting learned model, however, will most often avoid these obstacles. The reason for this strange behaviour can be explained by some geometric properties of the environment.

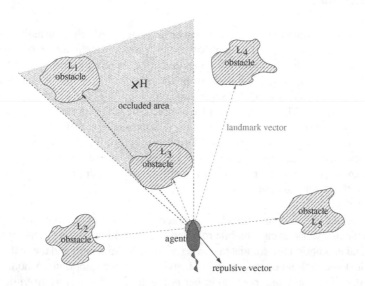

Fig. 3. Schematics of an agent homing in a complex environment with several landmarks which can at the same time be obstacles. The nearest obstacle is occluding other objects therefore producing a repulsive vector away from the obstacle. **H** indicates the home position.

In Figure 3, a scenario for an agent homing in an environment with several landmarks is plotted from a bird's eye view. The agent is moving straight towards the goal, however, the direct way is blocked by an obstacle. As soon as this obstacle is occluding another landmark, the landmark vector l_i^c which was formerly pointing from the current position of the agent in direction of the occluded landmark L_i, is now subtracted from the home vector without occlusions. For $L_m, \ldots, L_n$ being the occluded landmarks at the current position with $m < n$, we get a new homing vector h' from the difference of the two average landmark vectors a_c' and a_t:

$$h' = a'_c - a_t = \sum_{i=1}^{m} l_i^c - \sum_{i=1}^{n} l_i^t = h - \sum_{i=m+1}^{n} l_i^c$$

On average, the sum of the vectors pointing from the agent toward the occluded landmarks is pointing straight in the centre of the occluding landmark. As a consequence, the agent is repulsed from the landmark in front of it, allowing for any sidewards movement to take over, resulting in trajectories as can be found in Figure 2 right. This repulsion provides the same results as described in [8], where a force field method results in steering away from a surface. The problem of a dead spot, where the agent is repulsed exactly in the direction negative to its movement vector is avoided by noise in real robots. The agent is also automatically more strongly repulsed from close obstacles than from others further away, since the close ones are occluding a higher number of other landmarks.

3   Discussion

I have shown a situation, where a navigation strategy, which is usually considered to be on a higher level of complexity than basic obstacle avoidance behaviour, results in exactly this behaviour without having it explicitly encoded. We call this behaviour *emergent*, since it results from the agent-environment interaction, is not pre-programmed, and cannot be separated into a sub-module independent from the homing behaviour. This emergence of a simple behaviour (obstacle avoidance) from a more complex behaviour (visual homing) is unusual. In nature, evolution clearly does not come up with more complex behaviour before the most basic skills crucial for survival have been developed. In robotics, obstacle avoidance is usually one of the first behaviours to be implemented on a mobile robot in order to avoid damage to the property, the robot and others.

The effect of emergent obstacle avoidance during visual homing can be observed both in the real world using a mobile robot and in simulation, as long as the dynamics of vision and egomotion are simulated correctly. In both simulation and the real world, the dynamical interaction between agent and environment are important. A homing vector that is derived at only one position either points to a position away from the goal, or points directly towards the goal without taking any obstacles in between into account. Only through the dynamics arising from a constantly updated visual input caused by the movements of the agent within its environment can trajectories be produced which avoid obstacles and lead to the goal. The robot does not influence its environment substantially, but it influences its own sensory input through its movements in space.

What makes this example particularly interesting for *embodied AI* is its focus on the interaction with the environment. A navigating agent with sensors, motors and a brain[2] *has* to be embodied. It also shows that this concept cannot be restricted to mobile robots, but should also apply to simulated agents: If we call an agent that exploits the

---

[2] Note that having a brain is not necessarily a requirement for *EAI*, but for navigation. Animals evolved brains because they had to move.

interaction with its environment *embodied*, both mobile robots and simulated agents that fulfil this requirement should be called *embodied*.

The importance of real world interaction for the study of intelligence and intelligent behaviour in humans and other animals has been recognised a long time ago. This interpretation applied to the study or the design of artificial agents has recently been termed *embodied AI*. The aspects of the real world environment can vary. If we restrict *EAI* to aspects of the environment as we (humans) perceive them, we exclude many of the non-humanoid artificial agents. If we do not restrict it at all, then the concept of *EAI* becomes extremely vague, including even abstract algorithmic software such as a sorting algorithm which interacts with the memory stack. One problem with the *EAI* approach lies in the vagueness of the central concept of *embodiment*. This vagueness has hindered precise communication between researchers in the field so far.

As is clear from the example in section 2, there is no such thing as a dis-embodied navigating agent, and going one step further, there is no dis-embodied agent at all. Every agent, even a pure software agent, is defined by its interactions with the environment. One has to be careful not to restrict the definition of environment in this context too much. In conclusion, any interaction between an agent and an environment in their most open definition can be interesting for the study of intelligence.

References

1. Lambrinos, D., Möller, R., Labhart, T., Pfeifer, R., Wehner, R.: A mobile robot employing insect strategies for navigation. Robotics and Autonomous Systems, special issue on Biomimetic Robots **30** (2000) 39–64
2. Walter, W.G.: An imitation of life. Scientific American **182** (1950) 42–45
3. Holland, O.: The Grey Walter Online Archive. http://www.ias.uwe.ac.uk/extra_pages/gwonline.html, University of the West of England, Bristol (1996)
4. Hafner, V.V., Kunz, H., Pfeifer, R.: An Investigation into Obstacle Avoidance as an 'Emergent' Behaviour from two Different Perspectives. Proceedings of the EPSRC/BBSRC International Workshop on Biologically-Inspired Robotics: The Legacy of W. Grey Walter (2002) 166–173
5. Möller, R.: Insect visual homing strategies in a robot with analog processing. Biological Cybernetics, special issue: Navigation in Biological and Artificial Systems **83** (2000) 231–243
6. Hafner, V.V.: Adaptive Homing - Robotic Exploration Tours. Adaptive Behavior **9** (2001) 131–141
7. Hafner, V.V., Möller, R.: Learning of visual navigation strategies. In: Proceedings of the European Workshop on Learning Robots (EWLR-9), Prague (2001) 47–56
8. Reynolds, C.W.: Not bumping into things. Notes for the SIGGRAPH 88 course Developments in Physically-Based Modeling (1988)

Bayesian Modeling and Reasoning for Real World Robotics: Basics and Examples

David Bellot[1], Roland Siegwart[3], Pierre Bessière[2], Adriana Tapus[3],
Christophe Coué[2], and Julien Diard[2]

[1] Department of Statistics, University of California, Berkeley, 94720-3860, USA
[2] Gravir/IMAG/CNRS INRIA Rhône-Alpes ZIRST - 655 avenue de l'Europe
Montbonnot 38334 Saint Ismier France
[3] Swiss Federal Institute of Technology EPFL-STI-I2S-LSA1 CH-1015 Lausanne,
Switzerland

Abstract. Cognition and Reasoning with uncertain and partial knowledge is a challenge for autonomous mobile robotics. Previous robotics systems based on a purely logical or geometrical paradigm are limited in their ability to deal with partial or uncertain knowledge, adaptation to new environments and noisy sensors. Representing knowledge as a joint probability distribution increases the possibility for robotics systems to increase their quality of perception on their environment and helps them to take the right actions towards a more realistic and robust behavior. Dealing with uncertainty is thus a major challenge for robotics in a real and unconstrained environment. Here, we propose a new formalism and methodology called Bayesian Programming which aims at the design of efficient robotics systems evolving in a real and uncontrolled environment. The formalism will be exemplified and validated by two interesting experiments.

1 Incompleteness and Uncertainty in Robotics

One of the biggest challenge for autonomous mobile robotics is the navigation in unknown or partially known environments, when noisy sensors are used and where unexpected events happen. Even if recent research resulted in some very nice demonstrations of autonomous navigation in dynamic environments, we are still far from having concepts and algorithms that adapt to different environments and scale well with the complexity of the environment.

This paper suggests a generic approach based on the well-known Bayes theory, in order to progress toward cognitive systems that are able to reason in highly complex real-world environments. The proposed Bayesian framework is a generic approach for probabilistic reasoning. It combines probability distributions, established through *a priori* knowledge and learning, with Bayesian inference in order to make autonomous system capable of dealing with the uncertainty and incompleteness of the real world. *A priori* knowledge and models reduce significantly the complexity of the implementation. Thus, the probabilistic reasoning becomes more feasible for highly dynamic and complex environments.

F. Iida et al. (Eds.): Embodied Artificial Intelligence, LNAI 3139, pp. 186–201, 2004.

In classical robotics [1], the programmer of the robot has himself an abstract conception of the environment, described in geometrical, analytical and/or symbolic terms because the shape of objects, the map of the world, the laws of physics and the objects are known. Programming such a robot is a difficult task because the programmer needs to completely know the environment. The main example of this kind of robotics are the robots used to manufacture cars. Their environment is highly constrained and their behavior is usually described through a finite-state automaton. This is the usual answer to the problem of uncertainty: let the environment be as predictable as possible by controlling and constraining it. If the environment is open and if it cannot be constrained, or if the programmer aims at a more versatile robot, then the complexity of the program increases dramatically and lead to intractable models and representation of the real world. Therefore, it is necessary just to take into account a small part of the environment leading to a large number of hidden or unknown variables.

From an engineering point of view, an accurate control of both the environment and the tasks ensures that industrial robots work properly. However, this approach is no longer possible when the robot must act in an environment not specifically designed for it. The purpose of this chapter is to give an overview of a generic solution to this problem especially to present a versatile framework called Bayesian Programming (BP). Section 2 presents the Bayesian Programming paradigm. It establishes a common formalism and methodology that will be used throughout this chapter. The last section will be devoted to two complex examples in robotics. A solution based on Bayesian Programming will also be presented.

2   The Bayesian Programming Framework: A Generic Formalism

This section introduces the Bayesian Programming formalism. As mentioned in the introduction, when programming a robot, the programmer constructs an abstract representation of its environment, which is basically described in geometrical, analytical or symbolic terms. In a way, the programmer imposes to the robot, his or her own abstract conception of the environment. The difficulties appear when the robot needs to link these abstract concepts with the robot's raw signals (either obtained from the robot's sensors or being sent to the robot's actuators). The central origin of these difficulties is the irreducible incompleteness of the models. Probabilistic methodologies and techniques offer possible solutions to the incompleteness and uncertainty difficulties when programming a robot. The basic programming resources are probability distributions. The Bayesian Programming (BP) approach was originally proposed as a tool for robotic programming (see [2]), but nowadays used in a wider scope of applications: CAD systems [3], path planning [4] or medical diagnosis [5].

The Bayesian Programming formalism allows using a unique notation and structure to describe probabilistic knowledge and its use. The elements of a

Bayesian Program are illustrated in Figure 1. A BP is divided in two parts: a description and a question.

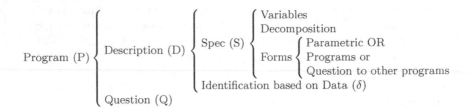

Fig. 1. Structure of a Bayesian program.

2.1 Description

The first component is a *declarative* component, where the user defines a *description*: it is a way to specify a joint distribution over a set of variables $\{X_1\ X_2\ \ldots\ X_n\}$, given a set of experimental data δ and preliminary knowledge π. The variables have to be relevant for the environment one would like to model. The joint distribution $P(X_1\ X_2\ \ldots\ X_n\mid\delta\ \pi)$ is decomposed into a product of simpler terms based on some conditional independence assumptions. This set of assumptions belongs to the set π of *a priori* knowledge. In order to complete the description, parametric forms (also belonging to π) and *a priori* distributions (numerical parameters of the so-called parametric forms) are given. If there are free parameters in the parametric forms, they have to be manually defined or fitted using a learning procedure on the set of experimental data δ.

The aim of a decomposition is to introduce some conditional independence assumptions between variables so that to decrease the complexity of the inference process or more generally to introduce *a priori* knowledge about the environment or the behavior of the robot. This kind of knowledge is provided by the programmer and represents either causal interactions [6] or structural relations between variables. For example, a first-order Markov assumption claims that the belief state of a variable X_t at time t is independent of its long-term past, given its short-term past. In other words X_t is independent of $X_{t-i}, \forall i > 1$ given X_{t-1}. Therefore the decomposition for such a simple system is $P(X_t\mid X_{t-1}).P(X_{t-1})$.

Variables represent facts about the environment or the robot. For example, a light sensor could be represented by a variable L where its probability distribution is assumed to be Gaussian, $L \sim \mathcal{N}(\mu, \sigma^2)$, and represent the intensity of light occurring at the sensor.

2.2 Question

Now, let assume that an environment can be described with the following set of variables $\mathcal{S} = \{A, B, C, D\}$. Our *a priori* (or prior) knowledge can be summarized by the statement "*C is independent of D given A and B*". No other particular knowledge about A and B is available. Therefore, an obvious decomposition would be $P(ABCD) = P(C|AB)P(D|AB)P(AB)$. This decomposition is not easy to use since the joint probability distribution over $\{AB\}$ has to be computed. The probability $P(AB)$ can be approximated using sampling techniques or can be decomposed into a simpler joint probability distribution using the so-called chain's rule: $P(AB) = P(A|B)P(B) = P(B|A)P(A)$. More formally, a question is obtained by partionning the initial set of variables into three distinct subsets: *Known*, *Searched* and *Unknown*. The first set denotes the set of observed variables. The second is the subset for which one wants to know the posterior joint probability distribution. And finally, the third subset contains unobserved or latent variables.

Using knowledge is answering the question. Answering the question is solving a Bayesian inference problem on the description in order to compute the posterior probability distribution described by the question. Therefore, a question in a Bayesian Program is the posterior probability distribution one is interested, given some measurements on the other variables. For example, let assume that we know some facts about B, but nothing about the other variables, say $B = b_1$. We would like what is the posterior distribution of D given $B = b_1$. The question is $P(D|B = b_1)$. Here we assume we have an algorithm to solve this Bayesian inference problem is available, and so by giving the description, π and δ, the probability distribution of $P(D|B = b_1)$ can be computed.

The general question $P(\mathcal{S}_{\setminus B}|B = b_1)$ is also known as the belief propagation problem [7]. This chapter is mainly concerned with modeling issues, and we assume the inference problem to be solved and implemented in an efficient way by an inference engine. The reader should be warned that Bayesian inference is not an obvious problem and inference algorithms are usually designed together with the model itself in order to obtain optimal results in terms of computational costs and accuracy. However, general algorithms are also available, based on messages and beliefs propagation [8], sampling techniques or variational approximations [9].

3   Complex Bayesian Programming for Robotics

This section presents two applications of Bayesian Programming. The first one is an extension of occupancy grids using *a priori* knowledge to perform target position and velocity in an urban traffic situation. The grids are combined with danger estimation to perform an elementary task of obstacle avoidance with an electric car. The second application is devoted to topological global localization by using sequences of features forming a global distinctive fingerprint. The topological representation gives a compact representation since only distinctive places within the environment are encoded.

3.1 Bayesian Programming for Multi-target Tracking: An Automotive Application

The ADAS Context. Unlike regular cruise control systems, Adaptive Cruise Control (ACC) systems use a range sensor to regulate the speed of the car while ensuring collision avoidance with the vehicle in front. ACC systems were introduced on the automotive market in 1999. Since then, surveys and experimental assessments have demonstrated the interest for this kind of systems. They are the first step towards the design of future Advanced Driver Assistance Systems (ADAS) that should help the driver in increasingly complex driving tasks. The use of today commercially available ACC systems is pretty much limited to motorways or urban expressways without crossings. The traffic situations encountered are rather simple and attention can be focused on a few, well defined detected objects (cars and trucks). Nonetheless, even in these relatively simple situations, these systems show a number of limitations: they are not very good at handling fixed obstacles and may generate false alarms; moreover, in some 'cut-in' situations, *i.e.* when the intrusion of an other vehicle or a pedestrian in the detection beam is too close to the vehicle, they may be unable to react appropriately.

A wider use of such systems requires to extend their range of operation to some more complex situations in dense traffic environments, around or inside urban areas. In such areas, traffic is characterized by lower speeds, tight curves, traffic signs, crossings and "fragile" traffic participants such as motorbikes, bicycles or pedestrians.

The Related Multi-Target Tracking Problem. A prerequisite to a reliable ADAS in such complex traffic situations is an estimation of dynamic characteristics of the traffic participants, such as position and velocity. This problem is basically a *Multi-target Tracking* problem. The objective is to collect *observations*, *i.e.* data from the sensor, on one or more *potential obstacles* in the environment of the vehicle, and then to estimate at each time step and as robustly as possible the obstacles position and velocity. Classical approach is to track the different objects independently, by maintaining a list of *tracks*, *i.e.* a list of currently known objects. The main difficulty of multi-target tracking is known as the *Data Association* problem. It includes observation-to-track association and track management problems. The goal of observation-to-track association is to decide whether a new sensor observation corresponds to an existing track or not. Then the goal of track maintenance is to decide the confirmation or the deletion of each existing track, and, if required, the creation of new tracks. A complete review of the tracking methods with one or more sensors can be found in [10].

Urban traffic scenarios are still a challenge in multi-target tracking area: the traditional data association problem is intractable in situations involving numerous appearances, disappearances and occlusions of a large number of rapidly maneuvering targets.

The approach presented here is a new approach for a robust perception and analysis of highly dynamic environments. This approach has been designed in order to avoid the data association problem previously mentioned. It is based on a probabilistic *grid representation of the obstacles state space*. As we consider the position and the velocity of the potential obstacles with respect to our vehicle, this grid is 4-dimensional. Then for each cell of the grid, the occupancy probability is estimated using sensor observations and some prior knowledge.

Estimation of the Occupancy Grid. The objective is to compute from the sensor observations the probability that each cell is full or empty. To avoid a combinatorial explosion of grid configuration, the cell states are estimated as *independent* random variables.

The occupancy grid framework was extensively used for mapping and localization. Of course, for an automotive application, it is impossible and useless to model the whole environment of the vehicle with a grid. Thus we will model only the near-front environment of our vehicle. As we want to estimate the relative position and the relative velocity of objects, each cell of our 4-D[1] grid corresponds to a position and a speed relative to the vehicle.

Figure 2 presents the Bayesian Program for the estimation of the occupancy probability of a cell. To simplify notations, a particular cell of the grid is denoted by a single variable X, despite the grid is 4-D. The number of sensor observations at time k is named N^k. One sensor data at time k is denoted by the variable Z_i^k, $i = 1 \ldots N_k$. The set of all sensor observations at time k is noted Z^k. The set of all sensor observations until time k is referred by the notation $Z_{1:k}$. A variable called the *matching* variable and noted M^k is added. Its goal is to specify which observation of the sensor is currently used to estimate the state of the cell.

Bayesian Occupancy Filter. To take into account the dynamic environment, and to be as robust as possible relatively to objects occlusions, it is necessary to take into account the *sensor observations history and the temporal consistency of the scene*. This is done by introducing a two-step mechanism in the occupancy grid estimation. This mechanism includes a prediction (history) and an estimation (new measurements) steps. This mechanism is derived from the *Bayes filtersq* approach [11] and it is called the *Bayesian Occupancy Filter* (BOF). Figure 3 shows the corresponding Bayesian Program.

Experimental Results. To test the estimation of occupancy grids both a simulator and the real Cycab vehicle were used. Figure 4 shows the first results of estimation and prediction steps, for a static scene. The upper left scheme depicts the situation: two static objects are present in front of the Cycab. These two objects are fixed. The Cycab is static too. Thus only 2-dimensional grids are depicted, corresponding to the object's position at a null speed. Figure 4b represents the occupancy grid, knowing only the first sensor observations. The gray level corresponds to the probability that a cell is occupied. In this case, the two objects are detected by the sensor. Consequently, two areas with high

---

[1] 2 dimensions for the x, y position and 2 dimensions for the $\dot{x}, \dot{y}$ velocities

$$P \begin{cases} D \begin{cases} \begin{array}{l} \text{Variables} \\ \quad X^k \ \ : \text{cell } X \text{ at time } k \qquad E_X^k : \exists \text{ an object in cell } X \\ \quad \mathcal{Z}^{1:k} : \text{sensor observations } M^k : \text{``matching'' variable.} \end{array} \\[2ex] \begin{array}{l} \text{Decomposition} \\ \quad P(X^k \ E_X^k \ \mathcal{Z}^{1:k}) = \\ \quad \begin{pmatrix} P(X^k)P(\mathcal{Z}^{1:k-1})P(E_X^k \mid X^k \mathcal{Z}^{1:k-1}) \\ P(M^k) \prod\limits_{s=1}^{N_s} P(Z_s^k \mid M^k \ E_X^k \ X^k) \end{pmatrix} \end{array} \\[3ex] \begin{array}{l} \text{Parametric Forms} \\ \quad P(X^k): \text{Uniform} \hspace{3em} P(\mathcal{Z}^{1:k-1}): \text{Unknown} \\ \quad P(E_X^k \mid X^k \ \mathcal{Z}^{1:k-1}): \text{Prediction step} \quad P(M^k): \text{Uniform} \\ \quad P(Z_s^k \mid [M^k = s] \ E_X^k \ X^k): \text{Sensor model} \\ \quad P(Z_s^k \mid M^k \neq s \ E_X^k \ X^k): \text{Uniform} \end{array} \end{cases} \\[2ex] \begin{array}{l} \text{Question :} \\ \quad P(E_X^k \mid X^k \ \mathcal{Z}^{1:k}) \end{array} \end{cases}$$

Fig. 2. Estimation Step at time k.

$$P \begin{cases} D \begin{cases} \begin{array}{l} \text{Variables} \\ \quad X^k, E_X^k, X^{k-1}, E_X^{k-1}, \mathcal{Z}^{1:k-1} : \text{same semantic as previously} \\ \quad U^{k-1} \hspace{6em} : \text{control of the Cycab at time } k-1 \end{array} \\[2ex] \begin{array}{l} \text{Decomposition} \\ \quad P(X_k \ E_X^k \ X^{k-1} \ E_X^{k-1} \ \mathcal{Z}^{1:k-1} \ U^{k-1}) = \\ \quad \begin{pmatrix} P(\mathcal{Z}_{1:k-1})P(U^{k-1})P(X^{k-1}) \ P(E_X^{k-1} \mid X^{k-1} \ \mathcal{Z}^{1:k-1}) \\ P(X^k \mid X^{k-1} \ U^{k-1}) \ P(E_X^k \mid E_X^{k-1} \ X^k \ X^{k-1}) \end{pmatrix} \end{array} \\[2ex] \begin{array}{l} \text{Parametric Forms} \\ \quad P(\mathcal{Z}^{1:k-1}) \hspace{5em} : \text{Unknown} \\ \quad P(U^{k-1}) \hspace{5.5em} : \text{Uniform} \\ \quad P(X^{k-1}) \hspace{5.5em} : \text{Uniform} \\ \quad P(X^k \mid X^{k-1}) \hspace{3.5em} : \text{dyn. model} \\ \quad P(E_X^k \mid E_X^{k-1} \ X^k \ X^{k-1}) : \text{Dirac.} \\ \quad P(E_X^{k-1} \mid X^{k-1} \ \mathcal{Z}^{1:k-1}) \ : \text{estim. at } k-1 \end{array} \end{cases} \\[2ex] \begin{array}{l} \text{Question :} \\ \quad P(E_X^k \mid X^k \ \mathcal{Z}^{1:k-1} \ U^{k-1}) \end{array} \end{cases}$$

Fig. 3. Prediction Step at time k.

occupancy probabilities are visible (dark gray areas). These probability values depend on the probability of detection, the probability of false alarm, and on the sensor precision. All these characteristics of the sensor are taken into account in the sensor model.

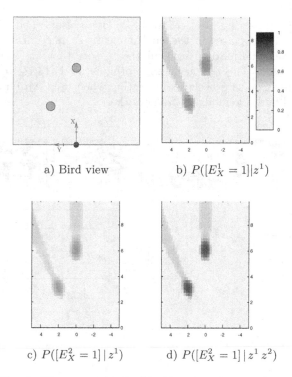

a) Bird view   b) $P([E_X^1 = 1]|z^1)$

c) $P([E_X^2 = 1]\,|\,z^1)$ d) $P([E_X^2 = 1]\,|\,z^1\,z^2)$

Fig. 4. First example of grid estimation, for a static scene.

The cells hidden by a sensor observation have not been observed. Thus we can not conclude about their occupancy. That explains the two areas of probability values close to 0.5. Thanks to this property of occupancy grids and to the prediction phase, the estimation of the grid is robust to temporary occlusion between moving objects. Finally, for cells located far from any sensor observation, the occupancy probability is low (plain grey areas).

To validate the approach in dynamic situations, an application involving an electric car has been implemented [12]. The car is longitudinally controlled in order to avoid obstacles. This basic behavior is obtained by combining the occupancy probability and the danger probability of each cell of the grid. Results of the experiments clearly show that this approach is able to prevent collisions even when moving obstacles (pedestrians for example) are temporally hidden (by a parked car for example).

3.2 Bayesian Programming for Topological Navigation with the Fingerprint Concept

Introduction. The topological approach yields a compact representation and allows high-level symbolic reasoning for map building and navigation. With this method we try to eliminate the perceptual aliasing (*i.e.* distinct locations within the environment appearing identical to the robot's sensors) and to improve the distinctiveness of the places in the environment. To maximize the reliability in navigation, the information from all the sensors that are available to the robot must be used. The notion of fingerprint is used [13,14] to characterize the environment. This is especially interesting when used within a topological localization and multiple modality framework.

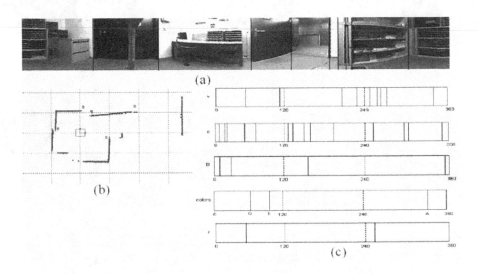

Fig. 5. Fingerprint generation. (a) panoramic image with the vertical edges 'v' and color patches detection, (b) laser scan with extracted corners 'c' and beacons 'b', (c) the first four images depict the position (0° to 359°) of the vertical edges, the corners, the beacons and the colors (G-green, E-light green, and A-red) respectively. The fifth image describes the correspondence between the vertical edge features and the corner features. By regrouping all this results together and by adding the empty space features, the final fingerprint is: *cbccbnfGcnEnvccncbcvncnnfvvvnccAcb*.

A fingerprint is a circular list of features, where the ordering matches the relative ordering of the features around the robot. We denote the fingerprint sequence using a list of characters, where each character represents the instance of a specific feature type. In our case we choose to extract color patches and vertical edges from visual information and corners and beacons from laser scanner. The letter 'v' is used to characterize an edge, the letters A, B, C, ..., P to represent hue bins, the letter 'c' to characterize a corner feature and the letter 'b' to characterize a beacon feature. Details about the visual features extraction can be found in [14,15] and laser scanner features extraction can be found in [16].

Fingerprint Generation. The fingerprint generation is done in three steps (see Figure 5). The extraction of the different features (*e.g.* vertical edges, corners, color patches, beacons) from the sensors is the first phase of the fingerprint generation. The order of the features, given by their angular positions $(0° \ldots 359°)$ is kept in an array. At this stage a new type of virtual feature 'f' is introduced, that reflects the correspondence between a corner and an edge. The ordering of the features in a fingerprint sequence is highly informative and for that reason the notion of angular distance between two consecutive features is added. This geometric information increases, once again, the distinctiveness between the fingerprints. Therefore, we introduced an additional type of feature, the empty space feature 'n', to reflect angular distance. Each 'n' covers the same angle of the scene (20 degrees). This insertion is the last step of the fingerprint generation [14].

Fingerprint Matching for Localization. The string-matching problem is not easy. Usually strings do not match exactly because the robot may not be exactly located on a map point and/or some changes in the environment or perception errors occurred. The standard algorithms are quite sensitive to insertion and deletion errors, which cause the string lengths to vary significantly. The approach adopted previously in the fingerprint approach for sequence matching is inspired by the minimum energy algorithm used in stereo-vision for finding pixels in two images that correspond to the same point of a scene [17]. More details can be found in [13,14]. Our current approach is a combination of the global alignment algorithm and the Bayesian formalism and it is described below.

Probabilistic fingerprint matching algorithm. The new approach comprises two steps. The first step is the phase of supervised learning where the robot inspects several locations, denoted by *Loc*. From each location $loc \in Loc$ the robot extracts the fingerprint data, as explained earlier, and stores it along with the name of the location in a database, denoted by the symbol π.

The second step is the phase of application, where we want the robot to localize itself in the environment. To answer at the question "Where am I?", the robot will extract the fingerprint fp of its current surroundings and solve the basic formula of probabilistic localization:

$$loc^* = \arg\max_{loc \in Loc} P(loc \mid fp\ \pi).$$

This means that if fingerprints are associated to each location, then the actual location of the robot may be recovered by comparing the fingerprint fp with the data of known locations and choosing the location loc^* which has the highest probability. In what follows we show how $P(loc \mid fp\ \pi)$ can be solved by applying Bayesian Programming.

Figure 6 shows the Bayesian Program used for the fingerprint matching. The features are denoted by: *Ve* the set of vertical edges and *Cp* the set of color patches extracted by the omni-directional camera; *Ex* the set of line extremities and *B* the set of beacons extracted from the data given by the laser scanner. For the fingerprint of a location, which is encoded as a circular string the notation *Fp* is used, and for the set of known (learned) locations the notation *Loc* is

$$P \begin{cases} D \begin{cases} \begin{array}{l} \text{Variables} \\ \quad Ve\text{: vertical edges} \qquad\qquad Cp\text{: color patches} \\ \quad Ex\text{: extremities} \qquad\qquad\; B\text{: beacons} \\ \quad Fp\; \text{: a fingerprint of a location } Loc\text{: the set of locations} \\[6pt] \text{Decomposition} \\ \quad P(Loc\; Ve\; Cp\; Ex\; B\; Fp\mid \pi) = \\ \qquad \begin{pmatrix} P(Loc\mid \pi)\, P(Ve\mid Loc\; \pi)\, P(Cp\mid Loc\; \pi) \\ P(Ex\mid Loc\; \pi)\, P(B\mid Loc\; \pi)\, P(Fp\mid Loc\; \pi) \end{pmatrix} \\[6pt] \text{Parametric Forms} \\ \quad P(Loc\mid \pi)\text{: Unif.} \\ \quad P(f\mid loc\; \pi)\text{: } |\sqrt[f]{\prod_{f_i \in f} p_{MOG(\theta_{f_i^{loc}})}(f_i)},\, \forall loc \in Loc \\ \quad \text{where } f \in (Ve, Cp, Ex, B) \\ \quad P(Fp\mid loc\; \pi)\text{: } \frac{1}{GlobalAlignment(Fp, fp_{loc})+1} \\ \quad \text{where } fp_{loc} \text{ is the fingerprint of the location } loc \end{array} \end{cases} \\[6pt] \begin{array}{l} \text{Question :} \\ \quad P(Loc\mid Ve\; Cp\; Ex\; B\; Fp\; \pi) \end{array} \end{cases}$$

Fig. 6. The fingerprint matching formalism written in BP

employed. Although the fingerprint string Fp, constructed over all the features (see [15]), adds some redundancy to the system, it introduces at the same time valuable information about the relative order of the features, which will improve the results. We assume that the variables Ve, Cp, Ex, B and Fp are independent from one another. We consider that the features (Ve, Cp, Ex, B) are dependent of the location and these dependencies lead to the decomposition described in the Bayesian Program (see Figure 6). From the result of the decomposition formula (see Figure 6) we can distinguish three different kinds of probability distributions:

- Since we have no *a priori* information about locations, we consider each location to be equally probable and consequently we express the probability of a location given all the prior knowledge as a uniform distribution.
- To determine the probability of one feature f, where $f \in \{$Ve, Cp, Ex, B$\}$, given the location and all the *a priori* knowledge, we suggest to express this probability as the likelihood of the new feature data f with respect to the distribution of the same feature as encountered at the given location during the learning phase. We calculate the distribution as a mixture of Gaussians (MOG) in angle space, optimizing the mixture parameters $\theta_{f_i^{loc}} = \{w_{f_i^{loc}}, \mu_{f_i^{loc}}, \sigma_{f_i^{loc}}\}$ (where $w_{f_i^{loc}}$ is the weight, $\mu_{f_i^{loc}}$ is the mean and $\sigma_{f_i^{loc}}$ is the standard deviation of the f_i-th mixture component), by making use of the Expectation Maximization (EM) algorithm [18]. Let us illustrate the $P(f = Ve \mid loc\; \pi)$ with an example. We start with a set of 13 occurrences of vertical edges and we calculate the MOG for it. We then generate a second set, this time with 18 occurrences, and evaluate the probability

$P(f = Ve \mid loc\ \pi)$ for both data sets with the same MOG parameters (see Figure 7a and Figure 7b). As expected, the resulting value is for the first data set significantly higher than for the second, since the parameters of the MOG were chosen to maximize the first set. Note how flexible this method is with respect to the number of features per set: A MOG can be generated from a set of any number of features, and it can be evaluated later for samples of arbitrary length.

– To calculate the probability of the fingerprint sequence given the location and all the prior knowledge: we will use the global alignment algorithm [19] used usually for the alignment of DNA sequences. Let GlobalAlignment(Fp, fp_{loc}) be a function yielding the minimal cost of the global alignment algorithm of two fingerprint strings.

Obviously, the three equations from the parametric forms will solve the basic question described in the Bayesian Program.

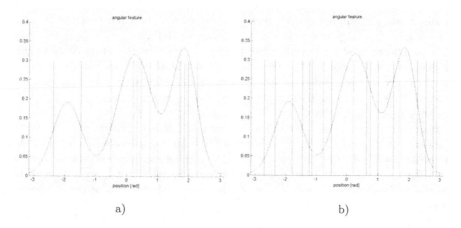

a)   b)

Fig. 7. (a)Evaluation of $P(f = Ve \mid loc\ \pi)$ for the original data set. (b)Evaluation of $P(f = Ve \mid loc\ \pi)$ for other data set, resulting in a smaller value, since the MOG is not optimal for this data.

Experimental Results. The approach has been tested in ten rooms, in a $50 \times 25\ m^2$ portion of our institute building. For the experiments, Donald Duck (see Figure 8), a fully autonomous mobile robot, has been used.

In order to validate the probabilistic fingerprint approach, for each of the ten rooms, fingerprints have been extracted. This experiment has been repeated ten times for each room. Eight times it was placed on a circle 40 cm to 70 cm of radius, yielding the training data, and two times inside the same circle, yielding the test data. For a given observation (fingerprint), a match is successful if the best match with the database (highest probability) corresponds to the correct

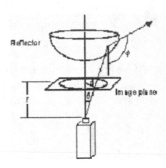

Fig. 8. System used for the experimentation: The fully autonomous robot Donald Duck and the panoramic vision system. The camera has a 640×480 pixel resolution and an equiangular mirror is used so that each pixel in the image covers the same view angle.

room. Since the number of occurrences of the beacon and color patch feature was too small to give significant results, they were omitted for the MOG calculations, but they were used for the fingerprint strings. The results yield a percentage of successful matches of 82.4%. The method presented does not always lead to a perfect success rate, but it still delivers valuable information for false-matched rooms. When the room is successfully matched, the probabilistic matching algorithm gives a high probability: 0.79 in average (between 0.62 and 0.89). Even if it detects the correct room with the second or third highest probability, a Bayesian localization approach, like for example a Partially Observable Markov Decision Process (POMDP) [20,21] can use this information in its observation function. An amelioration of the results can be expected with the augmentation of the number of components of Mixture of Gaussians (MOG) and of the number of observations of a feature [22].

4   Conclusion and Open Problems on Bayesian Programming

The main interest of Bayesian Programming is its ability to describe real-world models with partial and incomplete knowledge about the world. Bayesian Programming is a promising framework and a lot of exciting open problems still exists. To progress toward more robust and sophisticated robotics control systems, these problems need innovative and original solutions. Apart from robotics, those problems are common to other artificial intelligence related fields. It was shown before that it is impossible to completely represent an environment and the strength of Bayesian Programming is to deal with this incompleteness by transforming it into uncertainty. However, the more knowledge is used, the more accurate is the behavior of the robot. Therefore, the problem of making realistic and robust behaviors can be summarized as follow:

- how to make a well-adapted Bayesian Program?
- how to know that a program fit perfectly into a particular task?
- how to learn unknown parameters from real data and experiences?
- how to efficiently use a complex program with many variables and many probabilistic forms?

The answer to those questions is not obvious and leads to more general and exciting questions : learning and inference. How to learn a Bayesian Program instead of making it by hand and how to use the data provided by sensors in order to extract and learn a program? It is out of the scope of this paper to present details about state-of-the-art research on algorithm for Bayesian Programming, but we give here a few facts on this:

- inference is a NP-hard problem for a general Bayesian Program, but solutions exists for particular problems. For example, a state-space model Bayesian filter is usually dealt with using Kalman filter [23,24] or the Forward-Backward algorithm [25]. If the time series analysed by the filter is stationary gaussian, then Durbin-Levison approaches are technically efficient [26],
- inference on regular lattices of variables can be solved using suited algorithms. For instance, factorial hidden Markov models represent a complex stochastic process decomposed into several independent Markov chains given observations. The inference problem is intractable but the use of a variational approximation helps to overcome the computational cost of exact inference [27],
- probabilistic forms are usually discrete or gaussian. However, Bayesian Programming aims at representing whatever probability distributions where probabilistic forms are numerous or even unknown. Numerous approaches exists for dealing with other probabilistic forms, like Mixtures of Gaussians or exponential forms [28],
- complexity of probabilistic forms is sometime a bottleneck for robotics applications. Some techniques aims at reducing the memory footprint of those forms by approximating the distribution leading to a more efficient internal representation [29],
- making versatile programs is hard, but making small programs is quite easier. Does it exist a similar way as object software engineering to link and join small Bayesian Programs into a larger one. Several approaches have been developed: relational probabilistic models [30] or active learning [31] in the context of expert systems.

These techniques and approaches have been designed for particular purposes in the field of statistical learning and artificial intelligence and solve specific problems. They can be adapted to robotics and lead will to more efficient robots systems being able to deal with more complex environments as those of the real world.

References

1. T. Lozano-P rez, J. Jones, E. Mazer, and P. O'Donnell, *HANDEY, A Robot Task Planner*. Cambridge, Massachussets, USA: The MIT Press, 1992. ISBN 0-262-12172-7.
2. O. Lebeltel, P. Bessi re, J. Diard, and E. Mazer, "Bayesian robots programming," *Autonomous Robot*, vol. 16-1, pp. 49–79, 1 2004.
3. K. Mekhnacha, E.Mazer, and P. Bessi re, "A robotic cad system using a bayesian framework," in *Int. Conf. on Intelligent Robots and Systems*, vol. 3, (Takamatsu,Japan), pp. 1597–1604, October 2000.
4. C. Koike, C. Pradalier, P. Bessi re, and E. Mazer, "Proscriptive bayesian programming application for collision avoidance," in *Proc. of the IEEE-RSJ Int. Conf. on Intelligent Robots and Systems*, vol. 1, (Las Vegas, USA), pp. 394 – 399, October 2003.
5. D. Bellot, A. Boyer, and F. Charpillet, "Designing smart agent based telemedicine systems using dynamic bayesian networks: an application to kidney disease people," in *Proc. HealtCom 2002*, (Nancy, France), pp. 90–97, 2002.
6. J. Pearl, *Causality - Models, reasoning and inference*. Cambrige University Press, 2001.
7. I. Rish, *Efficient Reasoning in Graphical Models*. PhD thesis, University of California, Irvine, 1999.
8. M. I. Jordan, Z. Ghahramani, T. Jaakkola, and L. K. Saul, "An introduction to variational methods for graphical models," *Machine Learning*, vol. 37, no. 2, pp. 183–233, 1999.
9. D. J. MacKay, *Information Theory, Inference and Learning Algorithms*. Cambridge University Press, 2003.
10. S. Blackman and R. Popoli, *Design and Analysis of Modern Tracking Systems*. Artech House, 2000.
11. A. H. Jazwinsky, *Stochastic Processes and Filtering Theory*. New York : Academic Press, 1970.
12. C. Cou , C. Pradalier, and C. Laugier, "Bayesian programming for multi-target tracking: an automotive application," in *Proceedings of the International Conference on Field and Service Robotics*, (Lake Yamanaka, Japan), 07 2003.
13. P. Lamon, I. Nourbakhsh, B. Jensen, and R. Siegwart, "Deriving and matching image fingerprint sequences for mobile robot localization," in *Proceedings of the International Conference on Robotics and Automation*, vol. 2, (Seoul, Korea), pp. 1609–1614, May 2001.
14. P. Lamon, A. Tapus, E. Glauser, N. Tomatis, and R. Siegwart, "Environmental modeling with fingerprint sequences for topological global localization," in *Proceedings of the International Conference on Intelligent Robots and Systems*, vol. 4, (Las Vegas, USA), pp. 3781–3786, October 2003.
15. A. Martinelli, A. Tapus, and R. Siegwart, "Multi-resolution slam for real world navigation," in *11th International Symposium of Robotics Research*, (Siena, Italy), October 2003.
16. K. O. Arras and R. Siegwart, "Feature extraction and scene interpretation for map-based navigation and map building," in *Proceedings of the Symposium on Intelligent Systems and Advanced Manufacturing*, (Pittsburgh, USA), October 1997.
17. T. Kanade and Y. Ohta, "Stereo by intra- and inter- scanline search dynamic programming," *IEEE Transactions on pattern analysis and machine intelligence*, vol. PALMZ, March 1985.

18. J. A. Bilmes, "A gentle tutorial of the EM algorithm and its application to parameter estimation for gaussian mixture and hidden markov models," *ICSI-TR-97-021*, 1997.
19. S. Needleman and C. Wunsch, "A general method applicable to the search for similarities in the amino acid sequence of two proteins," *Journal Molecular Biology*, vol. 48, 1970.
20. A. R. Cassandra, L. Kaelbling, and J. Kurien, "Acting under uncertainty: Discrete bayesian models for mobile robot navigation," in *Proceedings of the International Conference on Robotics and Automation*, vol. 2, (Osaka, Japan), pp. 963 – 972, November 1996.
21. N. Tomatis, I. Nourbakhsh, and R. Siegwart, "Hybrid simultaneous localization and map building: a natural integration of topological and metric," *Robotics and Autonomous Systems*, vol. 44, pp. 3–14, 2003.
22. A. Tapus, S. Heinzer, and R. Siegwart, "Bayesian programming for topological global localization with fingerprints," in *International Conference on Robotics and Automation*, (New Orleans, USA), May 2004.
23. R. Kalman, "A new approach to linear filtering and prediction problems," *Journal of basic Engineering*, vol. 35, Mars 1960.
24. P. Smyth, D. Heckerman, and M. Jordan, "Probabilistic Independence Networks for Hidden Markov Probability Models," Tech. Rep. MSR-TR-96-03, Microsoft Research, June 1996.
25. L. Rabiner, "A tutorial on hidden Markov models and selected applications in speech recognition.," in *Proceedings of the IEEE*, vol. 77, pp. 257–285, 1989.
26. P. Brockwell and R. Davis, *Introduction to Time Series and Forecasting*. Springer, 2002.
27. Z. Ghahramani and M. Jordan, "Factorial hidden Markov models," MIT Computational Cognitive Science Report Technical Report 9502, MIT, 1995.
28. M. Jordan, "Graphical models," *Statistical Science (Special Issue on Bayesian Statistics)*, p. In press, 2002.
29. D. Bellot and P. Bessi re, "Approximate discrete probability distribution representation using a multi-resolution binary tree," in *ICTAI 2003*, (Sacramento, California, USA), 2003.
30. L. Getoor, N. Friedman, D. Koller, and B. Taskar., "Learning probabilistic models of relational structure," in *Eighteenth International Conference on Machine Learning (ICML)*, (Williams College), 06 2001.
31. S. Tong and D. Koller, "Active learning for structure in bayesian networks," in *Seventeenth International Joint Conference on Artificial Intelligence*, (Seattle, Washington), pp. 863–869, August 2001.

From Humanoid Embodiment to Theory of Mind

Yasuo Kuniyoshi[1], Yasuaki Yorozu[1], Yoshiyuki Ohmura[1], Koji Terada[1],
Takuya Otani[1], Akihiko Nagakubo[2], and Tomoyuki Yamamoto[3]

[1] The University of Tokyo, 7-3-1 Hongo, Bunkyo-ku, Tokyo 113-8656, Japan,
kuniyosh@isi.imi.i.u-tokyo.ac.jp
http://www.isi.imi.i.u-tokyo.ac.jp/
[2] National Institute for Advanced Industrial Science and Technology
[3] Japan Advanced Institute of Science and Technology

Abstract. We propose to investigate the foundations of communication
and symbolic behavior by means or a robotics approach, i.e. by studying
how these behaviors might emerge from the physical dynamics of an
agent and its sensory-motor interactions with the real world. In this
perspective, the human-robot interface problem can be viewed as one
of coupling the interaction dynamics of all agents. Through a number
of case studies we will show that within this interaction dynamics there
is sparse global structure, i.e. a structure that can be characterized by
only a small number of points in phase space, and that it is best to
interact with the agent, i.e. interfere with its dynamics, at these points.
We introduce a humanoid robot with the capability for dynamic fullbody movement. The preliminary results of two experiments, sitting and
standing up, are presented. Lastly, experiments with self exploratory
learning of embodiment and visual motor learning of neonatal imitation
abilities are introduced.

1 Introduction

Over the past decades there have been substantial research efforts devoted to developing human-robot interfaces. Recently, partly boosted by a drastic increase
in computational power, there has been a lot of progress in achieving skilled behavior, e.g. in visual tracking, face recognition, gesture recognition/production,
action understanding, speech recognition/synthesis, compliant motion, and real-
time hand-eye coordination.

However, the following principle-level issues seem to remain largely unexplored; What are the *essential* factors for assuring the *meaningfulness* of the
executed tasks? Task execution in the real world is always under unpredictable
perturbations. And the details of the execution *should* change in order to adapt
to various situations. Therefore appropriate control is crucial to assure that the
important conditions are not missed while being adaptive. So far finding and
defining such conditions and control laws are done by humans in a problem
specific mannar.

A symmetrical problem can be seen in recognition tasks. A well known
"pattern-to-symbol" problem has exactly the same structure in that the system

F. Iida et al. (Eds.): Embodied Artificial Intelligence, LNAI 3139, pp. 202–218, 2004.

is required to extract essential meaningful information from continous patterns with changing details due to real world situations even though they maintain the same meaning. So far pattern-to-symbol conversion rules are explicitly defined by humans in a problem-specific mannar.

These problems are naturally extended to human-machine interface issues. There, the "important conditions" above are not statically given, but dynamically commanded by humans. And machines should perceive human behavior with changing details and extract essential meaningful information. So far the interface protocols are defined by humans in a context-specific mannar. But real human behavior cannot be treated as static patterns.

In many cases the existing methodologies above are very fragile and that is why most of the robotic demos only work in laboratories. But we do not have general rules which tell us what the appropriate control laws, conversion rules, interface protocols are, in uncertain situations. Neither do we have a methodology for automatic acquisition of them. Are there any general principles that guide us finding solutions for these questions across many task domains and situations?

Open-Ended And Emergent Information Structure. Humanoid robotics necessitates an entirely different approach from traditional engineering. Because humanoids should have what has been called a "globally well-balanced functionality" [1], i.e. they should be able to perform a large set of tasks with very different requirements such as walking, grasping and manipulating objects, recognizing faces and, interacting with humans by means of natural language, optimal performance can no longer be defined. The strategies employed must be open-ended which means that they can continuously adapt to changing task demands and changing environments.

There is a considerable body of literature on the so-called symbol grounding problem. Rather than assuming that symbols are there as extant structures and need to be grounded, we suggest to view symbols as being emergent from a complex system-environment dynamics. The question to be tackled then is how, within this continuous stream of physical processes (stream of sensory stimulation from the visual, auditory, tactile, olfactory, and proprioceptive system, and motor actions of body and limbs), discrete entities can be clearly identified. If these entities are stable over certain variations in environmental conditions, they can be taken to designate what we might want to call low-level "symbols".

We start from the assumption that whenever we are engaged in sensory-motor interactions, some global information structure will emerge from the natural dynamics of the body-environment system[1].

By global information structure we mean that if we repeat the "same" task many times in different situations, we can find clearly identifiable features common to all of these instances. Details of these instances are always different as a result of adaptation to changing situations. But there are some persistent

---

[1] It is important to note that the structure does not exist before the interaction occurs. It emerges only when the interaction is taking place.

features by which we identify the category of each action and task. This is analogous to the mathematical concept of "structural stability", which means that the referred stucture is persistent even if the underlying manifold deviates from its original form in almost every point.

This information structure is sparse in the sense that within the continuous dynamics, it comprises only a certain finite and relatively small number of discrete states. This information structure forms the basis for agent-agent interaction, or to use the terminology introduced earlier, it forms the basis for the coupling of the dynamics of two agents. This coupling can be achieved, for example, by observation which in turn forms the basis of imitation learning.

Because of the discrete and persistent nature we can identify this global information structure with the notion of "symbols". Note that when we talk about information structure, we do not mean an extant memory structure of sorts, but a structure that emerges during the interaction. In this sense, the global information structure is the result of a self-organizing process.

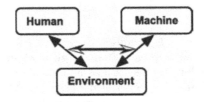

Fig. 1. Three Term Interaction Structure

Meaningful Three-Term Interaction. Three-term structure (Fig. 1) sets the foundation for the principle level discussion of the general interface problem. It consists of a human, a machine, and the environment. The human and the machine are individually involved in autonomous interaction (i.e. sensory-motor loops) through the environment[2]

The interface problem is formulated as interaction between the two interaction dynamics, e.g. some information (or sometimes, force) is transmitted from human-environment interaction and affect the machine-environment interaction.

Our interest is in how such inter-dynamics interaction (i.e. the interaction between the human-environment and the machine-environment interactions) is realized effectively, without destroying the current intra-dynamics (i.e. the behavior of each agent). In other words, how can autonomy and sociality be fused together consistently?

The concept of three-term interaction structure will become clear as we go through some examples.

Outline of the paper. We start by introducing the concept of a "three term interaction". This is followed by the presentation of two case studies. The first one

---

[2] An observer point of view may attribute certain "behaviors" to these interaction dynamics.

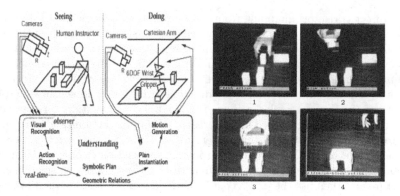

Fig. 2. Learning by Watching system (Left). A human performance of an arbitrary block stacking task is observed by a trinocular (stereo + zoom) camera system. Action units are identified in real time. The resulting symbolic action sequence is fed to a robot manipulator system which re-generates the learned actions to achieve the same task goal in a different workspace with a different initial block placements. Some snapshots of monitor display during task recognition (Right). Picking up the fourth pillar (top row) and placing the top plate (bottom row) in a "table" building task. The real time recognition results are displayed at the bottom; 1. reach, 2. pick, 3. pick, 4. place-on-block.

is about "learning by watching" where a robot equipped with a vision system observes a human performing a particular task. This case study deals with the problem of how to generate discrete states - symbols - from a continuous flow of visual stimulation. The second case study describes the full body dynamic motion of a simulated humanoid. We will demonstrate that the behavior of this robot is best influenced at certain critical points as defined by the sparse global information structure. Then, we introduce our current research platform which consists of a human-size humanoid robot with a large number of degrees of freedom, and we describe a real-time vision-based imitation experiment. Next, we discuss two experiments on body-schema acquisition and boot-strap learning of action categories based on self-exploration. Then we briefly introduce a system that learns to imitate the movements of another agent. The learning behavior is achieved by a neural mechanism capable of identifying attractor states in temporal sequences of high-dimensional sensory-motor data. Self-exploration designates sensory-motor activity that is not triggered by a particular interaction with other agents but is self-motivated. It is called "exploration" because, from an observer's perspective, this serves the purpose of "exploring" the sensory-motor potential of the body.

2   Qualitative Action Recognition – Case Study I

Kuniyoshi et al. [2] built an experimental system which recognizes pick and place sequences performed by a person in real time (Fig. 2).

A direct trajectory mapping from the human hand movement to the robot's movement will not result in performing the same task. This is because the initial placements of the blocks are different in teaching phase and task execution phase, The system must extract "symbolic" information from the continuos image data of human performance. Here, "symbolic" means that it is a unit of information which is reusable and invariant over different situations and sensory-motor interactions. In this particular system, the symbolic information corresponds to action units which are defined for observation of human actions and corresponding robot actions. Later in this paper, we will discuss how such action unit segmentations emerge from body-environment interaction dynamics. Ultimately a robot should be able to create its own action concepts based on such emergent structure and then use them to recognize other's actions. But for the time being, we investigate the perceptual process.

Kuniyoshi et al. [3] proposes the following principles of action recognition.

– Action recognition is detecting causal relationship connecting the subject of action, its motion, the target object, and its motion.
– Temporal segmentation of actions is done when the causal relationship changes.
– The causal relationship is affected by the ongoing context of overall task.

Mathematically, causality can be modelled as a consistent "dynamics", i.e. for a function F which is stable over a period of time, $x_{t+1} = F(x_t)$. The most important point is that a boundary of action emerges as the boundary between more than one F. And this boundary articulates symbolic action units, and important perceptual information is mainly collected at this boundary, according to past psychological experiments [4]. In other words, interfering multiple interaction dynamics define boundary structures which are the foundation of symbols. And the symbols are the units of information which maintains commonality over different situations and agents.

3   Controlling at the Boundaries of Intra-dynamics – Case Study II

Let us discuss about the idea of global structure within intra-dynamics, using an example of "rising" action [1,5]. Here we assume that a person is initially lying flat on the floor, then the person is asked to stand up quickly. One typical strategy would be to first swing up the legs, swing them down, and use the inertia of the legs to bring up the torso in a ballistic movement.

Figure 3 shows some snapshots from the motion capture data of a human subject. Hip and knee angles phase space trajectories are shown in Fig. 4.

What we([5]) found out was that the trajectory bundle from multiple trials have non-uniform structure. The trajectories converge at certain critical parts and diverges in other parts.

Based on further analysis and speculations, we proposed that the entire phase space for a dynamic whole body task has non-uniform structures with sparse

Fig. 3. Snapshots of motion capture data of a "Rising" action performed by a human subject.

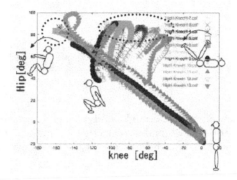

Fig. 4. Phase space trajectories of multiple trials of a "Rising" action. Trajectories converge at critical control points and diverges at ballistic movement parts.

critical points where the phase space trajectories converge or branch. This leads us to a novel strategy for controlling complex whole body dynamic actions. In the following, we describe one example of such a strategy.

Figure 6 is an output from a general purpose dynamics simulator ADAMS. The physical process of body motion including multiple contacts is faithfully simulated. The joint angles of the robot is manually fed by a human programmer in an open-loop manner (i.e. without feedback control). In some sense this is similar to teleoperation because there the operator specifies the desired sequence of positions of the robot.

Initially the robot is lying flat on the floor(a). Then the robot swings up its legs(b), and swings them down, which makes it rolling forward (a ballistic motion), passing a crouching position(c) (local maxima), until it hits the ground with its hands (d) (potential barrier). There it pauses. Then slowly rolls back by pushing against the ground by hands until it balances on its feet(e). Finally it slowly stands up(f). And walks several steps (not shown in the figure).

Our control strategy was extremely simplistic. The programmer just chose the "key poses" (corresponding to the snapshots in Fig. 6) intuitively based on his own motor imagery, i.e. based on what he imagined would be natural intermediate positions. Then he arranged them at appropriate timings with straightforward temporal interpolations which connect them. The control was entirely

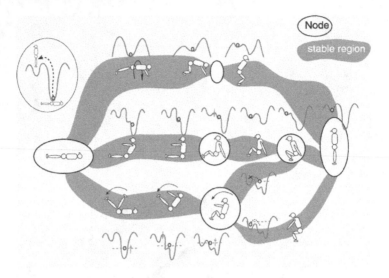

Fig. 5. Schematic view of non-uniform global-dynamics structure. Gray bands denote multiple phase space trajectory bundles connecting the initial state and the goal state. Circles denote converging or branching points on the trajectory bundles.

open-loop. We just let the system go and let the natural dynamics take over. And, lo and behold, it worked! Moreover, it was even robust against small perturbations in the postures and dimensional parameters of the body.

We can see the following important points from the above example.

1. While the system is captured in strong physical dynamics, e.g. the ballistic motion part ((b)-(c) in Fig. 6, and point A in Fig. 7), it is hard to change the trajectory of the system. By contrast, when the system is in a weaker dynamics, i.e. at a non-stable equilibrium point such as a crouching posture, the controller can bring the system into any of the adjacent stronger dynamics by a slight intervention, such as kicking or pushing lightly.

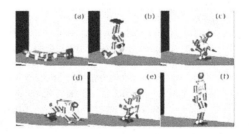

Fig. 6. "Rising" action. Output from a full dynamics simulator. (Created by A. Nagakubo)

Fig. 7. Types of intervention. Left: Intervention at boundary points. Right: Intervention by actively changing the dynamical landscape.

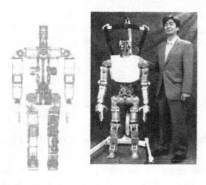

Limb	DOF
Arms	6 × 2
Torso	3
Neck	3
Eyes	3 × 2
Fingers	5 × 2
Legs	6 × 2
Total	46

Part	Mass
Arm Shoulder	3.8
Arm Link 2-4	5.2
Each Arm	9.0
Hand (each)	1.2
Each Leg	9.8
⇓	
Arm×2	18.0
Hand×2	2.4
Buttock	11.8
Leg×2	19.6
Waist&Torso	5.9
Neck	1.6
Full Body Total	59.3

Fig. 8. Our Humanoid: The schematic figure, photo, joint configuration, and mass(kg) distribution of the mechanism.

2. The reason why the naive pose mapping method succeeded, was because the global dynamics structures were very similar between the programmer's body and the robot's body (i.e. the dynamics of this task is mostly governed by the outer shape, posture and mass distribution of the body), and that the global structures were robust enough to tolerate small differences and perturbations. In order to exploit this property, the intervention was made only at the boundaries in the global structure in order to avoid destroying the useful global dynamics.

3. In the current example, the intervention was made by specifying key poses. Its effect is to change the global landscape of the dynamics when the system reaches certain key points (Fig. 7). It is important to note that these landscapes do not correspond statically to a specified pose. Because the robot is not statically fixed to the ground, the dynamical structure also depends on the current orientation, contact state, and motion of the body as well as the pose. Moreover, the pose making itself is affected by the dynamics, and here there is a cyclic dependency. This cyclic problem structure is important in the context of emergent structure, and we will discuss about this later.

Previous dynamical systems approaches exploited limit cycle attractors[6,7, 8]. This corresponds to our notion of "intra-dynamics", where intervention to the system dynamics can be kept to minimum yet its stability is maintained. This is because the strong attractor dynamics stabilizes itself. Our notion of "inter-dynamics" emphasizes a complimentary point to the above; switching from one global dynamics to another by sparse intervention at the boundary of global dynamics structure. If the intervention is made within the attractor structure, it may destroy the attractor because the intervention changes the underlying dynamics. Our claim is that the system gains more degrees of freedom at the boundaries between the attractor, and there the controller can take the system along/across the boundary structure to a desired direction. The general interface problem, mentioned earlier, can be understood well by the above notion of boundary structures.

4   Humanoid for Whole-Body Dynamic Actions

Figure 8 shows our current research platform. It has been developed at ETL and then at The Univ. of Tokyo[3].

Most of the design efforts have been devoted to achieve generality and openness, assuming no particular task or posture. The robot possesses 46 degrees of freedom, with the height and weight of an average Japanese person. It was intended to be an experimental platform to explore a novel principle of controlling complex embodied systems characterized by exploitation of natural physical dynamics of the body, and sparse control at the boundaries of global structure of interaction dynamics [9,10].

The idea of exploiting the physical interaction dynamics between a body and the environment is shared by a number of researchers, for example [11]. This is closely related to the principle of "cheap design" which states that intelligent agents exploit the intrinsic dynamics of the agent-environment interaction [12]. However, what we are interested in is the more general problem of identifying many different dynamics structures and navigating through them to achieve task goals, which is an open problem.

The design criteria is to keep the system's mobility range and strength as humanly as possible. Also the shape of the system is made to closely match that of a human. In order to achieve these objectives we also take into consideration compactness and modularity, while maintaining high power to weight ratio for the overall system. The details of the whole system will be presented in the following sections.

---

[3] Original mechanical design was done by A. Nagakubo. The overall system development was conducted by Y. Kuniyoshi. G. Cheng made significant contribution to the hardware debugging of the initial control system. Original control software was developed by G. Cheng which was later replaced by a different software package developed by Y. Ohmura.

4.1 Design Criteria

In order to pursue the research framework stated so far, we need a humanoid system whose embodiment is as close as possible to humans. However, with the current technology, it is unrealistic to mimic the complete organic structure of human body. Here, we are interested in overall mobility and interaction dynamics with the environment. Therefore we focus on realizing the similiar motor performance, mass distribution, and outer shape as the average human body. On the other hand we ignore internal differences in a way such as using motors instead of muscles and adopting metal structures instead of bones. We are well aware of importance of physical properties of such organic components. However, with the current state of the art, if we choose artificial muscles it will take many years before we achieve the overall mobility and the whole body motor performance comparable to humans. As such, our design is the result of strictly estimating the technical trade-offs and balances for building a whole-body humanoid given the current technological situation. Hence it is not ultimate and may change as new technology becomes available.

The above considerations lead us to a unique set of physical specifications [13]: 1) the joint torques must be able to support its own body weight, 2) the joints should be torque controllable and back drivable, 3) allow motion strategies to exploit inertia, such as, ballistic motion or dynamic motion, 4) the overall dimensions must be as close as possible to a small adult human it should have a smooth surface to allow arbitrary contact with surrounding objects; 6) the degrees of freedom and joint motion ranges should be close to humans to allow a broad range of motion.

In addition to these design considerations, the following criteria also needs to be taken into account, the overall system should be light in weight while keeping the requirement of high power. The overall mechanical system should be kept in a modular fashion, allowing ease of access and maintainability. Compactness will also need to be kept in order to keep overall proportion of the system's shape.

4.2 Experiments

Two preliminary experiments with our humanoid are presented.

Figure 9 shows a dynamic sitting up action[4]. Here the robot swings up its legs, swings them down and uses the inertia to bring up its torso. Rolling contact at the bottom is smooth with the specially designed outer cover. It plays a crucial role in this action.

Figure 10 shows simple dual-arm movement imitation[5]. The vision system uses color-based (e.g. skin-color) temporal differentiation and attention control based on simple knowledge of upright human posture, to detect the head and the arms. In other words, the robot will focus its attention on anything that it recognizes as an upright human Then it interprets the target posture with regard to multiple postural primitives. The result of this interpretation invokes

---

[4] This experiment was done by Ohmura
[5] This experiment was done by Otani and Ohmura

the humanoid's dual-arm movement, achieving the interpreted posture by means of balancing between the gravitational force and the directly controlled joint torque. As a result, the robot's arm movement appears quite natural.

In human interaction, smooth and autonomous transition between different interaction modes is also very important. Cheng and Kuniyoshi [14] built an integrated system which is capable of auditory target detection, visual target tracking, and real-time arm movement imitation, with a potential based attention mechanism that enables on the one hand smooth integration of these functionalities into the overall behavior, and on the other, transitions between them.

Fig. 9. Snapshots of a dynamic "Sitting up" action performed by our humanoid.

Fig. 10. Snapshots of a dual-arm movement imitation performed by our humanoid.

5   Self Exploratory Learning of Body Schema

This and the following sections present our recent attempts to develop a neural model for learning the global information structure of embodied interactions.

Our first experiment examines if simple spatio-temporal correlation can extract global strcture within the sensor data from embodied intraction. This could provide the first step towards acquisition of a body schema.

5.1   Acquisition of Topographic Somatosensory Map

Figure 11 illustrates one of our (Yorozu and Kuniyoshi) experiments with a simulated baby. The model of the body has 250 tactile sensors (actually, pressure sensors) distributed on its skin. All the limbs are driven by random signals in

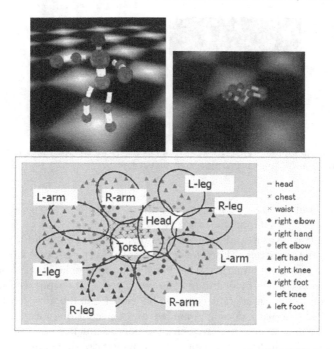

Fig. 11. The simulated baby body. Uupper left: Simulated baby body. Upper right: A snapshot during random movement. Bottom: Topographic somatosensory cluster map.

a simulated water, as if a fetus is moving inside the womb, and the tactile sensor data are collected. Then the spatio-temporal correlation is computed for all the pairs of sensing points. Then the sensor points are plotted on a 2D plane as a result of a self-organizing process in which the degree of pair-wise correlation is represented as a spring constant connecting the two points. And this is done for all pairwise relationships among the tactile points. The result shows a topographic structure corresponding to the physical topography of the body.

5.2   Bottom-Up Free Exploration of Sensory-Motor Patterns

In the previous example, the bodiliy movement was random, and the learning system passively observes the incoming tactile data. Pfeifer et al. [12] points out that active sensory-motor coordination is essential for learning to categorize the world. In the past models, the sensory-motor coordination strategy has been pre-defined. In this section we show an early attempt to let a robot acquire a repertoire of explorative behavior without any predefined behavioral primitives. We designed a very simple robot body in a dynamics simulation environment as shown in Fig. 12. It consists of a ball and a stick whose endpoint is connected by a motorized joint to the ball. Despite its simple structure, it can exhibit a rich variety of motion due to its physical dynamics which involves inertia, gravity

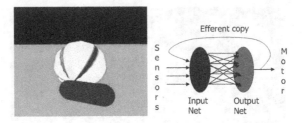

Fig. 12. A simple model for self exploratory sensory-motor learning. Left: Simulated body. Right: Sensory-motor learning architecture.

and rolling contact. Tactile sensors are distributed all over the surface of the robot, on both the ball and the stick.

A simple two-layer neural network as shown in Fig. 12 was adopted for learning sensory-motor patterns. Each layer is a hopfield type network ofneurons with non-monotonic output function [15]. The input layer takes tactile sensor data which then gets transformed through the two layers and output as a motor command for the joint. An efferent copy of the motor command is fed back to the input layer. The network does hebbian learning in real time, forming "trajectory attractors" representing repetitive sensory-motor sequence patterns. No reward or initial set of primitive behaviors are provided to the system. The system initially generates random motion due to the noise term. After that it continuously learns the sensory-motor patterns and as the learning proceeds and the synaptic connections are modified, the sensory-motor transformation immediately changes. Thus the system behaves in a boot-strap (self-referential) mannar in the sense that the learning modifies the explorative behavior, which then changes the sensory-motor patterns to be learned, affecting the learning process.

After several thousand steps of learning, the system found a few interesting behavior patterns. One example was lifting the bar and holding it still until the ball rolls due to the weight of the bar. Another example was repetitively swinging the bar and hit the ground so that the whole body "jumps". These are two of many possible ways to exploit the body-environment interaction dynamics to experience consistently repeated sensory-motor patterns. We interpret that the system discovered "bodily affordances" by acting on the body and self-guided along the motion patterns which emerge naturally from the physical properties of the body and the environment. This phenomenon may be similar to what Piaget named "second order circular reactions" [16], i.e. adaptively acquired behavior patterns in which the self action generates sensory stimuli that triggers the same action.

6   From Visual-Motor Learning to Neonatal Imitation

Imitation abilities of neonates discovered by Meltzoff and Moore [17] implies the existence of an innate mechanism in newborns for matching visual patterns of other persons' body movements to self movements. However, recent findings about rich exploratory movements of fetuses and extensive neural development during this period suggest a possibility of self exploratory learning during the fetus period and its effect on newborn imitation abilities.

Fig. 13. Visuo-Motor Learning Experiment. Left: Outlook of the robot. Right: View from the robot's eye.

As a synthetic approach to this hypothesis, we developed a visuo-motor neural learning system which consists of orientation selective visual movement representation, distributed arm movement representation, and a high-dimensional temporal sequence learning mechanism [18].

Initially, our robot generates random arm movements in front of its eye. At this point the robot has no knowledge about the relationship between its arm motor commands and the resulting visual patterns. A learning neural network takes the visual motion (i.e. optical flow patterns) and proprioceptive data (i.e. joint angles) mixed together as a high-dimensional sensory-motor vector. As the arm moves, the network learns the temporal sequence of the sensory-motor vectors as a "trajectory attractor" in its state space.

After the learning, a human comes in front of the robot showing an arm movement. If it is similar to the one experienced by the robot in the past, then the following will happen. The visual motion pattern of the human arm takes the network state into the pre-learned trajectory attractor. Since the trajectory goes through the sensory-motor state space, it automatically generates an arm movement. This results in an apparent imitative response.

7   Implications of Three-Term Interaction Dynamics Idea

As a summary and discussions, let us get back to the idea of the "three-term interaction structure", and discuss its relationship with the above presented examples, as well as implications to other standard interface problems, including dynamic motion control, teleoperation, behavior imitation (teaching by showing), cooperation, and communication.

Dynamic Motion Control: In whole body motion the robot-environment interaction is strongly governed by the physical dynamics as determined by gravity and inertia. Whenever there is this dominance of physical dynamics it is best to interfere as little as possible with the intrinsic whole-body dynamics. Stated differently, in such cases it is best to control the robot by sparse intervention. Rigid trajectory control is undesirable because this would often lead to interference with the system's natural dynamics.

Teleoperation: Modern teleoperation (i.e. remotely controlled by humans) systems have certain degrees of autonomous behavior capabilities. A human operator sometimes experiences difficulty, e.g. the system's behavior conflicts with the operator's intention. This is a typical example of a failure of interfacing two autonomous interactions. Moreover, as revealed by previous works on "shared control" and "shared autonomy" [19,20,21], it is desirable that the human intervention can be made at dynamically changing level of autonomous robot control at various times and situations through the task. The idea of dynamically choosing non-destructive intervention points in the ongoing intra-dynamics (i.e. autonomous behavior) of the system will be very useful.

Teaching by Showing/Behavior Imitation: As shown in the previous case study, essential information about the target task can be extracted at the boundaries of each actions. Remember also that interventions to a dynamic whole-body task can effectively be made at the boundaries of constituent actions. These two statements imply that in observational task learning or behavioral imitation, the essential information is extracted from, and exploited at, the action boundaries. In the development of communication between a mother and her child, communicative symbols initially emerge as private ones that are understood only by the mother and the child. Such symbols may have the property as action boundaries because the symbols emerge through mutual imitative interactions between the mother and the child, always accompanied by mutual efforts to obtain the common *interpretation* of each other's behavior [22].

Cooperation: It is straightforward to see the three-term structure in cooperative tasks. Two agents should effectively combine their individual task performances. A failure to properly coupling the two autonomous dynamics (i.e. each behavior) will result in conflicts and interferences. Then the task performance will be worse than doing it alone. Intervening other's task at action boundaries is effective. The "cooperation by observation"[23] scheme realizes the idea by invoking a helping behavior right at the time of action boundaries.

Communication: Generating a symbolic expression which best describes self actions, and interpreting a received expression and reflecting it on self behavior is the most basic form of effective communication. This is quite similar to the situation in behavioral imitation. Hence, effective situated communication may rely on proper identification of action boundaries.

Throughout the above discussions, an outstanding question is how to couple two independent interaction dynamics and create a meaningful inter-dynamics without destroying them.

Our hypotheses for the above problem are the following:

1. Each intra-dynamics has a sparse global structure which is quasi-stable and reproducible in its own context[6]. Such structure emerges as "boundaries", or "interference modes", of underlying dynamics. And these boundaries are "branching points" between one dynamics and the other.
2. External intervention/coupling can take place at these boundaries without destroying the underlying dynamics. This way an external process can take the system towards a desirable state from a global viewpoint.
3. A simplest mechanism for detecting/memorizing the global structure would be a spatio-temporal correlation network with adaptivity. The distributed correlation elements tune into the sensory-motor data flow. By introducing a competition mechanism, or lateral inhibition, the network will create clusters, each corresponding to different underlying dynamics. The boundaries between the clusters may correspond to the global structure.

8   Conclusions

So far we presented our hypotheses and basic learning mechanisms through several case studies. The core idea is that within each agent's sensory-motor interaction with the world, there is a natural global dynamical structure. This structure emerges from natural physical constraints or interactions between multiple sensory-motor flows. And the interaction with the agent should take place through an interface provided by this global structure, in order to effectively influence but not destroy. We also presented neural mechanisms which has basic capability to detect, learn, and use (to generate behaviors) such a global dynamical structure.

The above framework will eventually connect physical embodiment, sense of self, understanding other's actions, cooperation and communication. In short, it suggests a pathway from humanoid embodiment to theory of mind.

References

1. Kuniyoshi, Y., Nagakubo, A.: Humanoid as a research vehicle into flexible complex interaction. In: Proc. IEEE Int. Conf. Intelligent Robots and Systems. (1997) 811–819
2. Kuniyoshi, Y., Inaba, M., Inoue, H.: Learning by watching: Extracting reusable task knowledge from visual observation of human performance. IEEE Trans. Robotics and Automation 10 (1994)
3. Kuniyoshi, Y., Inoue, H.: Qualitative recognition of ongoing human action sequences. In: Proc. International Joint Conf. on Artificial Intelligence, Morgan Kaufmann (1993) 1600–1609
4. Newtson, D., et al.: The objective basis of behavior units. J. of Personality and Social Psychology 35 (1977) 847–862
5. Yamamoto, T., Kuniyoshi, Y.: Stability and controllability in a rising motion: a global dynamics approach. In: Proceedings of the International Conference on Intelligent Robots and Systems. (2002) 2467–2472

---

[6] The structure is not static, it may disappear and then re-appear.

6. Taga, G., Yamaguchi, Y., Shimizu, H.: Self-organized control of bipedal locomotion in unpredictable environment. Biol. Cybern **65** (1991) 147–159
7. Rizzi, A.A., Koditschek, D.E.: Further progress in robot juggling: The spatial two-juggle. In: Proc. IEEE Int. Conf. Robotics and Automation. (1993) 919–924
8. Miyakoshi, S., Taga, G., Kuniyoshi, Y., Nagakubo, A.: Three dimensional bipedal stepping motion using neural oscillators – towards humanoid motion in the real world. In: Proc. IEEE Int. Conf. Intelligent Robots and Systems. (1998)
9. Kuniyoshi, Y., Nagakubo, A.: Humanoid Interaction Approach: Exploring Meaningful Order in Complex Interactions. In: Proceedings of the International Conference on Complex Systems. (1997)
10. Kuniyoshi, Y., Nagakubo, A.: Humanoid As a Research Vehicle Into Flexible Complex Interaction. In: Proceedings of IEEE/RSJ International Conference on Intelligent Robots and Systems (IROS'97). (1997)
11. Raibert, M.H.: Legged Robots That Balance. The MIT Press (1986)
12. Pfeifer, R., Bongard, J., Iida, F.: New robotics: design principles for intelligent systems. Artificial Life (2004)
13. Nagakubo, A., Kuniyoshi, Y., Cheng, G.: Development of a High-Performance Upper-Body Humanoid System. In: Proceedings of IEEE/RSJ International Conference on Intelligent Robots and Systems (IROS'00). (2000)
14. Cheng, G., Kuniyoshi, Y.: Complex continuous meaningful humanoid interaction: A multi sensory-cue based approach. In: Proceedings of IEEE International Conference on Robotics and Automation, San Francisco, U.S.A. (2000) (to appear).
15. Morita, M.: Memory and learning of sequential patterns by nonmonotone neural networks. Neural Networks **9** (1996) 1477–1489
16. Piaget, J.: Play, Dreams and Imitation in Childhood. New York: W. W. Norton (1962)
17. Meltzoff, A.N., Moore, M.K.: Imitation of facial and manual gestures by human neonates. Science **198** (1977) 75–78
18. Kuniyoshi, Y., Yorozu, Y., Inaba, M., Inoue, H.: From visuo-motor self learning to early imitation – a neural architecture for humanoid learning. In: Proc. IEEE Int. Conf. on Robotics and Automation. (2003) 3132–3139
19. Sato, T., Hirai, S.: Language-aided robotic teleoperation system (larts) for advanced teleoperation. IEEE J. Robotics and Automation **3** (1987) 476–481
20. Lee, S.: Intelligent sensing and control for advanced teleoperation. IEEE Control Systems Magazine **13** (1993) 19–28
21. Brunner, B., Arbter, K., Hirzinger, G.: Task directed programming of sensor based robots. In: Proc. IEEE/RSJ/GI Int. Conf. Intelligent Robots and Systems (IROS). (1994) 1080–1087
22. Asao, T.: From Gestures to Language. Shin'yosha (1992) (In Japanese).
23. Kuniyoshi, Y.: Behavior matching by observation for multi-robot cooperation. In Giralt, G., Hirzinger, G., eds.: Robotics Research – The Seventh International Symposium. Springer (1996) 343–352

Robot Finger Design
for Developmental Tactile Interaction
Anthropomorphic Robotic Soft Fingertip
with Randomly Distributed Receptors

Koh Hosoda[1,2]

[1] Graduate School of Engineering, Osaka University
[2] Handai Frontier Research Center
hosoda@ams.eng.osaka-u.ac.jp

Abstract. The developmental approach enables us to build adaptive robots, and furthermore, to understand the essence of intelligence from the constructivist viewpoint. In this paper, a new design principle for tactile sensors is proposed to investigate and to utilize developmental processes of robots. Based on the design principle, an anthropomorphic fingertip is developed. The fingertip is made of soft material with randomly distributed receptors inside. The robot learns to acquire meaningful information such as the slip and the object texture from the outputs of receptors through interaction with the environment like a human does. Several experimental results are shown to demonstrate its sensing ability and applicability for the developmental approach.

1 Introduction

Robots are going out of the laboratories, and therefore, have to deal with an uncertain real-world environment in which environmental change is more than the designer can predict. Looking at biological systems, they might utilize the "developmental process" to deal with such a real environment. Applying the developmental process for designing robots provides us the comprehensive understanding of intelligence from the constructivist viewpoint [1], which makes it possible to construct adaptive robots.

It is very difficult to implement physical development on a robot as far as we do not use biological material. Instead, by implementing as many actuators and sensors as possible, we can study how the robot develops the connection between them through interaction with the environment. Although they have designed and developed robots that have many degrees of freedom such as humanoids, the variety and the number of sensors are still not sufficient.

A camera is the only sensor that has been utilized for the developmental research so far: it has basically so many pixels that can be used for image processing. By changing the image processing in a coarse–to–fine manner, for example, we can simulate the developmental changes and investigate its effect in

F. Iida et al. (Eds.): Embodied Artificial Intelligence, LNAI 3139, pp. 219–230, 2004.
© Springer-Verlag Berlin Heidelberg 2004

the process of learning [2,3]. For the other kinds of sensing modalities, however, there is no developmental study to the best of the author's knowledge.

In this paper, a new design principle for tactile sensors is proposed to enable us to investigate and to utilize developmental processes of robots: embedding as many receptors as possible in soft material randomly. Based on the design principle, an anthropomorphic fingertip is developed. The word *anthropomorphic* has two meanings: one is that the fingertip is made of soft material with randomly distributed receptors inside and like a human does, the robot learns to acquire meaningful information such as the slip and the object texture from the outputs of receptors through interaction with the environment. The other is that the structure of the fingertip is similar to that of a human's; it consists of a bone, a body, a skin layer, and randomly distributed receptors.

The remainder of this paper is organized as follows. First, an overview of the existing design of tactile sensors is explained. Then, we introduce the design of an anthropomorphic fingertip based on a new design principle that relies on learning ability of the robot. Following that, several experimental results are shown to demonstrate its sensing ability and applicability for the developmental approach.

2   Toward Adaptive Manipulation: Overview

A human being can manipulate various objects by fingers dextrously and adaptively. Although there has been an enormous number of studies on robot hands trying to reproduce such adaptive and dextrous manipulation [4], so far the performance is not satisfactory. One of the reasons is that these existing hands are basically designed and controlled so that the designers can understand the manipulation. Although it is easy for them to implement their knowledge to the robot, it gives certain constraints on design and control of the robot hand, and as a result, it prevents manipulation from being adaptive. If the robot would have an ability to develop manipulation by itself, it would be freed by such constraints and the resultant manipulation would be adaptive.

In order for a robot to learn and/or develop its own representation of manipulation in its own sensor spaces, it should have several different sensing modalities. Among such modalities, tactile sensing plays a great role to gather information about the object and contact conditions. Many kinds of tactile sensors are proposed (we can find a comprehensive survey in [5] until 1999). Sensors with distributed receptors are especially effective to observe detailed contact conditions for adaptive manipulation. Many attempts have been made to construct such sensors with pressure-conductive rubber [6], an optical position sensitive detector [7], capacitor arrays [8,9,10], a LC network [11], ultrasonic sensors [12], force sensing resistance [13], conductive fabric [14], and conductive gel [15].

Almost all robotic fingertips that have been developed so far have their sensing receptors only on their surfaces. One of the reasons is that the fingers are basically made of rigid materials such as metals, and the receptors cannot be embedded in a deeper part of the finger. Rigid fingertips make the control easy since the position of the manipulated object is easily calculated by configurations of the fingers. However, the fact that the receptors are only embedded on the surface limits the sensing ability of existing robotic fingers. We humans have many receptors (corpuscles) of several kinds broadly distributed in the finger. The receptors embedded in a deep part of the finger are able to acquire the information filtered by its material property whereas the ones in a shallow part are sensitive to high frequency transient phenomena. Therefore, it would be possible to obtain more useful information about the object by combining the sensory information at many different locations in the finger rather than just using receptors on the surface.

Several studies mentioned that receptors embedded in soft material could provide useful information about dynamic characteristics such as the slip and the friction coefficient [16,17,18,19,20]. Although it is promising to get more information about them by increasing the number of receptors at various depths, there have been very few studies on it. It is difficult for the designer to derive the translation from the raw signals to meaningful information if the positions of the receptors are not controlled and the property of material between them are not known. Only Shinoda and his colleagues discussed on randomly distributed receptors in a soft material [21,22] to the best of the author's knowledge. However, they only showed the characteristics of one receptor, and did not study the influence of the depth nor on the interplay of receptors.

3   Design of an Anthropomorphic Fingertip

3.1   Sensor Design That Relies on the Learning Ability

In order to translate raw signals into the meaningful information, the underlying structure provided by bodily, environmental, and task constraints is essential. For example, the electrical resistance of a strain gauge of a force sensor itself does not make any sense. If the robot knows the resistance-to-strain translation that is determined by the gauge material and structure and knows the strain-to-force translation determined by the sensor physical structure, it can translate the measured resistance into force.

A human designer usually calibrates the translation from the raw signals to meaningful information. He or she understands the constraints and implements knowledge about them as a *sensing model* for a robot. Then, the robot can behave properly even with a few receptors by compensating for a missing information with the model. Receptors of existing sensors are, therefore, placed regularly on relatively hard surface so that the designer can easily analyze the structure. As long as the task of the robot is simple, such a sensing model is functional. Recently, however, the task has become more complicated such as handling of the objects with various properties (e.g. material, size, mass, etc.), and the physical interaction between the finger and an object has also become complicated (e.g. grasping with slippage, finger gait, etc). Consequently, the realized behavior based on the human-designed sensing model is no longer robust against modelling errors and disturbances.

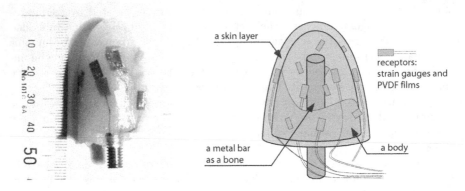

Fig. 1. A developed fingertip(left) and its cross sectional sketch(right): The fingertip consists of a metal bar, a body, and a skin layer inspired by the structure of the human finger. The body and the skin layer are made of different kinds of silicon rubber. Strain gauges and PVDF films are embedded randomly in the body and the skin layer as receptors.

Owing to the recent development, the learning function of a robot is now ready to be used for many applications. If the robot can acquire the sensing model through experience, the receptors can be distributed randomly in or on soft material. The softness of a tactile sensor provides not only stability of grasping and protection against strong impact forces, but also more sensing abilities than hard sensors. It would even be possible for the robot to have a sensing ability that is excluded by the human designed sensors whose receptors are placed regularly. In this sense, the learning ability will change the design principle for tactile sensors. This paper describes a new design principle for tactile sensors that relies on the learning ability: **embedding as many receptors as possible randomly in soft material**. The word "many" means not only the number but also the variations of receptors.

It is obvious that the variety of receptors provides more sensing abilities. Even with receptors of the same kind, the robot can get different information from them in different depths since material existing between receptors play a role of a low-pass filter. In this sense, embedding many receptors provides not only redundancy, but also variety of sensing abilities. Another important point is randomness: non-uniform and anisotropic sensor structure potentially provides information that is excluded by the human design bias, that is, the uniformity of the sensor structure.

3.2   Structure of the Finger

By following the design principle explained above, an anthropomorphic fingertip is developed (Figure 1). The fingertip consists of a metal bar that plays a role of a bone, a body, and a skin layer inspired by the structure of the human finger. The silicon used for the skin layer is slightly harder than that for the

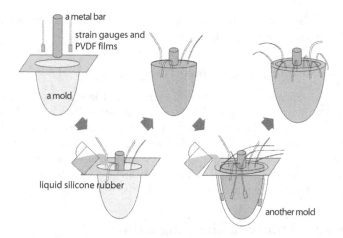

Fig. 2. A procedure to build a soft fingertip:a metal bar and several receptors, strain gauges and PVDF films are inserted into a mold, and silicon rubber is cast into it. This mold is then inserted into another mold that is slightly bigger. The additional receptors are implemented in this layer, then silicon rubber is again cast.

body. Strain gauges and PVDF (polyvinylidene fluoride) films are embedded randomly both in the body and in the skin layer as receptors. A PVDF film is sensitive to the strain velocity by using the piezo effect, whereas a strain gauge measures the static strain. In the human skin, there are also several corpuscles that are sensitive to the change of the strain (Meissner's corpuscle and Vater-Pacini corpuscle), and to the static strain (Merkel's disk and Ruffini ending). Since these receptors are embedded randomly, the robot has to learn to acquire meaningful information such as the slip and the object texture from the outputs of receptors through the interaction with the environment like a human does.

Figure 1 (left) shows a complete soft fingertip. Its diameter and length are 2[cm] and 9[cm], respectively. This finger has 6 strain gauges and 6 PVDF films both in the body and the skin layer, which results in totally 24 receptors. As mentioned above, the positions and the orientations of these receptors are not determined, i.e. the designer or the robot cannot know the geometries of the receptors beforehand.

We expect that the receptors of the same kind embedded in different positions would be able to measure different physical properties. A strain gauge embedded near the skin surface is expected to sense the local static strain between the skin and the object surface whereas a gauge embedded near the bone is expected to sense the total force exerted to the finger and is expected to be insensitive to the local texture of the object. A PVDF film senses the strain velocity, which means that it is more sensitive to the transient and the rapid strain changes (or stick-slip motions) than the strain gauges whereas it cannot sense the static strain. The silicon existing between two PVDF films is expected to function as a low-pass filter, therefore the difference between the signals is expected to represent the local stick-slip interaction.

3.3   Procedure to Make a Fingertip

Figure 2 shows the procedure to make the fingertip. First, a metal bar and several receptors, strain gauges and PVDF films are inserted into a mold, and silicon rubber is cast into it. The mold is put into the vacuum to remove bubbles, and is baked in the oven to be solid. It is then inserted into another mold that is slightly bigger. The additional receptors are implemented in this layer. Another kind of liquid silicon rubber that is harder than the previous one is cast, and the mold is put into the vacuum and is baked in the oven again.

4   Sensing Ability of the Fingertip

To investigate sensing ability of the anthropomorphic fingertip, it is mounted on a robotic finger (Figure 3), and rubbed on four different materials: wood, paper, cork, and vinyl. The finger is not force-controlled but position-controlled along a pre-determined trajectory.

Fig. 3. The robot finger used for experiments: An anthropomorphic fingertip is attached at the tip of a robot finger.

The data from the PVDF films are obtained from the rubbing experiments. Figure 4 shows variance of signals originating from a PVDF film in the skin layer and that from another PVDF film in the body layer. In the figure, stars, crosses, oblique crosses, and squares represent the data obtained during rubbing vinyl, cork, paper, and wood, respectively. Since variance ellipsoids depicted in this figure do not overlap each other, we can conclude that these four materials are distinguishable by combining the outputs of these two receptors. It is important to note that, as illustrated in this figure, the paper, the cork, and the vinyl cannot be identified only from the film in the skin layer. The same holds for the wood, the vinyl, and the cork measured by the film in the body.

Since the finger is not force-controlled and the height of the surface is also not precisely controlled, the contact force is not constant through the rubbing process. This could be the main reason of the relatively large variance in the data points obtained from the same material. We expect that, if the finger is precisely controlled, the variance should be smaller so that one receptor is sufficient to identify the material. However, even without such a precise control, it is shown that the distributed receptors are able to distinguish the different materials. From the viewpoint of the developmental process of the robot, this characteristics would be particularly important since the robot would not be able to perform the precise position and force control from the beginning.

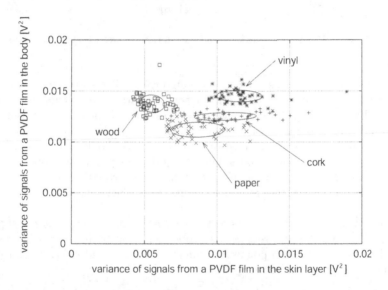

Fig. 4. The results from rubbing experiments: variance of signals of a PVDF film in the skin layer and that in the body layer are plotted. In the figure, stars, crosses, oblique crosses, and squares represent obtained data during rubbing vinyl, cork, paper, and wood, respectively. A ellipsoid represents the variance ellipsoid for each material. The paper, the cork, and the vinyl cannot be identified only from the film in the skin layer. The same holds for the wood, the vinyl, and the cork measured by the film in the body.

5   Toward Development: Representation of the Slip

Situatedness is one of the most essential properties for a robot to be truly autonomous [1]: an autonomous robot should have the perspective based on its own sensory system. It is very difficult, however, to transfer the knowledge of a

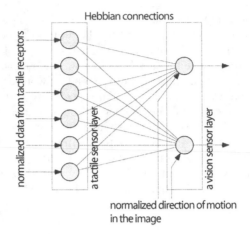

Fig. 5. A Hebbian network to find correlation between tactile and vision

designer into the control system of a situated robot. Therefore, the robot should be able to "develop" itself by using the sensory data from its own receptors.

At the beginning of developmental process, the robot get the sensory data flows but cannot see any correlation among them. If they come from the same physical interaction with the environment, the flows should have a certain relation between them because of the underlying dynamics, and the robot would be able to find it as a correlation after learning. The correlation should be robot's own representation of the interaction acquired through the developmental process.

From this perspective, in this section, we investigate a case study of slippage. The problem addressed here is how the robot can identify the correlation between the data from the anthropomorphic finger and the visual sensory information during the slippery interaction.

5.1   Network to Acquire Representation of the Slip

A simple Hebbian network is used to find correlation (Figure 5). There are two layers: a tactile receptor layer and a vision sensor layer. Two neurons in the vision layer are activated by the displacement of the image target in the image plane along x and y–axes, respectively. The 6 strain gauges in the skin layer are used, which are connected to the 6 neurons in the tactile receptor layer (PVDF films are not used in this experiment).

Since the relation between the vision sensor and the tactile receptors is not calibrated before learning, the weights between the neurons are initially 0. While the finger is touching an object and rubbing on it, the vision sensor can observe both the object and the fingertip. Therefore, when the tip slips, it is observed by the vision sensor as the difference of displacements of the object and the finger tip in the image plane. Simultaneously, the strain information can be obtained by the tactile receptors. If the direction of one tactile receptor happens to be along the slip direction in the image plane, the connection between the neurons is strengthened according to the Hebbian rule. Over time, the connection between a vision neuron and a tactile neuron in a corresponding direction has certain amount.

After some learning trials, the direction of a slip can be sensed by the tactile receptors as well as the vision sensor. This provides the system redundancy [1]. That is, even if the vision sensor cannot catch the slip information, for example, because of occlusion, the slip can be detected from the network and vice versa. An interesting aspect of this approach is the complementary nature of vision and tactile sensors concerning the sensitivity. Since the vision sensor is a non-contact sensor, the sensitivity (the minimum observable amount of a slip) changes according to the distance between the eye and the object whereas that of a tactile receptor does not change so much since the distance between the object and the receptor does not change. Therefore, at the beginning of learning when the vision sensor is mainly used, the sense of the slip is strongly affected by the position and orientation of the object. After learning, tactile receptors provide complemental information, and therefore, the sense will be insensitive to the object position and orientation. This process is supposed to be finding invariance in the observation.

In Figures 6 and 7, a broken line and a solid line represent normalized movement in the image plane along x axis (-1, 0, and 1 mean moving to $-x$ direction, stopping, and moving to $+x$ direction, respectively) and the sum of the outputs from the tactile sensor layer to the corresponding vision neuron, respectively. The continuous movement of the object is observed as pulses in the vision not as continuous signal because of the quantization of pixels.

Figure 6 shows the result from the first learning trial. The tactile output is 0 at the beginning since the the weights between the neurons are initially 0. There is almost no correlation between the outputs of the vision sensor and the tactile layer since the learning is not enough.

After 260 learning trials, the network obtains the correlation between them (Figure 7). The output of the tactile layer predicts the movement in vision, that is, at first the activation of the tactile sensors becomes larger, and then that of the vision is activated since the visual image is quantized.

6 Discussion

This paper has described a new design principle for the tactile sensor to investigate and to utilize the developmental processes of robots. The ability of the anthropomorphic fingertip that can discriminate vinyl, cork, paper, and wood is provided by its softness and placement of receptors. The slippery interaction is also mapped onto the multi-modal sensory space consisting of vision and tactile as a set of distributed weights through robot's interaction with the environment.

Since the receptors are randomly distributed in the soft fingertip, the designer cannot map the physical phenomenon with the receptor outputs explicitly.

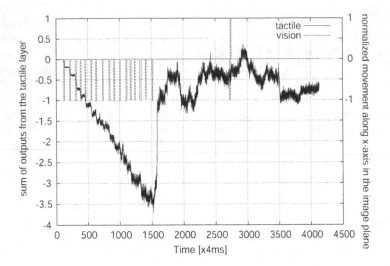

Fig. 6. The outputs of the vision neuron of x-direction and sum of corresponding tactile neurons of the first learning trial : the tactile output is 0 at the beginning since the weights between the neurons are initially 0. There is almost no correlation between the outputs of the vision sensor and the tactile layer since the learning is not sufficient.

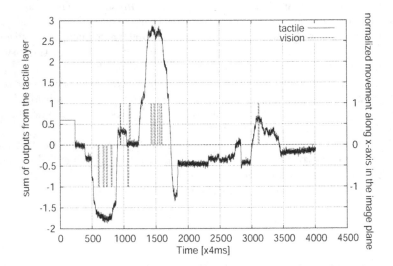

Fig. 7. The outputs of the vision neuron of x-direction and sum of corresponding tactile neurons of the 260th learning trial : the output of the tactile layer predicts the movement in vision, that is, at first the activation of the tactile sensor becomes larger, and then that of the vision is activated.

Therefore, the robot has to learn the mapping through its own experience, and to organize the outputs of receptors. The learning and organizing process is one of most important developmental aspects of an autonomous robot. In this sense, this design principle will shed a light on the developmental study of robots.

In the first experiment of the object discrimination, the category is given by the designer. However, the category should be obtained by the robot itself based on its behavioral result. It does not have to discriminate the objects as far as the probability of achieving a given task (e.g. manipulating or grasping an object) does not change. This is very important point since the developmental process of the robot must be triggered by the internal motivation of the agent.

We expect that the study of an anthropomorphic fingertip could also provide an additional insight to the developmental process of human manipulation. Although it is still under a developing stage, the representation of the slip discussed in this paper pointed out some interesting issues toward future. Particularly, some interesting issues include (a) what kind of mechanism is effective to acquire such distributed representation and (b) how we can utilize such a representation for the adaptive behaviors.

Acknowledgement. I would like to thank Mr. Yasunori Tada for his effort to produce the anthropomorphic fingertip and to perform an enormous number of experiments. I would also like to thank my colleague Minoru Asada for valuable discussions. Finally, I would also like to express my gratitude to the anonymous reviewers and Fumiya Iida for their valuable comments.

References

1. Pfeifer, R., Scheier, C.: Understanding Intelligence. The MIT Press (1999)
2. Nagai, Y., Asada, M., Hosoda, K.: Developmental learning model for joint attention. In: Proc. of the RSJ/IEEE Int. Conf. on Intelligent Robots and Systems. (2002) 932–937
3. Dominguez, M., Jacobs, R.A.: Developmental constaints aid the acquisition of binocular disparity sensitivities. Neural Computation **15** (2003) 161–182
4. Bicchi, A., Kumar, V.: Robotic grasping and contact: A review. In: Proc. of the IEEE Int. Conf. on Robotics and Automation. (2000) 348–353
5. Lee, M.H., Nicholls, H.R.: Tactile sensing for mechatronics – a state of the art survey. Mechatronics **9** (1999) 1–31
6. Shimojo, M., Ishikawa, M., Kanayama, K.: A flexible high resolution tactile imager with video signal output. In: Proc. of the IEEE Int. Conf. on Robotics and Automation. (1991) 384–391
7. Maekawa, H., Tanie, K., Komoriya, K., Kaneko, M., Horiguchi, C., Sugawara, T.: Development of a finger-shaped tactile sensor and its evaluation by active touch. In: Proc. of the IEEE Int. Conf. on Robotics and Automation. (1992) 1327–1334
8. Fearing, R.S.: Tactile sensing mechanisms. The International Journal of Robotics Research **9** (1990) 3–23
9. Johnston, D., Zhang, P., Hollerbach, J., Jacobsen, S.: A full tactile sensing suite for dextrous robot hands and use in contact force control. In: Proc. of the IEEE Int. Conf. on Robotics and Automation. (1996) 3222–3227

10. Hakozaki, M., Shinoda, H.: Digital tactile sensing elements communicating through conductive skin layers. In: Proc. of the IEEE Int. Conf. on Robotics and Automation. (2002) 3813–3817
11. Nilsson, M.: Tactile sensors and other distributed sensors with minimal wiring complexity. IEEE/ASME Trans. on Mechatronics **5** (2000) 253–257
12. Hutchings, B.L., Grahn, A.R., Petersen, R.J.: Multiple-layer corss-field ultrasonic tactile sensor. In: Proc. of the IEEE Int. Conf. on Robotics and Automation. (1994) 2522–2528
13. Lazzarini, R., Magni, R., Dario, P.: A tactile array sensor layered in an artificial skin. In: Proc. of the RSJ/IEEE Int. Conf. on Intelligent Robots and Systems. (1995) 114–119
14. Inaba, M., Hoshino, Y., Nagasaka, K., Ninomiya, T., Kagami, S., Inoue, H.: A fullbody tactile sensor suit using electrically conductive fabric and strings. In: Proc. of the RSJ/IEEE Int. Conf. on Intelligent Robots and Systems. (1996) 450–457
15. Tajima, R., Kagami, S., Inaba, M., Inoue, H.: Development of soft and distributed tactile sensors and the application to a humanoid robot. Advanced Robotics **16** (2002) 381–397
16. Shinoda, H., Uehara, M., Ando, S.: A tacile sensor using three-dimensional structure. In: Proc. of the IEEE Int. Conf. on Robotics and Automation. (1993) 435–441
17. Yamada, D., Maeno, T., Yamada, Y.: Artificial finger skin having ridges and distributed tactile sensors used for grasp force control. In: Proc. of the RSJ/IEEE Int. Conf. on Intelligent Robots and Systems. (2001) 686–691
18. Yamada, Y., Maeno, T., Fujimoto, I., Morizono, T., Umetani, Y.: Identification of incipient slip phenomena based on the circuit output signals of pvdf film strips embedded in artificial finger ridges. In: Proc. of SICE 2002. (2002) 3272–3277
19. Nakamura, K., Shinoda, H.: A tactile sensor instantaneously evaluating friction coefficients. In: Proc. of the 11th Int. Conf. on Solid-State Sensors and Actuators. (2001) 1430–1433
20. Yamada, K., Goto, K., Nakajima, Y., Koshida, N., Shinoda, H.: Wire-free tactile sensing element based on optical connection. In: Proc. of the 19th Sensor Symposium. (2002) 433–436
21. Hakozaki, M., Nakamura, K., Shinoda, H.: Telemetric artificial skin for soft robot. In: Proceedings of TRANSDUCERS '99. (1999) 844–847
22. Shinoda, H., Oasa, H.: Passive wireless sensing element for sensitive skin. In: Proc. of the RSJ/IEEE Int. Conf. on Intelligent Robots and Systems. (2000) 1516–1521

The Autotelic Principle

Luc Steels

VUB AI Lab - Brussels
Sony Computer Science Laboratory - Paris
steels@arti.vub.ac.be

Abstract. The dominant motivational paradigm in embodied AI so far is based on the classical behaviorist approach of reward and punishment. The paper introduces a new principle based on 'flow theory'. This new, 'autotelic', principle proposes that agents can become self-motivated if their target is to balance challenges and skills. The paper presents an operational version of this principle and argues that it enables a developing robot to self-regulate its development.

1 Introduction

The design and implementation of self-developing robots has become a focal point of recent efforts in robotics and AI research [21]. It builds further on the work of developmental psychologists, who have a long history of studying 'epigenetic' or 'ontogenetic' development [6], [8]. A lot of research in developmental robotics focuses on finding powerful learning mechanisms that can run continuously in open-ended environments [11]. This paper turns to a more global issue: How can the developmental process as a whole be orchestrated.

The problem of regulating development is very challenging for three reasons. (1) Certain things often cannot be learned before other things are mastered, so the developmental process must be scaffolded somehow, to enable bootstrapping from simple to complex. Thus, it is not possible to learn fine-grained control of grasping if there is no ability to identify and track the objects that need to be grasped. (2) In a complex agent, each component depends on others, either to provide input or to produce appropriate feedback. But if there are many subcomponents, each developing at their own pace, regulating global development becomes a non-trivial issue. (3) An agent may reach a level of performance which is adequate with respect to a given environment but which is nevertheless a local maximum in the sense that a richer interaction can be achieved by further exploration and development. So a big challenge is to avoid that the agent gets stuck in development, even if this means a decreased performance in the short run.

Some researchers have proposed that nothing special needs to be done to orchestrate the developmental process, because the development of one skill naturally creates new opportunities for the development of other skills in a changing ontogenetic landscape [19]. For example, once the arm can be controlled, it is possible to start exploring the uses of the hand. Although it is obviously the

F. Iida et al. (Eds.): Embodied Artificial Intelligence, LNAI 3139, pp. 231–242, 2004.

case that one opportunity may lead to the next, it is now generally recognised that more needs to be done, particularly to avoid that the agent remains in local maxima which do not exploit the full capacity of what is possible. Three approaches have already been discussed in the literature.

- Scaling of input complexity

  A first group of researchers has proposed that development can be organised by regulating the complexity of the external environment. This way the agent can build up capacity in a simple environment before tackling additional challenges. Usually a small subtask is isolated and the agent is trained for that specific subtask with prepared scaffolded data [1], [5]. In more sophisticated applications, several stages and subcompetences are identified and input data is carefully prepared to pull the agent through each stage. [20].
- Scaffolding of reward function

  Other researchers have proposed to scaffold the reward function, i.e. to give external feedback to the agent which makes sure that simpler and foundational skills are learned before more complex skills are tackled and that the stakes are increased as soon as steady performance has been reached [21]. In the case of language development for example, we could envision first a high reward for producing single word sentences, then a higher reward for multiple-word sentences, then a higher reward for constructing grammatical phrases with increased complexity.
- Staging of resources

  Yet another approach is to stage the resources available to the agent in a kind of 'maturational schedule'. For example, Elman [5] has shown that a recurrent neural network can be trained first with a small 'look back' window, then this window is progressively increased to take more of past input into account. Such an approach has been shown to give better performance compared to one where the full complexity of internal resources is available from the beginning.

All these approaches are valuable and have shown to yield interesting results. Moreover they are not completely devoid of naturalness because in the case of infants, caregivers often scaffold the environment or 'up the ante' to push the infant to higher competence. However these approaches assume a very strong intervention by 'trainers' and/or a careful a priori design of developmental scenarios. The real world always presents itself with the same complexity to the learner and it is therefore artificial to constrain it. It would be much more desirable if the agent could develop independently and autonomously in an open-ended environment by actively self-regulating his own development.

This is precisely the goal of the research reported in this paper: a general principle is proposed by which a complex agent could self-regulate its-build up of skills and knowledge without the need for the intervention of a designer to scaffold the environment, stage the reward functions, or bring resources progressively on-line in a maturational schedule. The main idea is to introduce a new motivational principle gleaned from recent work in humanistic psychology. This principle is introduced in the next section of the paper. Further sections present an operationalisation of this principle. We have already conducted various experiments to exercise the principle in the context of grounded language development, [18]. The results are encouraging and will be reported in more detail in forthcoming papers.

2   Motivation and Flow

Reinforcement Learning

Most models in psychology and neuroscience are still rooted in the behaviorist framework of reward and punishment, originallly coming from the work of Skinner and his associates [16]. Also a lot of autonomous robotics work, particularly under the banner of reinforcement learning, is implicitly based on the same approach. This theory makes four major assumptions.

First, it assumes that the overall goal of the organism is to keep its critical parameters for survival within viable bounds [12]. The challenge of a developing organism is to acquire the necessary behaviors so that such a viable state is maintained, or to adapt the behaviors if the environment changes.

Second, it argues that certain behaviors get rewarded, for example with food or other means that give direct pleasure, and others are punished, for example through the inducement of corporal pain. Rewards reinforce specific behaviors because they inform the organism that they are beneficial, in other words that a viable state can be reached and maintained. Punishment signals that the behaviors that were enacted need to be abandoned or new knowledge and skills need to be acquired. In natural circumstances, reward and punishment is generated by the environment.

Third, it proposes that organisms start with a repertoire of reflex behaviors and an innate value system. New behaviors are shaped by reward and punishment. When a trainer or educator hands out the reward or punishment, she can push development in specific directions and the trainer's value system may become progressively internalised by the trainee.

Fourth, classical behaviorism proposes that this reinforcement framework is an adequate theory of motivation, in the sense that the main purpose of the organism is to seek reward and avoid punishment, and so all the rest (acquisition of new behaviors and internalisation of a value system) follows.

Flow Theory

More recently, a complementary motivational theory has been proposed in psychology, which points to a richer notion of motivation. This theory was originally developed by the humanistic psychologist Csikszenmihalyi, based on studying the activities of painters, rock climbers, surgeons, and other people who showed to be deeply involved in some very complex activity, often for the sake of doing it, i.e. without direct reward in the form of financial or status compensation [2].

He called these activities autotelic. "Autotelic" signifies that the motivational driving force ("telos") comes from the individual herself ("auto") instead of from an external source, administered by rewards and punishments.

Autotelic activities induce a strong form of enjoyment which has been characterised as "flow". The word "flow" is a common sense word and so there is a risk to interpret it too broadly. Csikszenmihalyi intends a restricted usage, being a state which often occurs as a side effect of autotelic activities:

> People concentrate their attention on a limited stimulus field, forget personal problems, lose their sense of time and of themselves, feel competent and in control, and have a sense of harmony and union with their surroundings. (...) a person enjoys what he or she is doing and ceases to worry about whether the activity will be productive and whether it will be rewarded. o.c. p. 182.

Because the activity is enjoyable, the person who experiences this enjoyment seeks it again, and therefore it becomes self-motivated. Moreover due to the high concentration and the strong self-motivation, learning takes place very fast. The learner is eager to find the necessary sources and tools herself and spends time on the acquisition of skills, even if they are not exciting in themselves, as long as they contribute to the autotelic activity.

Given this description, it is quite obvious that many people will have experienced some form of flow in their life, and that children in particular enter into flow experiences quite often, particularly during play. Flow is sometimes associated with the ultimate high experience of the rock climber that has finally managed to climb Mount Everest, but that is an exceptional situation. Flow - as defined here - is much more common and can just as well happen in every-day experiences like playing with children or engaging in a long term love relationship.

It is also important to distinguish flow from directly pleasurable activities like going down a roller coaster. A key difference is that the activity must in itself be challenging - otherwise there is no feeling of satisfaction after difficulties have been surmounted. Moreover there must be a steady progression in the nature and particularly the level of the challenge. This is the reason why child rearing can be so enjoyable and fascinating. A child keeps developing all the time - which is what makes the interaction fun - and that creates continuously new challenges for the parent to figure out what she is thinking, what she might want to do or not do, and so on. The rock climber can also scale up the level of difficulty with which rocks are being climbed or the kinds of rocks that are tackled. Similarly, the musician can first play easy pieces and then steadily move up. she can first play with other amateur musicians and then play with better and better musicians. The performance can be first for a few friends, but then for a larger and larger unknown audience.

An obvious key question is: What makes activities autotelic? Here comes Csikszenmihalyi's most important contribution, I believe. He argues that it lies in a balance between high challenge, generated through the activity and perceived as meaningful to the individual, and the skill required to cope with this challenge:

Common to all these forms of autotelic involvement is a matching of personal skills against a range of physical or symbolic opportunities for action that represent meaningful challenges to the individual. o.c. p. 181

When the challenge is too high for the available skill, in other words the opportunity for action is so bewildering that no clear course can be seen, and when there is at the same time no hope to develop appropriate skills by learning, anxiety sets in and the person gets paralysed and eventually may develop symptoms of withdrawal and depression. When the challenge is too low for the available skill, boredom sets in and the long term reaction may be equally negative. The optimal regime is somewhere between the two, when there is a match of challenge and skill. It follows that it is important for the individual to be able to decrease challenge when it is too high so as to get an opportunity to increase skills, but it is equally important that the individual can increase challenge when the skill has become higher than required to cope with the challenge, or that the environment generates new opportunities for the individual to grow.

Let us now see how these intuitive ideas can be operationalised into a design principle that can be implemented on physical self-developing robots.

3   Operationalising the Autotelic Principle

A cognitive agent is a physically embodied organism embedded in an environment in which there is a steady stream of sensori-motor inputs and a steady stream of decisions for action which translate into motor commands or internal state changes, such as switch goals or move to another location in the world. The key challenge for the agent is to survive in this environment and hence choose the right action based on an interpretation of the current situation.

We assume that the agent is organised in terms of a number of sub-agencies further called components. Each component establishes an input-output mapping based on knowledge and/or skill. For example, a segmentation component takes a camera bitmap and produces a list of segments using some segmentation algorithm. Each component requires a set of resources (memory, computer time) and makes use of knowledge or skill that is typically adapted or learned. For example, the segmentation algorithm may progressively build up a database of the shapes or movement trajectories of the objects in the environment so that segmentation can be done more quickly or more reliably.

A realistic system needs of course many components. For example, in the case of an embodied agent interacting through language with another agent [17], we need components for grounding world models through vision, speech and gesture recognition, speech and gesture production, selection of a topic, conceptualisation of what to say, lexicon lookup, grammatical parsing and production, interpretation of semantic structures, dialog management, etc.

The autotelic principle suggests that the balancing of skill and challenge should be the fundamental motivational driving force of the agent. This implies (1) that each component must be parameterised so that challenge levels can be self-adjusted based on self-monitoring of performance, (2) that each component must have the ability to increase skill to cope with new challenge, and (3) that there is a global dynamics regulating the adjustement (both increase and decrease) of challenge levels. The reward function of the total agent is the degree of balance between challenge and skill for each of its components. The increase in complexity of the agent's behavior (and hence the kinds of tasks and environmental complexity it can handle) will be an emergent side effect of the system's effort to keep this balance between challenge and skill.

The following subsections provide more detail on each of these aspects.

3.1   Parameterisation of Components

Each component of the system must be parameterised to reflect different challenge levels. The nature of the parameterisation obviously depends on the task that the component must achieve and on the nature of the algorithms that are used. For example, suppose that a robot has a subsystem that has the task of moving an object using vision and hand/arm motor control. One parameterisation of such a component could concern the precision with which the object is to be moved: Is pushing it aside in a broad gesture enough, or should the object be picked up and put down carefully in a precise location. Another parameter is the nature of the object: Is it of a simple uniform shape or does it contain handles or other structures that need to be recognised and used to manipulate the object. Another parameter concerns the weight of the object. A heavy weight might require the agent to adopt a specific posture so as not to get out of balance.

Formally, we associate with each component c_i a parameter vector $\langle p_{i,1}, ... p_{i,n} \rangle$. The set of all parameters for all m components in the agent forms a multi-dimensional parameter space P. At any point in time, the agent s adopts a particular configuration of these parameters. $p(s,t) in P$.

The problem of self-regulation in development can now be seen as a search process in a multi-dimensional parameter space to maintain optimal (or acceptable) performance. The performance is determined by a cost function $C : P-> R$ where R is a real number between 0.0 and 1.0. Formally, the goal is to find a configuration $p(s,t)$ such that: $C(p(s,t)) = C_{opt}$ where C_{opt} is the optimum cost.

Given this formulation, many techniques from optimisation theory (such as the Simplex algorithm, combinatorial optimisation, simulated annealing, evolutionary programming, etc.) become relevant. There can be little doubt that we are dealing with an NP-hard problem because the parameter space for any realistic developmental system is typically very large. So we must expect approximations, sub-optimal performance, and the use of heuristics. Moreover, the goal of the developmental system is not to reach a stable state, but to keep exploring the parameter landscape so as to maintain a balance of challenge and skill. In other words, as soon as a stable state is reached there should be a force to pull the system out of equilibrium again (see next section).

3.2   Monitoring Performance (The Cost Function)

Next, each component must have a subprocess to monitor the performance of that component. Various types of monitors can normally be formulated easily for a particular component. Performance data is collected over a certain window of time, known as the observation window, and values are typically averaged and then compared to desired performance levels.

Thus there can be various performance measures related to the nature of the task that a component is trying to achieve. For example, a component in a language production system concerned with lexicon lookup can monitor how far the lexicon can cover all the meanings that are required to be expressed and how far the words that were chosen have been understood by the hearer. The optimal levels for these performance measures must be defined, and they are often related to challenge parameters. For example in lexicon lookup, one challenge is to keep the ratio between the number of words used and the number of predicates covered low (pushing the system to create words with complex meanings), another challenge is to increase the certainty with which a certain word has a certain meaning (pushing the system to seek disambiguated words as much as possible). In these cases, the monitored value reflects how far actual performance deviates from the desired performance level.

Without loss of generality, we assume that monitors yield a real value in the range $[0, 1]$ with 1 being optimal performance for a specific dimension. We associate with each component c_i and with the total agent c_T a monitor vector $\langle m_{i,1}, ..., m_{i,n} \rangle$. The set of all monitors for all m components in the system (and the total) forms a multi-dimensional space and system performance in response to a given stream of environmental stimuli traces a trajectory in this space. The performance of the agent at a time t, denoted as $M(s, t)$, is the averaged sum of the performance of all monitors for all components actively used by the agent. We can then define the cost function as $C(P(s,t)) = 1.0 - M(s,t)$, so that $C_{opt} = 0.0$.

3.3   Learning and Skill Levels

When the developing system is attempting to establish its global input-output mapping by chaining the mappings of each of its subcomponents, various failures may occur. Moreover, even if a mapping could be established, there may be a negative feedback signal later. Each component of the agent should be equiped with mechanisms to try and repair these failures. It is not important in the present context what kind of mechanisms are used. They could range from methods to increase needed resources (for example increase the memory available to a component), simple learning mechanisms (such as various forms of neural networks), or sophisticated symbolic machine learning techniques.

It is necessary for the agent to internally measure characteristics of the skill level of each component so that the system can track whether there is any significant change. For example, the amount of memory required by a component, the number of rules learned, the number of nodes or links in a network, etc. can all be quantified so that their evolution during development can be followed. Each component c_i has therefore an associated skill vector $\langle s_{i,1}, ..., s_{i,n} \rangle$ which measures knowledge and skill levels.

It follows that each component c_i (and the agent as a whole) has an associated triple $c_i = \langle P_{i,j}, M_{i,k}, S_{i,l} \rangle$, where $P_{i,j} = \langle p_{i,1}, ...p_{i,n} \rangle$ is the challenge parameter vector, $M_{i,k} = \langle m_{i,1}, ..., m_{i,k} \rangle$ is the monitor vector and $S_{i,l} = \langle s_{i,1}, ..., s_{i,l} \rangle$ is the skill vector.

4   Self-Regulation

Assuming that all components of the developing agent are designed this way, we can now focus on the global behavior of the agent, and particular the strategy to regulate challenge levels for a smooth, progressive self-development. The global system is an instance of combinatorial optimisation and hence has the same structure as well-known optimisation algorithms such as simulated annealing [9], in which a configuration of parameters needs to be found which gives optimal performance. There are two complications compared to traditional optimisation tasks: (1) The cost of a parameter configuration cannot simply be computed by applying a simple function (as in the travelling sales man for example, where cost is basically the length of a path) but must be derived from monitoring actual performance of the system over a particular period of time, including enought time to achieve the acquisition of the necessary skills to reach a certain performance level, (2) this monitoring period must include enough time for the system to acquire the necessary skills to reach a certain performance level. It is to be noted that the objective is not to get optimal performance, but rather to explore the landscape of possibilities in such a way that a higher degree of complexity is reached.

Optimization algorithms typically combine iterated improvement, in which there are small-scale changes to a configuration in order to find optimal parameter settings in a hill-climbing process, and randomisation, in which there is a change in a parameter which may initially cause a decrease in performance but helps the system to get out of a local minimum. Both aspects are present in the algorithms that we propose in the form of two alternating phases: (1) a phase in which challenge parameters are clamped until a steady performance level is reached after increases in skill levels through learning or resource allocation, this is called the operation phase, (2) a phase in which the challenge parameters are changed either because a steady performance could be reached, and so the skill level is getting too high for the challenge posed, or because performance could not be reached, and so the challenge is too high for the skill level. This is called the shake-up phase.

In our experiments to date we have found that the system should start with the lowest challenge levels possible for all components (instead of starting with a random configuration) so as to build up steadily in a bottom-up fashion.

4.1   Operation Phase

The operation phase assumes that the challenge parameters are set at certain levels. The agent exercises its components and monitors the performance of each. A component becomes active when its various inputs are available. In case of failures, each component is assumed to have a set of processes (called 'repairs') that can be used to fix the failure. For example, if a grasp action failed, the categorisation component receives a negative feedback signal and must extend its categorial repertoire to distinguish a new situation. Some of the repairs just involve the addition of additional resources, such as more memory or more processing cycles, others may require more sophisticated forms of learning. Because there are many possible failures in a given run and many possible repairs, some choice must be made about which repairs will be tried and how many.

The operation phase can be algorithmically described as follows:

Procedure Operation Phase

1. Select all executable components, i.e. components for which inputs are available, and activate them.
2. Monitor performance of these components (Could the input-output mapping be established? In how far does it satisfy the criteria set by current challenge parameters?)
3. If a component fails, extract a list of possible repairs and add them to the 'possible repair list'.
4. Consider the next series of components (go to step 1). If no more components can be executed, go to step 5.
5. Given a set of repair on the 'possible repair list'. First filter out those repairs that were executed on the same input stimuli, but failed. If there are repairs left, select the one(s) with the highest estimated effectiveness and execute it. If there are no more repairs restart from 0. A possible variation which considerable speeds up development is to restart the execution of components with the same input stimuli in order to attempt a solution.

4.2   Shake Up Phase

The goal of the shake up phase is to adjust the challenge parameters. In most combinatorial optimisation algorithms (such as simulated annealing) this is done in a random fashion by selecting arbitrarily a parameter and changing it. However, given the size of the parameter space for developmental systems of realistic complexity, such a weak search method does not give adequate results. Instead it is necessary to adapt parameters in a more structured way and maximally exploit available heuristics.

The shake up phase takes place after the system has sufficient experience with a given parameter setting. Sufficient experience means that the average performance level of each of the components and of the total agent does no longer change significantly during specific observation windows, and that there is no longer any significant increase in skill.

Two situations can now occur:

1. Performance does not reach anywhere near the desired levels. This means that the challenge levels are too high and that learning is no longer improving performance. We call this the A-state (where A comes from Anxiety).
2. Performance is consistently at a very high level. This means that operation of all components becomes routine and there is a potential for increased challenge. We call this the B-state (where B comes from Boredom).

Depending on the specific state, specific actions can be performed. Moreover a fine-grained analysis of these states is possible because performance for one component can be very high whereas that of another one can be very low. So changes to parameters should be heuristically guided by taking into account which components are in the A-state and which ones are in the B-state.

Another source of heuristic information is the dependency of components on each other. If a component is in the A-state, then this can be due to the complexity of the output coming from components that feed into it.

The final source of heuristic information is performance on the previous parameter configuration. Because optimisation algorithms are known to require an iterated approach towards optimal configurations, it is necessary to locally explore the parameter space hill climbing towards an adequate solution.

Procedure for the A-state

The goal of this procedure is to decide which challenge parameters to decrease.

1. If the previous parameter configuration had a better performance than the current one, then first switch back to the earlier configuration before making any change.
2. Select all components which are in the A-state. Either one of the challenge parameters must be decreased or else, one of the components feeding into it must be signalled to decrease the complexity of its output, by a recursive application of step 2. This step generates a set of possible choices for parameter adaptation. These choices can be heuristically ordered based on the performance of the components involved.

3. Choose one or more parameters, enact the change, and go back to the Operation Phase.

Procedure for the B-state

The goal of this procedure is to decide which challenge parameters to increase. There is a steady performance with the given parameters but there is perhaps an opportunity for further increase in skill. Note that this phase correspondence to the 'randomisation' phase of many combinatorial optimisation algorithms, although the parameter change is not completely random.

1. Collect all components which are in the B-state. The possibilities are: to increase one of the challenge parameters of this component, or else to recursively collect one of the challenge parameters that give input to this component. The possibilities can again be heuristically ordered based on the performance of the components involved.

2. Choose one or more of these parameters, enact the change, and go back to the Operation Phase.

Note that a record must be kept of parameter configurations and their associated performance in order to backtrack if needed.

We also know from our experiments that a conservative strategy (where only one repair is executed in the Operation Phase, and one parameter is changed in the Shake Up Phase) is much more desirable than drastic and rapid change.

From the viewpoint of optimization theory, the need for this shake-up process is not surprising. Optimization algorithms like simulated annealing typically combine iterated improvement, in which there are small-scale changes to a configuration in order to find optimal parameter settings in a hill-climbing process, and randomization, in which there is a change in a parameter which may initially cause a decrease in performance but helps the system to get out of a local minimum [14]. In fact it is only because of randomization that these local minima, i.e. situations where a stable but suboptimal solution is reached, can be avoided. Given that a complex cognitive agent is exploring a vast parameter space, the problem is an NP-hard (i.e. nondeterministic polynomial time-hard) problem as defined according

5   Conclusions

The paper has focused on the problem how an agent can self-regulate his own developmental process. It has proposed the autotelic principle, as a way to go beyond the classical reinforcement learning framework initiated by behaviorist psychology. There must be (i) ways to monitor performance and change in knowledge, skill, or resource use, (ii) ways to control the challenge level for the different components of the agent, and (iii) a general mechanism that self-adjusts challenge levels or shakes the system up to push the agent towards new heights. The motivational structure of the agent continuously tries to strike a balance between the highest possible level of challenge and skill.

Acknowledgement. I am indebted to Frederic Kaplan for many discussions on the architecture of agents.

References

1. Cohen, L. et al.: A constructivist model of infant cognition. Cognitive Development, 17[2002] 1323-1343.
2. Csikszentmihalyi, M.: Beyond Boredom and Anxiety: Experiencing Flow in Work and Play. Cambridge University Press, Cambridge. [1978]
3. Csikszentmihalyi, M.: Flow. The Psychology of Optimal Experience. Harper and Row, New York. [1990]
4. Csikszentmihalyi, M. and I. Selega (Editors): Optimal Experience : Psychological Studies of Flow in Consciousness. Cambridge University Press, Cambridge. [2001]

5. Elman, J.: Learning and development in neural networks: The importance of starting small. Cognition, v. 48. p. 71-89 [1993]

6. Elman, J. L., Bates, E. A., Johnson, M. H., Karmiloff-Smith, A., Parisi, D., Plunkett, K.: Rethinking innateness: A connectionist perspective on development. Cambridge, MA: MIT Press. [1996]

7. Hopfield, J.: Neural Networks and Physical Systems with Emergent Collective Computational Abilities. Proceedings of the National Academy of Sciences, USA, volume 79 [1982] pages 2554 to 2558.

8. Johnson, M.: The Infant Brain. In: Tokoro, M. and L. Steels (eds.) The Future of Learning Vol 1. IOS Press, Amsterdam. [2003] p. 101-116.

9. Kirkpatrick, S., Gerlatt, C. D. Jr., and Vecchi, M.P.: Optimization by Simulated Annealing, Science 220. [1983] 671-680.

10. Matthews, J.: Art Education as a form of child abuse. Lecture for the National Institute of Education Singapore. [1993]

11. Kaplan, F. and PY Oudeyer: A generic engine for open-ended sensory-motor development. Epigenetic Robotics Workshop, Genoa. [2004]

12. McFarland, D. and M. Boesser.: Intelligent Behavior in Animals and Robots. MIT Press, Cambridge Ma. [1993]

13. Newell, A.: Unified Theories of Cognition. Harvard University Press, Cambridge, MA. [1990]

14. Papadimitrious, C. and K. Steiglitz: Combinatorial Optimization: Algorithms and Complexity. Dover, New York. [1998]

15. Steels, L. and R. Brooks (eds.): The Artificial life Route to Artificial Intelligence. Building Situated Embodied Agents. Lawrence Erlbaum, New Haven. [1995]

16. Skinner, B.F.: Science and Human Behavior. New York: Macmillan. [1953]

17. Steels, L.: Language games for Autonomous Agents. IEEE Intelligent Systems. Sept/Oct Issues [2001]

18. Steels, L.: Evolving grounded communication for robots. Trends in Cognitive Science. Volume 7, Issue 7, July [2003] pp. 308-312.

19. Thelen, B. and L. Smith: A dynamic systems approach to cognition and development. MIT press, Cambridge Ma. [1994]

20. Uchibe, E, M. Asada and K. Hosoda: Environmental complexity control for vision-based learning mobile robot. In: Proc. of IEEE Int. Conf on Robotics and Automation. [1998] p. 1865-1870.

21. Weng, J., McClelland, J., Pentland, A., Sporns, O., Stockman, I., Sur, M., and Thelen, E.: Autonomous mental development by robots and animals. Science, 291. [2001] p. 599-600.

Toward a Cognitive System Algebra: Application to Facial Expression Learning and Imitation

Philippe Gaussier[1], Ken Prepin[1,2], and Jacqueline Nadel[2]

[1] Neuro-cybernetic team, Image and Signal processing Lab., UMR CNRS 8051
Cergy Pontoise University / ENSEA, 6 av du Ponceau, 95014 Cergy , France
gaussier@ensea.fr
http://www.etis.ensea.fr/~neurocyber
[2] UMR CNRS 7593, Hopital la Pitié Salpétrière, Paris, France

Abstract. In this paper, we try to demonstrate the capability of a very simple architecture to learn to recognize and reproduce facial expressions without the innate capability to recognize the facial expressions of others. In the first part, the main properties of an algebra useful to describe architectures devoted to the control of autonomous and embodied "intelligent" systems are described. Next, we propose a very simple architecture and study the conditions for a stable behavior learning. We show the solution relies on the importance of the interactions with another system/agent knowing already a set of emotional expressions. A condition for the learning stability of the proposed architecture is derived. The teacher agent must act as a mirror of the baby agent (and not as a classical teacher). In conclusion, we discuss the limitations of the proposed formalism and encourage people to imagine more powerful theoretical frameworks in order to compare and analyze the different "intelligent" systems that could be developed.

1 Introduction

Nowadays hardware and software technologies allow to build more and more complex artifacts. Unfortunately, we are almost unable to compare two control architectures proposed to solve one given problem. Of course, one can try an experimental comparison on a given benchmark but the results focus on the optimality regarding the benchmark (how to deal with really unknown or unpredictable events?). We should be able to analyze, compare and predict in a formal way the behaviors of different control architectures. For instance, we must be able to decide if two architectures belong or not to the same family and can be reduced to a single architecture.

On another level, new design principles are proposed to create more "intelligent" systems [1] but there is no real formalization of these principles. The only way to correctly understand and use them is to have a long explanation build on examples showing cases of success stories (examples of good robotic

F. Iida et al. (Eds.): Embodied Artificial Intelligence, LNAI 3139, pp. 243–258, 2004.

architectures). Hence, we have good intuitions about what to do or not to do to build a control architecture but it remains difficult to deal with really complex systems. Our situation can be compared to the period before Galileo when people knew objects fall but were unable to relate that to the concept of mass and acceleration in order to predict what will happen in new experiments. We urgently need tools to analyze both natural and artificial intelligent systems. Previous works have focused on mathematical tools to formalize pure behaviorist or reactive systems [2]. People have also tried with no real success to measure the complexity (in terms of fractal dimension for instance) of very simple behaviors like an obstacle avoidance [3]. The most interesting tools are dedicated to specific part of our global problem such as learning (see NN literature), dynamical systems [4] or some game theory aspects [5]. Yet, it remains difficult to overstep the old frame of the cybernetics [6,7]. Finding the fundamental variables and parameters regarding some particular cognitive capabilities will be a a long and difficult work but we believe this should be related to the invariant properties of cognitive mechanisms and to the variation laws linking learning and embodiment.

In the present paper, we would like to show that a mathematical formalism used previously to represent for instance a control architecture dedicated to the visual homing [8], can also be used to build a simple theoretical model of the development of the capability to express and recognize more and more complex facial expressions. We will try to discuss, using this mathematical formalism, which are the basic mechanisms necessary to allow a naive agent to acquire the capability to understand/read the facial emotions of a teacher agent and to mimic them (so as to become a teacher and to allow turn taking in an emotion expression game). We will try to show that a newborn do not need a hardwired mechanism of emotion recognition to begin to interact in emotional games with adults. At last, we will discuss the drawback of the proposed formalism and try to propose directions for future researches since this work is at its very beginning.

2   Basic Formalism of a Cognitive System

We summarize here the basis of our mathematical formalism. Figure 1 shows a typical control architecture for what we will call a cognitive[1] system (CS). The input and output of a CS are represented by vectors in the "bracket" notation[2]. An input or output vector x (column vector of size m) is noted $|x\rangle$ with $|x\rangle \in R^{+m}$[3] while its transposed vector is noted $\langle x|$. Hence $\langle x|x\rangle$ is a scalar rep-

---

[1] The term cognitive must be understood here in the sense of the study of particular cognitive capabilities and not as a positive a priori for any kind of cognitivist approach.

[2] The formalism is inspired from Hilbert space used in quantum mechanics. Nevertheless, in our case it is not an Hilbert space since the operator is not linear...

[3] We consider the components of the different input/output vectors can only be positive/activated or null/inactivated. Negative activities are banned to avoid positive effects when combined with a negative weight matrix.

resenting the square of $|x\rangle$ norm. The multiplication of a vector $|x\rangle$ by a matrix A is $|y\rangle = A|x\rangle$ with $|y\rangle \in R^n$ for a matrix A of size $n \times m$.

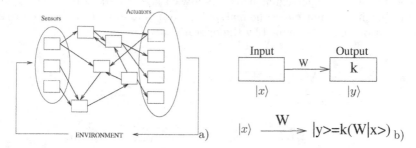

Fig. 1. a) Typical architecture that can be manipulated by our formalism. b) Graphical representation and formalism of the connection between 2 boxes in a CS.

A CS is supposed to be made of several elements or nodes or boxes associated with input information, intermediate processes and output (command of actions). We can consider that any element of a CS filters an input vector according to a matrix of weights W and a non-linear operator $\mathbf{k}$. This operator represents the way to use the W matrix and the pattern of interactions between the elements of the same block. It can be a simple scalar product (or distance measure) or even a more complex operator such as an "If...then...else..." treatment (hard decision making), a pattern of lateral interactions in the case of a competitive structure, a recurrent feedback in the case of a dynamical system, a shifting mechanism, a mechanism to control a focus of the attention... Hence, we can consider these elements as "neurons" even if they can be more complex algorithmic elements in other programming languages. For instance, in the case of a simple WTA[4] box, we can write the WTA output $|y\rangle$ is $wta(A|x\rangle)$ with $|y\rangle = (0,...,y_j,...0)$ and $j = ArgMax(q_i)$ and $q_i = \langle A_i|x\rangle$. In the case of a Kohonen map, $|y\rangle = koh(A|x\rangle)$, the main difference is the way the output is computed: $q_i = \sum_j |A_{ij} - x_j|$. To be more precise, we should write $|y\rangle = koh(A, |x\rangle)$. Because, in the general case, an operator can have an arbitrary number of input groups, we will consider the recognition of an input is performed according to the type of its associated weight matrix. For instance, "one to one" input/output connections represented by the general identity weight matrix I is considered as the signature of a reflex pathway (because there is almost no interest to consider "one to one" learnable links). Basically, we distinguish 2 main types of connectivity according to their learning capabilities (learning possible or not): the "one to one" links (see fig. 2a) and the "one to many" connections (see fig. 2b) which are used for pattern matching processes, categorization... or all the other possible filtering. "One to many" connections will be represented in general by a A. In the case of a complex competitive and conditioning structure

---

[4] Winner Takes All.

with 1 unconditional (US) and 2 conditional (CS) inputs, we should write for instance $|y\rangle = c(A_1, |CS_1\rangle, A_2, |CS_2\rangle, I, |US\rangle)$. To avoid too many commas in the operator expression, we simply write $|y\rangle = c(A_1|CS_1\rangle, A_2|CS_2\rangle, I|US\rangle)$[5]. This allows to be sure a particular matrix is always associated to the correct input vector but it does not mean the matrix has to be multiplied by the vector (this computation choice is defined by the operator itself).

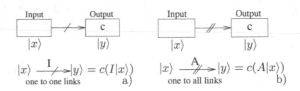

$$|x\rangle \xrightarrow{\ I\ } |y\rangle = c(I|x\rangle) \qquad\qquad |x\rangle \xrightarrow{\ A\ } |y\rangle = c(A|x\rangle)$$
one to one links   a)   one to all links   b)

Fig. 2. Arrows with one stroke represent "one to one" reflex connections (one input connected to one output in an injective manner). Arrows with labels and 2 parallel strokes represent "one to many" modifiable connections between input and output nodes. a) Unconditional "one to one" connections (used as a reflex link) between two groups. Upper image is the graphical representation and lower image is the formal notation. b) "One to many" connections with a competitive group representing the categorization of the input stimulus at the level of the output group.

The main difference with classical automata networks is that most of our operators can adapt or learn online new input/output associations according to their associated learning rule. For instance, in the case of a classical Kohonen rule, we should write $\frac{dA_{ij}}{dt} = koh\_learning\,(|y\rangle, |x\rangle)$. Hence, 2 equations have to be written for each elementary box: one for the computation of the system output and another one for the weight adaptation (modification of the box memory). In the following, it will be crucial to remember our operators represent 2 different functions and flow of information moving in opposite directions. The first one will allow to transform sensorial information in an output code while the second one will act on the group memory in order to maintain a certain equilibrium defined by the learning rule [9].

In this paper, we will not discuss the interest or defaults of particular learning rules. Learning rules such as Least Mean Square algorithm (LMS - used for conditioning) or Hebb rule variants or any competitive rule will be sufficient for our demonstration since they are able to stabilize their associated weight matrices in the case the system is in a simple behavioral attractor or "perception state" (here simple means fixed point attractor).

Definition 1. *The perception Per can be seen as a scalar function ψ representing an attraction basin of the agent behavior. It can be seen as a sensorimotor invariant of the system (a kind of energy measure). Hence, the percep-*

---

[5] In previous papers, it was possible to write $|y\rangle = c(A_1|CS_1\rangle + A_2|CS_2\rangle + I|US\rangle)$ but many reviewers complained about the risk of misunderstanding the meaning of the operator +.

tion can only be defined for an active system and is dependent of the system dynamical capabilities (kind of body, sensors and actuators). We will write: $Per(\boldsymbol{p}) = -\langle Ac|\boldsymbol{p}\rangle = -\int_{\boldsymbol{p}+\delta\boldsymbol{p}} Ac \, d\boldsymbol{r}$ where $|\boldsymbol{p}\rangle >$ describes the position of the system in the considered environment[6].

This corresponds to our intuition of the recognition as an attraction basin. We will say a system is in a **stable state of perception** if it is able to maintain itself in the associated attraction basin. Hence, learning to recognize an object (from visual, tactile, auditory... informations) can be seen as learning to maintain the system in a particular dynamical attraction basin [10]. More illustrations and justifications of this definition can be found in [11].

So when studying a control architecture, we will not need to take into account all the details of its implementation. We will have to focus on the global architecture and the way its elements are able to shape the behavior (building attractor basins).

3   Formal Simplification Rules

Now, the problem is to be able to simplify a CS architecture in another one (either simpler to analyze and to understand the architecture or more complex to provide more degrees of freedom to increase the architecture performances). Two architectures will be considered as equivalent if they have the same behavioral attractors (or perception state as defined previously). This means we cannot study a control architecture alone. The interactions with the environment must be taken into account. After the learning of a first behavior, the dynamics of the interactions with the environment (the perception state) is supposed to be stabilized. In the present formalism, two types of diagram simplifications will be considered. Simplifications of the first type can be performed at any time and leave the fundamental properties of the system completely unchanged (these are very restrictive simplification rules). Those of the second type only apply after learning stabilization (if learning is possible!). They allow strong simplifications but the resulting system is no more completely equivalent to the departure system (the new system will be less robust, less efficient and less precise for instance). At the opposite, the same formalism can be used to complexify an architecture in order to increase the efficiency of a given set of cognitive capabilities (increase of the system elasticity, robustness, precision...).

We present now a first example of simplification rule based on the existence of unconditional and reflex links. If we consider a linear chain of unconditional links between competitive structures of the same size such as "Winner Take All" (WTA), the intermediate competitive boxes are useless since they replicate on their output their input information. Hence we can write for instance that if we have: $|b\rangle = c(I|a\rangle)$ and $|d\rangle = c(I|b\rangle)$ then $|d\rangle = c(I|c(I|a\rangle))$ which should be equal to $|d\rangle = c(I|a\rangle)$ because a cascade of competitions leads to an isomorphism

---

[6] In naive cases, $|\boldsymbol{p}\rangle$ can be expressed in Cartesian coordinates or in any pertinent parameter space useful to describe more complex cases.

between the different output vectors which become equivalent to each other after the self organization of the different groups. So we can deduce the following rule $\mathbf{c}(\mathbf{I}|\mathbf{c}(.)) = \mathbf{c}(.)$. Other static simplification rules can be built in the same way [9]. Other simplifications can be used to represent the effect of learning. Except for robustness, these simplifications can be introduced to compare different control architectures (or to build more complex controllers). We will suppose that the system is in a stable state of perception or interaction with its environment. That is to say, it exists a time period where the system remains almost unchanged (internal modification must not have an effect on the system behavior). To be in a stable state, the environment properties must be quite constant. We postulate that for a given time interval, the learned configuration will be stable enough so that the simplifications can be applied (but they remain only valid for this time interval). Fig. 3 shows an intuitive representation of the evolution of a system behavior through time. The system behavior can evolve to adapt itself to an

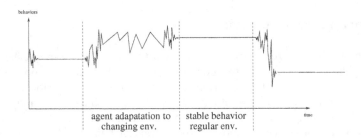

Fig. 3. Intuitive representation of what is a stable behavior allowing formal simplifications of the system.

environment variation (or to the variation of an internal signal). In this case, it moves from a stable state to an unstable state or transition phase. It is only during the stable phases that the following simplifications can be considered as valid. Hence, we have to highlight a "before learning state" and an "after learning state" since some of the simplifications can be made at any time while some others must necessarily be made in the "after learning state".

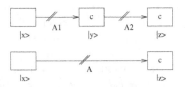

Fig. 4. A cascade of competitive or unsupervised classification structures can be simplified in a single competitive or classification box with a possible loss of performance but without a change in the main properties of the architecture.

A very simple example of such a simplification is the case of strict self organized learning group or competitive boxes (c operator) push-pully connected, fig. 4. We have $|y\rangle = c(A_1|x\rangle)$ and $|z\rangle = c(A_2|y\rangle)$ with A_1 and A_2 the matrices to learn the relevant input configurations. So $|z\rangle = \mathbf{c}(\mathbf{A_2}|\mathbf{c}(\mathbf{A_1}|\mathbf{x}\rangle)) = \mathbf{c}(\mathbf{A}|\mathbf{x}\rangle)$ since it is always possible to create a bijection between the activation of a given neuron in a first group and the activation of another neuron in a second group. Both sets of neurons can be considered as equivalents.

A more interesting case corresponds to the conditioning learning. The conditioning network (fig. 5 a) should be equivalent "after learning" to the simple network shown fig. 5 b and can be translated by the following equation: $c(I|US\rangle, A|CS\rangle) \approx c(A|CS\rangle)$ where $|US\rangle$ represents the unconditional stimulus and $|CS\rangle$ the conditional stimulus. The simplification "before learning" considers only the reflex pathway: $c(I|US\rangle, A|CS\rangle) \approx c(I|US\rangle)$ (functioning is equivalent in a short time delay but there is no possible adaptation) whereas the other simplification represents the equivalent NN in the "after learning" situation: not equivalent if the environment changes too much and leads the agent to be inadapted.

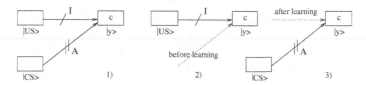

Fig. 5. Image 1 is the graphical representation of a conditioning learning $|y\rangle = c(I|US\rangle, A|CS\rangle)$. Image 2 is the graphical representation of the equivalent network before learning and Image 3 after learning $|y\rangle = c(A|CS\rangle)$.

We have shown in [9] that maximizing the dimensionality (rank) of the perception matrix $\sum_P |Ac\rangle\langle S|$ can be equivalent to the mean square error minimization performed when trying to optimize the conditioning learning between the action proposed by the conditional link and the action proposed by the unconditional link (where $|Ac\rangle$ represents the action vector (here $|y\rangle$) and $|S\rangle$ the sensorial input (here $|CS\rangle$)). Hence, learning can be seen as an optimization of the tensor representing the perception. In other words, we can say the proposed simplification rules are relevant if the system is adapted to its environment or if the system perceives its environment correctly according to the capabilities of its own control architecture (learning capabilities). We can notice that $Per = \sum_P \langle Ac|S\rangle = tr(\sum_P |Ac\rangle\langle S|)$ while the "complexity" of the system behavior can be estimated from $rank((\sum_P |Ac\rangle\langle S|)$.

4 Application to Social Interactions Learning

In this section, our goal is to show how our formalism can be applied to analyze a very simple control architecture and justify some psychological models (see [12] for a discussion on the importance of an emotional system in autonomous agents). At the opposite to the classical pattern recognition approach, we will show that an online dynamical action/perception approach between two interacting systems has very important properties. The system we will consider is composed of two identical agents (same architecture) interacting in a neutral environment (see fig. 6).

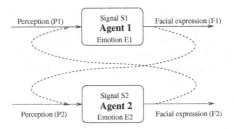

Fig. 6. The bidirectional dynamical system we are studying. Both agents face each other. Agent 1 is considered as a newborn and agent 2 as an adult mimicking the newborn facial expressions. Both agents are driven by internal signals which can induce the feeling of particular emotions.

One agent is defined as an adult with perfect emotion recognition capabilities and also the perfect capability to express an intentional emotion[7]. The second agent will be considered as a newborn without any previous learning on the social role of emotions. First, we will determine the conditions for a stable interaction and show that in this case learning to associate the recognition of a given facial expression with the agent own "emotions" is a behavioral attractor of the global system. Our agents receive some visual signals (P_i perception of agent i). They can learn and recognize them ($|R_i\rangle$ activity). Hence, the perception of a face displaying a particular expression should trigger the activation of a corresponding node in R_i. This mechanism can use an unsupervised pattern matching technics such as any winner take all mechanism (WTA, ART network, Kohonen map...).

$$|R_i\rangle = c\left(A_{i1}|P_i\rangle\right) \tag{1}$$

c represents a competitive mechanism allowing for instance to select a winner among all the vector components. To simplify, this winning component is put to 1 and the other ones to 0 (any other non linear competition rule could be applied

----

7 The problem of the dynamical development of two identical agents in a more free interaction game will be studied in a following paper.

and should not change our reasoning). A_{i1} represents the weights of the neurons in the recognition group of the agent i allowing a direct pattern matching. Our agents are also affected by the perception of their internal milieu (hunger, fear etc.). We will call S_i the internal signals linked to physiological inputs such as fear, hunger... "Emotion" recognition E_i depends on the internal milieu. The recognition of a particular internal state will be called an emotional state E_i. We suppose also E_i depends on the visual recognition R_i of the visual signal P_i. At last, the agents can express a motor command F_i corresponding to a facial expression. If one agent can act as an adult, it must have the ability to "feel" the emotion recognized on someone else's face (empathy). At least, one connection between the visual recognition and the group of neuron representing its emotional state must exist. In order to display emotional state, we must also suppose there is a connection from the internal signals to the control of the facial expression. The connection can be direct or through another group devoted to the representation of emotions. For sake of homogeneity, we will consider that the internal signal activates through an unconditional link the emotion recognition group which activates through an unconditional connection the display of a facial expression (hence it is equivalent to a direct activation of F_i by S_i - see [9] for a formal analysis of this kind of properties). Hence, the sum of both flows of information can be formalized as follow:

$$|E_i\rangle = c\left(I|S_i\rangle, A_{i3}|R_i\rangle\right) \qquad (2)$$

At last, we can also suppose the teacher agent can display a facial expression without "feeling" it (just by a mimicking behavior obtain form the recognition of the other facial expression). The motor output of the teacher facial expression then depends on both facial expression recognition and the will to express a particular emotion:

$$|F_i\rangle = c\left(I|E_i\rangle, A_{i2}|R_i\rangle\right) \qquad (3)$$

Fig. 7 represents the network associated to the 3 previous equations describing our candidate architecture. In a more realistic architecture, some intermediate links allowing the inhibition of one or another pathway could be added but it is out of the scope of the present paper, which aims at illustrating what can be done with our formalism on a very simple example.

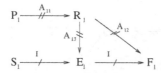

Fig. 7. Schematic representation of an agent that can display and recognize "emotions"(notations see fig. 2). Arrows with one stroke represent "one to one" reflex connections. Arrows with labels and 2 parallel strokes represent "one to all" modifiable connections.

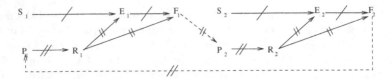

Fig. 8. Schematic representation of the global network representing the interaction between 2 identical emotional agents. The dashed links represent the connections from the display of a facial expression to the other agent perception system (effect of the environment).

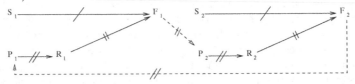

Fig. 9. Schematic representation of the simplified network representing the interaction between 2 identical emotional agents (modification of fig. 8)

4.1 Condition for Learning Stability

First, we can study the minimal conditions allowing the building of a global behavioral attractor (learning to imitate and to understand facial expressions). Fig. 8 represents the complete system with both agents in interaction. It is considered as a virtual net that can be studied in the same way than an isolated architecture thus allowing to deal at the same time with the agent "intelligence" and with the effects of the embodiment and/or the dynamics of the action/perception loops.

The following simplifications apply before learning and concern only the unconditional links (see in the previous section the simplification of a conditioning structure before learning). We simply consider the activation of S can induce a reflex activation of a stereotyped facial expression F before (and after) the learning of the correct set of conditioning. The resulting network is shown fig. 9.

Next, the linear chains of "one to many" modifiable connections and their associated competitive learning structures can also be simplified since $c(\mathbf{A}|\mathbf{c}(.)) \equiv \mathbf{c}(.)$. We finally obtain the network shown fig. 10 a).

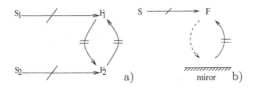

Fig. 10. a) Final simplification of the network representing the interaction between 2 identical emotional agents (modification of fig. 9). b) Minimal architecture allowing the agent to learn "internal state"-"facial expression" associations.

It is much simpler on fig. 10 to see the condition of the learning stability. Since, the chosen simplifications allow to obtain a virtual network with learnable bidirectional connections between F_1 and F_2, a condition for the learning stability is that these connection weights remain stable. If S_1 and S_2 are independent, learning cannot be stable since S_1 and S_2 are connected through unconditional links to F_1 and F_2 respectively. The only way to stabilize learning is to suppose S_1 and S_2 are congruent. Otherwise a lot of "energy" is lost to adapt continuously the connections between F_1 and F_2 (see [9] for more details). Because, the agent representing the baby must not be explicitly supervised, a simple solution is to consider the agent representing the parent is nothing more than a mirror[8]. We obtain the network shown in fig. 10 b) where the architecture allows the system to learn the "internal state"-"facial expression" associations. Hence, we show that from our initial control architecture, learning is only possible if the teacher/parent agent imitates the baby agent. The roles are switched according to the classical point of view of AI and learning theory. This shows how taking account the dynamics of interactions between two agents can change our way of thinking learning and more generally cognition problems.

4.2   Learning the Emotional Value of Facial Expressions

These first simplifications bring us to the conclusion that learning stabilization is possible if the teacher/parent agent acts as an imitator of the baby agent. Now, we will suppose these conditions are respected. From the initial equations of the system, we will derive another set of simplifications in order to prove the beginner (or naive) agent can learn to associate the visual facial expression displayed by the teacher agent to the correct emotional state. We suppose the agent 1 perceptive input P_1 is the result of a linear projection of the facial expression (output) of the agent 2 and vice versa. We will write $|P_1\rangle = B_1|F_2\rangle$ and $P_2 = B_2|F_1\rangle$. Hence, $|R_1\rangle = c(A_{11}|P_1\rangle) = c(A_{11}.B_1|F_2\rangle) = c(A'_{11}|F_2\rangle)$ (with $A'_{11} = A_{11}.B_1$). We can then replace in this new expression of R_1, $|F_2\rangle$ by the result of the computation of the second agent (using eq. 3). We obtain:

$$|R_1\rangle = c\left(A'_{11}|c\left(I|E_2\rangle, A_{23}|R_2\rangle\right)\right)$$
$$= c\left(A'_{11}|c\left(I|E_2\rangle, A_{23}|c\left(A_{21}|P_2\rangle\right)\right)\right)$$

On the other side, we have $|P_2\rangle = B_2|F_1\rangle$ so:

$$|R_1\rangle = c\left(A'_{11}|c\left(I|E_2\rangle, A_{23}|c\left(A_{21}\cdot B_2|F_1\rangle\right)\right)\right)$$
$$= c\left(A'_{11}|c\left(I|E_2\rangle, A_{23}|c\left(A'_{21}|F_1\rangle\right)\right)\right) \tag{4}$$

---

[8] Obviously, another possible solution is that the second agent tries to deceive the first agent. If the second agent displays an "unhappy face" every time the first agent displays an "happy face" and vice versa an incorrect learning is possible. Fortunately, the probability of such a learning is very low if the first agent interacts with several independent agents (no conspiracy!). Yet, we can predict that a baby interacting with a depressed mother (low probability of "happy face") will have some difficulties to create an unbiased repertory for the recognition of other's emotional states.

A'_{21} is defined as the matrix resulting from $A_{21} \cdot B_2$. The equation 4 can be represented by the virtual[9] network shown fig. 11. Intuitively, the network means the visual recognition in the first agent depends on the emotional state of the second agent and should also be a function of the facial expression of agent 1.

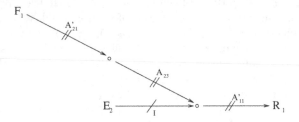

Fig. 11. Virtual net associated to eq. 4.

All the preceding simplifications could be made at any time (here, it is before learning). The following simplification can be done only after learning (and need the learning stability condition i.e. the second agent is a mirror of the first one). If the obtention of learning is possible (the error between F_1 and E_2 can be minimized in the mean square sense), conditioning learning in eq. 4 should result in:

$$I|E_2\rangle \approx A_{23}.c\left(A'_{21}|F_1\rangle\right) \qquad (5)$$

if both architectures are identical, since there is no influence of learning on this simplification, we obtain by symmetry:

$$|E_1\rangle \approx A_{13}.c\left(A'_{11}|F_2\rangle\right) \qquad (6)$$

Then, we can simplify eq. 4.

$$|R_1\rangle \approx c\left(A'_{11}|c\left(A_{23}|c\left(A'_{21}|F_1\rangle\right)\right)\right)$$
$$\approx c\left(A'_{123}|F_1\rangle\right) \qquad (7)$$

(we also have $|R_1\rangle \approx c\left(A'_{12}|E_2\rangle\right)$ but we won't use it.) Eq. 7 can be interpreted as the fact the activity of agent 1 visual face recognition is a function of its own facial expression. If we replace the value of F_1 obtained from eq. 3 in eq. 7, we obtain:

$$|R_1\rangle \approx c\left(A'_{123}|c\left(I|E_1\rangle, A_{13}|R_1\rangle\right)\right) \qquad (8)$$

Here again $I|E_1\rangle$ is the reflex link and $A_{13}|R_1\rangle$ the conditional information. The conditional link can learn to provide the same results as the reflex link. If E_1 can be associated to R_1 then we obtain:

---

[9] This network is virtual since it mixes together parts of networks belonging to two different agents.

$$|R_1\rangle \approx c\left(A'_{123}|c\left(I|E_1\right)\right)\right)$$
$$\text{and } |R_1\rangle \approx c\left(A'_{123}|E_1\rangle\right) \tag{9}$$

This result shows the activity of the face recognition system is a direct function of the agent emotional state (R_1 can be deduce from E_1). In conjunction with the relation linking E_1 to R_1 (eq. 2) we can deduce the agent 1 (baby) has learned to associate the visual recognition of the tested facial expressions to its own internal feeling (E_1). The agent has learned how to connect the felt but unseen movements of self with the seen but unfelt movements of the other. It could be generalized to other movements since we showed in [13,14,15] that a simple sensori-motor system is sufficient to trigger low level imitations.

5   Discussion and Perspectives

In this paper, we have applied a formalism proposed in [9] to simplify an "intelligent" system and to analyze some of its properties. We have shown a very simple architecture can learn the bidirectional association between an internal "emotion" and its associated facial expression. To demonstrate this feature, we have proved first that learning is only possible if one of the agents acts as a mirror of another. We have proposed a theoretical model that can be used as a tool not only to understand artificial emotional brains but also natural emotional brains. Let us consider a newborn. She expresses internal states of pleasure, discomfort, disgust, etc, but she is not aware of what she expresses. Within our theoretical framework, we can expect that she will learn main associations between what she expresses and what she experiences her partners' mirroring of her own expressions. Seeing what she feels will allow the infant to associate her internal state with an external signal (i.e. her facial expression mirrored by someone else). Empirical studies of mother-infant communication support this view. For instance, two-month-old infants facing a non contingent televised mother who mirrors their facial expressions with a delay become wary, show discomfort and stop imitating the mother's facial expressions (see [16]). The primary need of mirroring is also demonstrated by the progressive disappearance of facial expressions in infants born blind. Another prospective benefit of the model is to give a simple developmental explanation of how facial expressions come to inform the growing infant about external events through the facial reading of what those events trigger in others [17]. Finally the model leads to suggest a main distinction between two processes of emotional matching: matching a facial emotion without sharing the emotion expressed: in this case there is a decoupling (see [18]) between what is felt and what is shown, thus it is pure imitation, and matching a facial emotion with emotional sharing, that is to say feeling what the other expresses through the process of mirroring, a definition of empathy (see [19]). More complex architectures could be built on the basis of the studied model. For instance, adding feedback connections from the proprioceptive signal linked to the control of the facial expressions onto the recognition of an internal emotional state would allow the agent to "feel" a little bit more happy when he is smiling. Fig. 12 shows what could be such an architecture.

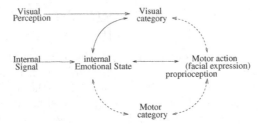

Fig. 12. Schematic representation of an agent that can show and recognize emotional states with feedbacks connections from action/proprioception to the different internal categorization structures.

In our lab., the formalism developed in this paper is used as a programming language allowing to describe architectures for visual object recognition, visual navigation, planning, visuo-motor control of an arm and imitation games... Yet, even if the size of our networks becomes more and more important, their intrinsic complexity remains low since it was our goal to prove complex dynamical behaviors could emerge from relatively simple architectures. At the opposite, Sporns et al. [20] study complex networks in terms of their structure and dynamics. They show highly complex networks have distinct structural characteristics such as clustered connectivity and short wiring length similar to those of large-scale networks of the cerebral cortex. Hence, future works will have to answer the following questions: Which are the minimal structures that cannot be simplified (or which intrinsic property is lost when an important simplification is made)? Which kind of really different operators have to be considered? And at a higher level, we will have also to understand how to manage different learning time constants and how to represent the body/controller codevelopment.

To answer these questions, it becomes necessary to test our formalism on other architectures and other problems to analyze precisely its limitations and to propose a more powerful framework. Another problem is that the demonstration proposed in this paper was based on the fact it was possible to isolate fixed point dynamics and to study one of them isolatedly. For more complex dynamical systems, we believe the same approach could be used (i.e. isolate the different dynamical regime and propose simplifications for each of them). Nevertheless, we believe the proposed approach could be directly applied to problems of the same kind of complexity level such as the problem of joint attention learning (see for instance [21] where a robot controller based on a principle close to the one developed in our own architecture is proposed).

To sum up, we have shown it is possible to develop theoretical tools taking into account the interactions with the environment in order to compare and analyze different control architectures. In this context, the more a system is embodied, the less it need explicit learning since it is well adapted to its function: it relies on the physical plasticity of its physical architecture (see for instance how our mechanical anatomy simplifies the wide variety of tasks we have to solve). The need for learning can be seen as the impossibility of a perfect embodiment according to a given ecological niche (need of a physical compromise between the requirement of the different behaviors). Hence, in an algebra of embodied cognitive systems, we will have to distinguish between 3 levels of cognition: an infra level of cognition linked to the physical properties of the body, the individual level of cognition (the control architecture for one isolated agent) and the social level dealing with the social interactions between agents. At each level, the measure of the system elasticity or adaptation capability might be performed to characterize the embodiment of the system (see [9] for a tool to compare the complexity of different control architectures in term of an energy measure). In conclusion, we believe the difficulty of a formal analysis of cognitive systems is much more a problem of choosing the correct postulates and axioms than the lack of mathematical tools to deal with the intrinsic complexity of the existing systems.

Acknowledgements. This work is supported by the CNRS "ACI Computational Neurosciences" and "ACI Time and Brain" and CNRS team project on "imitation in robotics and development".

References

[1] Pfeifer, R., Scheier, C.: Understanding intelligence. MIT press (1999)

[2] Steels, L.: A case study in the behavior-oriented design of autonomous agents. In: SAB'94. (1994) 445–451

[3] Smithers, T.: On quantitative performance measures of robot behaviour. Robotics and Autonomous Systems **15** (1995) 107–133

[4] Schöner, G., Dose, M., Engels, C.: Dynamics of behavior: theory and applications for autonomous robot architectures. Robotics and Autonomous System **16** (1995) 213–245

[5] Ikegami, T.: Ecology of evolutionary game strategies. In: ECAL 93. (1993) 527–536

[6] Wiener, N.: CYBERNETICS or Control and Communication in the Animal and the Machine. MIT Press (1948, 1961)

[7] Ashby, W.: Design for a brain. London: Chapman and Hall (1960)

[8] Gaussier, P., Zrehen, S.: Perac: A neural architecture to control artificial animals. Robotics and Autonomous System **16** (1995) 291–320

[9] Gaussier, P.: Toward a cognitive system algebra: A perception/action perspective. In: European Workshop on Learning Robots (EWLR), http://www-etis.ensea.fr/~neurocyber /EWRL2001_gaussier.pdf (2001) 88–100

[10] Gibson, J.: The Ecological Approach to Visual Perception. Houghton Mifflin, Boston (1979)

[11] Gaussier, P., Baccon, J., Prepin, K., Nadel, J., Hafemeister, L.: Formalization of recognition, affordances and learning in isolated or interacting animats. In: to appear in SAB04 (From Animal to Animat). (2004)

[12] Canamero, L.: Emotions and adaptation in autonomous agents: A design perspective. Cybernetics and Systems **32** (2001) 507–529

[13] Gaussier, P., Moga, S., Quoy, M., Banquet, J.: From perception-action loops to imitation processes: a bottom-up approach of learning by imitation. Applied Artificial Intelligence **12** (1998) 701–727

[14] Andry, P., Gaussier, P., Moga, S., Banquet, J., Nadel, J.: Learning and communication in imitation: An autonomous robot perspective. IEEE transactions on Systems, Man and Cybernetics, Part A **31** (2001) 431–444

[15] Andry, P., P.Gaussier, Nadel, J.: From sensorimotor coordination to low level imitation. In: Second international workshop on epigenetic robotics. (2002) 7–15

[16] Nadel, J., Revel, A., Andry, P., Gaussier, P.: Toward communication: first imitations in infants, low-functioning children with autism and robots. Interaction Studies **5** (2004) 45–75

[17] Feinman, S.: Social referencing and the social construction of the reality in infancy. Plenum Press, New York (1992)

[18] Scherer, K.: Emotion as a multicomponent process. Rev. Person. Soc. Psychol. **5** (1984) 37–63

[19] Decety, J., Chaminade, T.: Neural correlates of feeling sympathy. Neuropsychologia **42** (2002) 127–138

[20] Sporns, O., Tononi, G.: Classes of networks connectivity and dynamics. Complexity **7** (2002) 28–38

[21] Nagai, Y., Hosoda, Asada, M.: How does an infant acquire the ability of joint attention?: A constructive approach. In: Proceedings of the 3rd International Workshop on Epigenetic Robotics. (2003) 91–98

Maximizing Learning Progress: An Internal Reward System for Development

Frédéric Kaplan and Pierre-Yves Oudeyer

Developmental Robotics Group
Sony Computer Science Laboratory Paris
6 rue Amyot, 75005 Paris, France
{kaplan, py}@csl.sony.fr

Abstract. This chapter presents a generic internal reward system that drives an agent to increase the complexity of its behavior. This reward system does not reinforce a predefined task. Its purpose is to drive the agent to progress in learning given its embodiment and the environment in which it is placed. The dynamics created by such a system are studied first in a simple environment and then in the context of active vision.

1 Introduction

Models of natural or artificial autonomous agents usually imply the existence of primary motivational principles that govern the behavior of the agent. Biologists typically describe most of the processes of metabolical regulations as the results of homeostatic mechanisms maintaining internal "variables" such as hunger, thirst or temperature into desirable limits. Behaviorists have argued that behavior was driven by the search for a set of potential rewards corresponding to primary and conditioned reinforcers [1]. This kind of model has lead to reinforcement learning architectures in which a robot learns to behave in order to receive artificial "rewards"[2]. In return, recent results seem to suggest a convergence between the reinforcement learning theory and neurophysiological data from the midbrain dopamine system in particular [3,4,5].

Rewards have traditionally been viewed as external stimulations provided by the environment or by an experimenter. This chapter discusses the opportunity of considering the existence of an *internal* reward system that would act as a driving force for learning: a kind of "epistemic hunger". Several authors have argued that human behavior is probably driven by a principle of this sort (e.g. Korand Lorenz's "neophily" [6], Csikszentmihalyi's "flow experiences" [7]). But very few precise models exist to explain the mechanisms underlying such a curiosity principle.

In our vision, curiosity is tightly linked with prediction capabilities and learning experiences [8]. "Learning situations" occur when an agent encounters a situation which is not yet entirely predictable based on its current capabilities but learnable using its algorithms for adaptation. In models which view homeostasis as a major drive for living creatures, such kind of situations may be seen as perturbations [9]. If the goal of the agent is to constantly try to minimize its

F. Iida et al. (Eds.): Embodied Artificial Intelligence, LNAI 3139, pp. 259–270, 2004.
© Springer-Verlag Berlin Heidelberg 2004

error in prediction, learning is simply a way to reach again a form of equilibrium state (e.g. [10]). Taking an opposite point of view, several authors have suggested that in order to learn efficiently agents should focus on "novel", "surprising" or "unexpected" situations. This would mean that a "curious" agent should focus on situations for which it does not yet have adequate prediction capabilities (e.g [11] in field of developmental robotics or [12] in the field of "active learning").

The view presented in this chapter is different, but in the line of Schmidhuber's theoretical machine learning work [13]. Curiosity is defined neither as a pressure to minimize errors in prediction, nor as a tendency to focus on the most "surprising" situations, but on the contrary as a drive that pushes the agent to lose interest in both predictable and unpredictable areas, to concentrate on situations that *maximize learning progress.*

The next section presents an engine that can generate behaviors based on this reward principle. As the reward system is generic, this engine can be associated with any input-output device. However the resulting behavior highly depends on the embodiment of the device, that is on the physical structure of the device and on the particular implementation of the prediction systems used by the engine. In order to understand the learning dynamics created by this reward system, this chapter focuses first on a simple embodiment. Experiments with a more complex active vision system are then described.

2   A Generic Engine

2.1   Technical Description

Input, Output, Internal rewards. An agent can be viewed as a plant consisting of an input-output device and an engine controlling it. At any time t, the engine receives a vector $S(t)$ of input signals (either internal or external to the agent) and can send a vector $M(t)$ of control signals corresponding to its actions on the environment or on internal parameters. The set of internal reward received at time t is contained in a vector $R(t)$. The purpose of the engine is to maximize the amount of rewards received in a given time frame (possibly infinite).[1]

The complete situation (sensory-motor and rewards) is summarized in a vector $SMR(t)$. The behavior of the engine consists in determining $M(t)$ based on $S(t)$ and on previous sensory-motor-reward situations $SMR(t-1), SMR(t-2), \ldots$ Given the constraints provided by its embodiment and the environment in which it is placed, the engine develops in an unsupervised manner.[2]

---

[1] In this paper we only consider the case of a reward vector of dimension 1, but we describe the engine in its general form which can deal with more than one reward function.

[2] It is to be understood that, in the present paper, when the expression "sensory-motor" is used the word "motor" does not necessarily entail physical motion. The term "motor" refers more generally to any control signal having a potential effect either on the environment of the agent or on the agent itself: control of physical actuators, activation of sensory devices, change on internal parameters.

An important point is that the engine receives inputs and produces control commands without any a priori information about what they "externally" mean.

Predictors. Predictors form the most important part of the engine. They are responsible to anticipate future sensory-motor evolutions and expected rewards. Their function can be implemented as a single predictor Π that tries to predict future situations.

$$\Pi(SMR(t)) \rightarrow SMR(t+1) \tag{1}$$

However, in practice, it is often more efficient to implement this global predictor through three specialized *prediction devices*: Π_m, Π_s, Π_r. The three devices take the current situation $SMR(t)$ as an input and try to predict the future motor situation $M(t+1)$, the future sensory situation $S(t+1)$ and the future state of the reward vector $R(t+1)$, respectively. At each time step, once $SMR(t)$ is defined, the three devices learn in order to increase their prediction accuracy. Each predictor adapts differently depending on its implementation. The prediction devices can be implemented in different manners, for instance:

- A recurrent Elman neural network with a hidden layer / context layer. Because this network is recurrent it can predict its output based on the value of the sensory-motor vectors several time steps before t [14].
- A prototype-based prediction system that learns prototypic transitions and extrapolates the result for unknown regions.[3]
- A system using Hidden Markov Models [15].
- A mixture of experts like the one described in [16] and [17]

The performances of the prediction devices are crucial for the system, but the architecture of the engine does not assume anything about the kind of devices used. The choice of a particular technique and its implementation are considered to be part of the embodiment of the device.

Reward system. At each time step t, the reward system computes the current values of $R(t)$ based on internal computation on the architecture. The system we describe below is based on "maximizing learning progress". At any time t, the system can evaluate the current error for predicting sensory effect of a given command. It is the distance between the predicted sensory vector and its actual values.

$$\Pi_s(SMR(t-1)) \rightarrow S'(t), e(t) = distance(S'(t), S(t)) \tag{2}$$

---

[3] It takes the form of a set of vectors associating a static sensory-motor context $SMR(t-1)$ with the predicted vector ($M(t),S(t)$ or $R(t)$). New prototypes are regularly learned in order to cover most of the sensory-motor-reward space. Predictions are made by combining the results of the k closest prototypes. k is typically taken as size(SMR(t)) $+1$. This prediction system is faster and more adaptive than the Elman network, but may prove less efficient for complex sensory-motor-reward trajectories.

We define the "learning progress" $p(t)$ as the reduction of the error $e(t)$. In the case on an increase of e(t), progress is zero.

$$p(t) = \begin{cases} e(t-1) - e(t) & : \quad e(t) < e(t-1) \\ 0 & : \quad e(t) \geq e(t-1) \end{cases} \tag{3}$$

In the case when "learning progress" is the only variable to maximize, the vector $R(t)$ is of dimension 1 :

$$R(t) = \{p(t)\} \tag{4}$$

Maximizing learning progress forces the agent to move away from predictable trajectories in order to receive rewards when returning to predicted ones. This is very different from minimizing or maximizing the error $e(t)$. Minimizing the error involves carrying out the actions whose effects are the most easy to predict. This leads to the specialization of the agent into a very small sensory-motor domain, that it will try to master perfectly. With such a reward system, the diversity of the behavior of the agent tends to be rapidly reduced. Maximizing the error involves carrying out the actions whose effects are the most difficult to predict. This can lead to good results in some cases. But in the case where part of the sensory-motor space is very difficult to predict, this strategy is likely to result in destructive learning as the agent will not go back to predictable trajectories. Experiments with these different kind of reward functions are discussed in [8].

Action selection. The action selection module chooses the output commands that are expected to lead to the maximum rewards between t and t+T. Several techniques taken from the reinforcement learning literature can be used to solve these kind of problems[2]. In our system, the process can be separated into four phases:

1. Generation : The module constructs a set of possible commands $\{mi\}$. For some applications this phase can be trivial, but more elaborate computations may be required when dealing with complex actuators. As an example of a simple case: if the current value of an actuator control signal, m_0, is 0.7 then the controller may randomly shift the current value so as to produce candidate values such as 0.55, 0.67, 0.8, 0.75, for m_0.
2. Anticipation : With the help of the predictors, by using the prediction devices in a recurrent manner, the module simulates the possible evolution $\{SMR_{mi}\}$ over T time steps. The module combines the result of both Π_m and Π_s to predict future sensory-motor situations and uses Π_r to predict the evolution of the reward vector $R(t)$.
3. Evaluation : For each evolution $\{SMR_{mi}\}$ an expected reward r_{mi} is computed as the sum of all the future expected rewards.

$$r_{mi}(t) = \sum_{j=t}^{t+T} ||R(j)|| \tag{5}$$

4. Selection : The motor command $\{mi\}$ corresponding to the highest expected reward is chosen.

The action selection module monitors how well the system is behaving by computing the average reward $< ar(t) >$ over K timesteps. If $< ar(t) >$ is below a given threshold η the system acts randomly instead of using the anticipation process. This allows the discovery of opportunities for learning by chance and then to exploit them.

3 The Light Switching System: A Very Simple Embodiment

In order to better understand how the architecture works, this section describes first the behavior of the system for a very simple embodiment. Let's assume that the device is equipped with two sensors. The first one, $pos(t)$ is its position between 0 and 1. The second $light(t)$ can only take two values 0 and 1 and corresponds to the presence or absence of a light in the environment. This light is switched on when the agent occupies a position between 0.89 and 0.91, otherwise it is zero. The system is equipped with a single actuator $nextpos$ corresponding to the next position the system should go to. It tries to maximizes its "learning progress" $p(t)$. A step by step evolution of the system in this simple situation is described in a detailed manner in the appendix.

Π_s, Π_m, Π_r are three prototype-based predictors with a maximum capacity of 500 prototypes. For these simulations $T = 2, K = 1$. To understand the dynamics produced by the engine more easily, we set $\eta = 0$. This means that the system always attempts to use the action selection mechanism to maximize rewards.

Measures. In order to evaluate the behavioral effect of the action selection mechanism based on maximizing learning progress, we need to systematically compare the simulation results with results obtained for an agent that learns choosing random actions. Despite its simplicity, the "random" action strategy can be efficient for learning about unknown environments and discover sensory-motor contingencies.

The first question is whether the system manages to reach its goal : maximizing learning progress. We define the cumulative progress $P(t)$ as the integration over time of $p(t)$.

$$P(t) = \sum_{j=0}^{t} p(t) \tag{6}$$

To evaluate the performance of the action selection mechanism, we define the comparative progress ratio $C_P(t)$ as :

$$C_P(t) = \frac{P_{MAXPROGRESS}(t)}{P_{RANDOM}(t)} \tag{7}$$

In the context of this environment, the second question is whether the action selection mechanism based on progress leads to a different behavior towards the light. We define the number of times the light has been switch on as :

$$L(t) = \sum_{j=0}^{t} light(t) \tag{8}$$

To evaluate the difference in the behavior, we define the comparative ratio :

$$C_L(t) = \frac{L_{MAXPROGRESS}(t)}{L_{RANDOM}(t)} \tag{9}$$

Simulation results. Figure 1 shows the evolution of $C_P(t)$ and $C_L(t)$ for 10 000 time steps. In the beginning of the evolution, the random strategy outperforms the action selection mechanism. In a situation where very few learning situations are present, choosing actions based on anticipated progress is a worse strategy than choosing actions at random.

Fig. 1. Evolution of the comparative progress $C_P(t)$ and comparative light ratio $C_L(t)$ for 10 000 time steps

At $t = 579$, the agent using the action selection mechanism switches the light on by chance. From that moment, its starts progressing rapidly. The strategy outperforms the random one around $t = 1000$ and will continue since then.

The $C_L(t)$ curves follows a similar evolution as $C_P(t)$. Random action choices lead first to more situations where the light is on. But as soon as the possibility to switch on the light is "discovered" by chance by the agent, the action selection mechanism exploits that source of learning opportunities by switching the light on much more often than in the random case.

Such behavior is not complex in itself. It would have been easy to define a system to reward the agent when the light is switched on and off. The external behavior of such an agent may have been similar to the one of this experiment. The difference is that our agent had no a priori bias towards that particular stimulus. In this case, the light switching behavior is an "emergent property" resulting of a generic internal reward principle.

We can define γ, the point of transition where $C_P(t)$ becomes superior to 1. Once γ is reached, the action selection strategy outperforms the random strategy. In this simulation γ corresponds also to the transition when the agent starts switching on the light more often than by random moves.

Figure 1 shows a dramatic increase of $C_P(t)$ just after the "discovery of the light". But after γ, $C_P(t)$ reaches a stationary regime where the progress performances are not very different from the one obtained with the random strategy. A similar pattern is observed with the light ratio $C_L(t)$. The light is maximally switched on just after γ. After this initial drive for learning how to switch the light on or off, the system reaches a kind of *habituation* phase. There are no more opportunities for learning. Progress is not zero because there are still some residual errors in prediction to be improved. But the learning progresses as fast as with the random strategy.

4 Embodiment in an Active Vision System

The embodiment we present in this section shares some similarity with an active vision system described by Marocco and Floreano [18,19]. But the latter uses an evolutionary robotics paradigm: population of robots are evolved and the best individuals are selected according to a predefined fitness function. We use this embodiment in a developmental perspective.

The system is equipped with a squared retina using RxR perceptual cells. It can move this retina in an image and zoom in and out. Based on the zooming factor, the retina averages the color of the image in order to produce a single value for each cell of the retina. With such a system, it is possible to rapidly scan the patterns present in the overall image and zoom in to perceive some details more accurately. The system has to learn how to "act" on the image by moving and zooming the retina in order to get the higher reward as defined by its reward system.

More precisely, for a given image snapshot $I(t)$, the sensory vector $S(t)$ contains the renormalized grayscale value of the RxR pixels of the retina, its current position $(X(t),Y(t))$ and zoom factor $Z(t)$. The motor vector $M(t)$ contains the values for the three possible actions the device can perform: changing the X and Y values and the zooming factor Z.[4]

---

[4] It is not sure that the global position information $X(t)$, $Y(t)$, $Z(t)$ are necessary for the system to work as theoretically they can be deduced from the temporal integration of $D_X(t)$, $D_Y(t)$, $D_Z(t)$. However, in practice, it was difficult for the system to

$$S(t) = \begin{vmatrix} Pix_{1,1}(t) \\ Pix_{2,1}(t) \\ ... \\ Pix_{R,R}(t) \\ X(t) \\ Y(t) \\ Z(t) \end{vmatrix}, M(t) = \begin{vmatrix} D_X(t) \\ D_Y(t) \\ D_Z(t) \end{vmatrix} \tag{10}$$

The system is presented with a white image where a grey circle is drawn in the down right corner (Figure 2). This situation shares a lot of similarities with the light switching environment previously studied: the environment is uniform except in a small zone and at a given time t, the agent only perceives a small part of it. But the sensory-motor know-how to be developed to master the retina is much more complex.

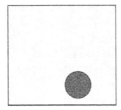

Fig. 2. Image used for the active vision experiment

The system is equipped with a 5x5 retina, so S(t) is of size $5*5+3 = 28$ and SMR(t) of size $28+3+1 = 32$. As in the previous experiments, Π_s, Π_m, Π_r are three prototype-based predictors with a maximum capacity of 500 prototypes, $T = 2$ and $K = 1$. Three simulations where conducted : one with $\eta_1 = 0.01$, one with $\eta_2 = 0$ and one with actions chosen randomly. $C_{P1}(t)$ and $C_{P2}(t)$ are defined as previously. By analogy with the light example, we can define $l(t)$ which has value 1 when the center of retina is inside the gray circle and 0 otherwise. $C_{L1}(t)$ and $C_{L2}(t)$ defined as previously measure the relative focus on that part of the image in comparison with a retina governed by random commands.

Figure 3 shows the evolution of $C_{P1}(t)$, $C_{L1}(t)$, $C_{P2}(t)$ and $C_{L2}(t)$. With $\eta_1 = 0.01$, the system discovers rapidly that to focus on the grey circle lead to more learning progress. $C_{P1}(t)$ and $C_{L1}(t)$ show similar features than the ones observed for the light switching problem : (1) the action selection mechanism outperforms clearly the random strategy, (2) once the "interesting" part of the environment is discovered the agent focuses mainly on it for a while, (3) eventually an habituation phase is observed. However with $\eta_2 = 0$, the performances are worse than random. A study of the trajectory shows that the agent focuses on corner instead of focusing on the grey circle. As the sensory-motor-reward

discover how the "borders" of the image constrained the retina's movements, without using this positional information.

space is larger then in the previous system we are confronted with a classical exploration/exploitation trade-off. An optimal strategy may consist in an adaptive system evaluating $C_P(t)$, increasing η when $C_P(t) < 1$, reducing it when $C_P(t) > 1$.

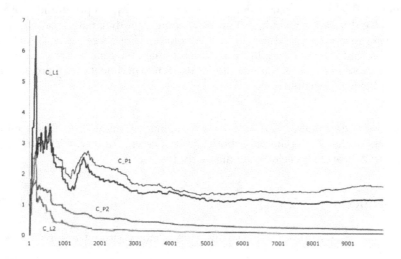

Fig. 3. Evolution of the comparative progress $C_P1(t)(\eta = 0.01)$ and $C_P2(t)(\eta = 0)$ and comparative focus on the grey circle $C_L1(t)(\eta = 0.01)$ and $C_L2(t)(\eta = 0.0)$ for 10 000 time steps

5   Conclusions

An agent motivated by maximizing learning progress constructs its behavior in order to go from unpredictable situations to predictable ones. Instead of focusing on situations that it predicts well (minimizing prediction error) or on situations it does not predict at all (maximizing prediction error) it focuses on the frontier that separates mastered know-how from unmastered know-how. This strategy enables the development of more complex behaviors in a given environment. In the two experiments described in this chapter, the agent discovers a feature of its environment that is only visible when specific actions are performed. As soon as the agent discovers this feature, it tries to learn about the sensory effects of its actions in this context. Once its learning progress ceases to increase, the agent stops focusing on this sensory-motor trajectory as it is now part of well predicted situations. This focusing behavior can be seen as a novel behavior mastered by the agent.

One major challenge for developmental robotics is to build a single architecture that would enable a robot to develop novel behaviors of increasing complexity. Can the same mechanisms enable a robot to autonomously learn how to avoid obstacles, discover how to use objects or structure its interactions with other robots? We believe reward systems like the one presented in this paper can play a key role to build robots capable of open-ended development. However, the reward system in itself is not sufficient. The development of complex behaviors results from the interplay between generic motivational principles, particular embodiment (including a particular physical structure, perceptual and motor apparatus and learning techniques) and environmental dynamics (see for instance the experiments discussed in [20] in this volume). For these reasons, further research in that direction should focus on exploring how generic principles such as the one presented in this chapter can be used in experiments with grounded and situated robotic agents.

Acknowledgements. The authors would like to thank Luc Steels and the members of the developmental robotics team (Verena Hafner, Claire d'Este and Andrew Whyte) for precious comments on this work.

References

1. Skinner, B.: The Behavior of Organisms. Appleton Century Crofs, New York, NY. (1938)
2. Sutton, R., Barto, A.: Reinforcement learning: an introduction. MIT Press, Cambridge, MA. (1998)
3. Montague, P., Dayan, P., Sejnowski, T.: A framework for mesencephalic dopamine systems based on predictive hebbian learning. Journal of Neuroscience **16** (1996) 1936–1947
4. Schultz, W., Dayan, P., Montague, P.: A neural substrate of prediction and reward. Science **275** (1997) 1593–1599
5. Doya, K.: Metalearning and neuromodulation. Neural Networks **15** (2002)
6. Lorenz, K.: Vom Weltbild des Verhaltensforschers. dtv, Munchen (1968)
7. Csikszenthmihalyi, M.: Flow-the psychology of optimal experience. Harper Perennial (1991)
8. Kaplan, F., Oudeyer, P.Y.: Motivational principles for visual know-how development. In Prince, C., Berthouze, L., Kozima, H., Bullock, D., Stojanov, G., Balkenius, C., eds.: Proceedings of the 3rd international workshop on Epigenetic Robotics : Modeling cognitive development in robotic systems. Number 101, Lund University Cognitive Studies (2003) 73–80
9. Varela, F., Thompson, E., Rosch, E.: The embodied mind : Cognitive science and human experience. MIT Press, Cambridge, MA (1991)
10. Andry, P., Gaussier, P., Moga, S., Banquet, J., Nadel, J.: Learning and communication in imitation: an autonomous robot perspective. IEEE Transaction on Systems, Man and Cybernetics, Part A : Systems and Humans **31** (2001) 431–444
11. Huang, X., Weng, J.: Novelty and reinforcement learning in the value system of developmental robots. In: Proceedings of the 2nd international workshop on Epigenetic Robotics - Lund University Cognitive Studies 94. (2002) 47–55
12. Thrun, S.: Exploration in active learning. In Arbib, M., ed.: Handbook of Brain Science and Neural Networks. MIT Press (1995)

13. Schmidhuber, J.: Curious model-building control systems. In: Proceeding International Joint Conference on Neural Networks. Volume 2., Singapore, IEEE (1991) 1458–1463
14. Elman, J.: Finding structure in time. Cognitive Science **14** (1990) 179–211
15. Rabiner, L., Juang, B.: An introduction to hidden markov models. IEEE Acoutics, Speech and Signal Processing Magazine **3** (1986) 4–16
16. Tani, J., Nolfi, S.: Learning to perceive the world as articulated : An approach for hiearchical learning in sensory-motor systems. Neural Network **12** (1999) 1131–1141
17. Jordan, M., Jacobs, R.: Hierarchical mixtures of experts and the em algorithm. Neural Computation **6** (1994) 181–214
18. Kato, T., Floreano, D.: An evolutionary active-vision system. In: Proceedings of the congress on evolutionary computation (CEC01), IEEE Press (2001)
19. Marocco, D., Floreano, D.: Active vision and feature selection in evolutionary behavioral systems. In Hallam, B., Floreano, D., Hallam, J., Hayes, G., Meyer, J.A., eds.: From Animals to Animats 7, Cambridge, MA., MIT Press (2002)
20. Lungarella, M., Berthouze, L.: Robot bouncing : On the interaction between body and environmental dynamics. this volume (2004)

Appendix

This appendix describes a step by step evolution of the system for the light switching problem. In this example the parameter η is set to 0.01. Let's assume that the system has been active for t time steps and that during that period the light stayed at value 0. For the sake of simplicity, we will make the unrealistic assumption that the predictors learn rapidly and "perfectly" based on the situations they encountered. So we will assume that the predictors have learned to predict that $pos(t) = nextpos(t-1)$ and that $light(t) = 0$.

Time t. At time t the situation is the following.

$$e(t-1) = 0, SMR(t-1) = \begin{vmatrix} pos(t) = 0.4 \\ light(t) = 0 \\ nextpos(t) = 0.2 \\ p(t) = 0 \end{vmatrix}, SMR(t) = \begin{vmatrix} pos(t) = 0.2 \\ light(t) = 0 \\ nextpos(t) = ? \\ p(t) = ? \end{vmatrix} \quad (11)$$

The "perfect" predictor for sensory information, Π_s, gave the following predictions:

$$\Pi_s(SMR(t-1)) \rightarrow \begin{vmatrix} position(t) = 0.2 \\ light(t) = 0 \end{vmatrix} \quad (12)$$

So $e(t) = 0$ and $p(t) = e(t-1) - e(t) = 0$.

The role of the action selection module is to determine $nextposition(t)$. The average progress, close to zero, is currently inferior to η. Acting randomly, the system chooses $nextposition(t) = 0.9$. The vector $SMR(t)$ is now completed. The three predictors can learn by comparing $SMR(t)$ with the values predicted based on $SMR(t-1)$.

Time t+1. As the agent moved to 0.9, the light was switched on. Consequently, the situation at time t+1 is the following, and Π_s could not predict that evolution.

$$SMR(t+1) = \begin{vmatrix} pos(t+1) = 0.9 \\ light(t+1) = 1 \\ nextpos(t+1) =? \\ p(t+1) =? \end{vmatrix}, \Pi_s(SMR(t)) \rightarrow \begin{vmatrix} pos(t+1) = 0.9 \\ light(t+1) = 0 \end{vmatrix} \qquad (13)$$

So $e(t+1) = (0+1)/2 = 0.5$ and as progress is negative $p(t+1) = 0$. Next action is random : $nextpos(t+1) = 0.7$. Predictors learn.

Time t+2. Situation at $t+2$ is the following

$$SMR(t+2) = \begin{vmatrix} pos(t+2) = 0.7 \\ light(t+2) = 0 \\ nextpos(t+2) =? \\ p(t+2) =? \end{vmatrix} \qquad (14)$$

Let's assume that the predictor has predicted correctly that the light will be switched off after that move.

$$\Pi_s(SMR(t+1)) \rightarrow \begin{vmatrix} pos(t+2) = 0.7 \\ light(t+2) = 0 \end{vmatrix} \qquad (15)$$

So $e(t+2) = 0$ and $p(t+2) = 0.5 - 0 = 0.5$. Because $p(t+2) > \eta$, next action will be chosen through anticipation. The system creates a set of possible values for $nextpos$ and tries to predict their effects in terms of rewards. If the system only looks one step ahead no reward can be anticipated. But if the system looks at least two steps ahead, it can anticipate that choosing $nextpos$ near 0.9 will lead to a situation similar to the one experienced at time $t + 1$. Using predictor Π_m to simulate what it would do next in such a situation and Π_r to evaluate the associated expected reward a total anticipation of the situation at t+3 is possible.

$$\Pi(\begin{vmatrix} pos(t+2) = 0.7 \\ light(t+2) = 0 \\ nextpos(t+2) = 0.91(tried) \\ p(t+2) = 0.5 \end{vmatrix}) \rightarrow \begin{vmatrix} pos(t+3) = 0.91(\Pi_s) \\ light(t+3) = 1(\Pi_s) \\ nextpos(t+3) = 0.7(\Pi_m) \\ p(t+3) = 0(\Pi_r) \end{vmatrix} \qquad (16)$$

The system can then use Π_r to anticipate the expected reward at $t + 4$ (e.g. $p(t + 4) = 0.5$). Based on this anticipation the system will decide to move to 0.91 in order to experience again the transition that lead to an increase of $p(t)$. This example illustrates one possible way for the system to optimize the learning progress $p(t)$. The variable $p(t)$ is positive when the system moves from an unpredicted situation to a predictable one. This means that the system is rewarded when it "returns" to known situations. But to be "returned", the system must first leave the situations it anticipates well.

You Did It on Purpose!
Towards Intentional Embodied Agents

Bart Jansen, Bart de Boer, and Tony Belpaeme

Artificial Intelligence Lab
Vrije Universiteit Brussel, Pleinlaan 2, 1050 Brussel, Belgium
{bartj,bartb,tony}@arti.vub.ac.be
http://arti.vub.ac.be

Abstract. The paper describes a road-map towards intentional behavior in artificial systems. We catch the developmental path in two dimensions, a social and an intentional dimension. Starting out with a babbling phase, development continues over an exploratory phase without social interactions and a phase in which action-level imitation is used. The pinnacle of development is the intentional imitation of goals. An experiment, together with preliminary results, is presented for each developmental phase.

1 Introduction

In recent years, a lot of attention has gone to the construction of *intentional agents*, *i.e.* agents that *purposefully* interact with their environment and with other agents, *e.g.* [1]. The field has witnessed several *ad hoc* solutions towards implementing intention in agents. Especially in the field of software agents several formalisms have been defined to enable agents to maintain beliefs about actions and their possible outcomes [2]. However, little attention has been paid to the developmental path that agents should take before arriving at full-fledged intentional behavior, and the importance of embodiment and grounding in this all.

In this paper we outline a possible developmental path for mastering the difficult task of intentional imitation. We discern four steps: motor babbling, imitation of gestures, individual exploration and intentional exploration [3]. We argue that embodiment is an important factor in all four steps. To substantiate the theory, we present computer simulations and results for the first two steps and outline possible computer simulations for the last two.

2 Social and Intentional Dimensions of Behavioral Development

When studying behavioral development and learning, it is easy to be baffled by the complexity of the subject. Many different types of behavior are encountered.

F. Iida et al. (Eds.): Embodied Artificial Intelligence, LNAI 3139, pp. 271–277, 2004.
© Springer-Verlag Berlin Heidelberg 2004

One encounters such apparently disparate phenomena as motor babbling, mimesis (when an individual mimics another individual without paying attention to the goal of the behavior), exploration of the world and imitation of object use. All of these phenomena are associated with different stages of development, but at the same time many of them occur in parallel. Although at first sight these phenomena may appear to have little in common, we argue that in fact they can be analyzed in terms of two developmental parameters: a social dimension and an intentional dimension.

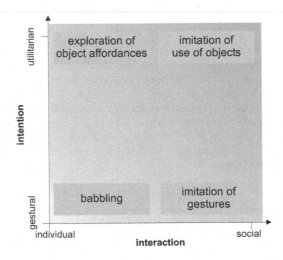

Fig. 1. Social and intentional dimensions of behavioral development.

The social and intentional dimensions can be imagined as forming an abstract two-dimensional space (see figure 1). At the zero ends of both dimensions there is behavior without intention and without social content. As one moves away from zero, behavior becomes more intentional and more social, respectively. Although it is clear that this proposed space is continuous, in the sense that behavior can be more or less social or intentional, for the sake of argument it is useful to define four corners in this space. We argue that the above-mentioned four examples of developmental behavior correspond to the four corners of this space. Babbling and motor babbling fall at the zero-point. These behaviors are neither social nor intentional. They are used for exploring the possible space of movements and articulations and possibly for finding a mapping between visual perception, proprioception and action. Imitation of gestures is social in that it requires at least two agents and in that it requires the imitating agent to map the other agent's actions onto its own. However, by definition, imitation of gestures is not concerned with the intent of the other agent's actions. Agents only mimic each other, this is why this type of imitation is also called mimesis [4]. Exploration of the world, on the other hand is intentional, in that the agent doing the exploration wants to figure out how its actions change the state of the world. As the agent is perfectly able to do this on its own, such behavior is not social. During the exploration phase in which affordances of objects are learned, the environment is of crucial importance. It provides situational constraints to which every action is bound [5]. The agent now does not only learn the properties of its own effectors, it learns the effect of its actions on the environment. In doing this, it acquires a representation of the physical properties of the environment. Finally, the combination of intentional and social behavior can result in imitation of object use, where by definition the agent's aim is to copy the other agent's intentions. Other social and intentional behaviors, such as constructing a theory of mind or trying to manipulate another agent are also possible. The success of an imitative attempt can be measured by observing the behavior of the robots. Measuring the success in constructing a theory of mind for instance requires inspection of the agents' internal memories. Imitation on robots was therefore heavily investigated, although no systems capable of intentional imitation were built.

3 Embodiment and Behavioral Development

Development in animals and humans is for a large part defined by constraints and changes in the body and brain structures and therefore embodiment is obviously important when studying development.

The importance of development for embodied systems is familiar to all researchers in robotics. It is almost always impossible to program a robot (a prototypical example of an embodied computer system) exactly to perform a certain behavior. It turns out that robots perform much better when they are able to adapt their behaviors to at least some extent. A capability of development is therefore a crucial aspect of an embodied system. However, as we will argue below, it turns out that when studying development it is also crucial to take into account the embodiment of the developing agent. Examples of the kinds of development we are referring to are the four different kinds of behavior mentioned above. Studying such behaviors without taking into account that they take place in an agent that has specific sensors and actuators and that has to operate in a specific environment becomes such an abstract exercise that it is almost meaningless. The kinds of behaviors an agent can perform, the events it can detect and the ease with which it can manipulate its environment, all depend on the embodiment of the agent. Therefore the mechanisms necessary for development as well as the developmental trajectories an agent can follow depend on the embodiment. But from a more practical point of view, embodiment is important when studying development on robotic agents. If one wants to train robots through imitation, for example, it is not a priori clear whether the particular embodiment selected is able to imitate a pre-defined set of behaviors. When studying imitation in robotic agents, it is therefore perhaps better to have the agents develop a repertoire of basic behaviors beforehand, for example through imitation games [6,7,8]. Such a repertoire of basic behaviors could then in principle be used to construct more intentional compound behaviors. In any case, the embodiment selected for an agent limits the behaviors it can acquire and the ways in which it can manipulate its environment, and therefore has a strong influence on the agent's development.

4   Experiments

For the four examples of developmental learning specific computer simulations of populations of embodied agents will be proposed. Concrete results are shown for motor babbling experiments and experiments on the imitation of gestures.

Fig. 2. The robot arm and stereovision system simulated in the experiments.

Motor Babbling
An individual agent learns a model of the forward and inverse kinematics of its effector by motor babbling. The agent executes random motor commands and observes how its effector behaves. We have experimented with Locally Weighted Learning [9], combined with selection and organization of stored points. In figure 3 on the left, the average error in the model of the inverse kinematics and the number of points stored are shown. It can clearly be seen that with very few samples good predictions can be made.

Imitation Game
We have investigated how agents can construct a shared repertoire of actions. We propose a multi agent system in which the interactions consist of imitation games. In an imitation game two agents are randomly selected from the population. Both agents are randomly assigned the roles of *initiator* or *imitator*. A single imitation game causes the repertoires of both agents to become more similar. By repeating the game over multiple combinations of randomly selected agents, the repertoires of all agents become shared. In each game, the imitator tries to imitate the action the imitator executed. The initiator decides on the success of the game and sends binary feedback to the imitator, such that the imitator can adapt its repertoire.

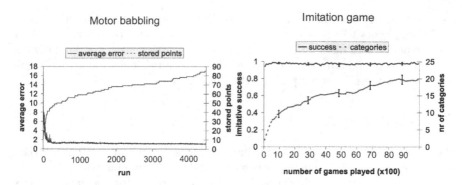

Fig. 3. On the left: the average error in the learnt model of the inverse kinematics and the number of associations stored. On the right: imitative success and number of categories for 10000 imitation games played by two agents.

The imitation game was first proposed by de Boer [6] in the context of vowel systems. He showed how a shared and human-like vowel system can emerge in a population of agents. In later experiments, the imitation game was used to show how a population of robotic agents develops a shared repertoire of actions. In this experiment, agents are equipped with a (simulated) 6-DOF robot arm with a gripper and a (simulated) stereo vision system, see figure 2. The actions performed by the agents are simple gestures with the robot arm. In figure 3 on the right results are shown for a population of two agents playing 10000 imitation games. While the imitative success remains high, the agents succeed in constructing a shared repertoire of actions. Detailed results are available in [7,8].

Individual Exploration

We plan to set up a concrete experiment in which a robotic agent learns how its movements can cause changes in its environment. Starting from the assumption that the agent already has a model of the kinematics of its actuator, we want to investigate how the agent can learn to modify its world. In future experiments, agents will explore their environment by performing random actions. Agents will assess the effect their actions have on the environment. Actions are rewarded proportionally to the observed external effect, such that agents learn to perform actions that require the least movement to cause a maximum effect.

Intentional-Social Exploration

In individual exploration, agents learnt that some of their actions cause changes in the environment and learn to prefer actions that have a maximal effect. This new skill can bootstrap the difficult task of intentional imitation. As the initiator performs a sequence of actions in order to obtain a certain goal, the imitator can deduce the first agent's goal by comparing the observed effect of the initiator's actions with the learnt associations between own actions and environmental effects. In this paradigm of goal-level imitation, both agents will thus learn to pursue the same goal. However, they might be performing different actions, because actions were learnt on an individual basis and because agents can have different embodiments and thus require other actions for accomplishing the same goal.

5   Discussion

When studying the developmental path towards intentional behavior, striking parallels are noted between motor behavior and vocal communication. Both start out with an initial "babbling" phase in which the articulatory devices—effectors or vocal tract—are explored and its inverse-kinematics are learned. Following this phase, agents explore the interaction with the environment. In motor behavior this is expressed as an exploration of the physical properties of objects and their behavior when being manipulated. In vocal communication this is less explicit, but it could exhibit itself for example in the fascination of children with echoing sounds. At the same time, through gestural imitation, the agent can achieve a representation which is influenced by observing other agents. In motor behavior, this is the phase where agents mimic the actions without picking up the purpose of the action. While in vocal communication, the agents acquire a repertoire of sounds from their peers yet without learning the meaning associated with each sound. The last phase is marked by learning to use behavior intentionally. Objects are manipulated with a clear goal in mind, and vocalizations are uttered with goal-like purpose. This marks the phase where actions are connected with meaning.

Acknowledgments. Bart Jansen is sponsored by a grant from the Institute for the Promotion of Innovation by Science and Technology in Flanders (IWT). Tony Belpaeme is a postdoctoral fellow of the Fund for Scientific Research - Flanders (Belgium) (FWO Vlaanderen). Experiments on motor babbling were performed by Steven Teerlinck.

References

1. Malle, B.F., Moses, L.J., Baldwin, D.A.: Intentions and Intentionality, Foundations of Social Cognition. The MIT Press, London (2001)
2. Wooldridge, M., Jennings, N.: Intelligent agents: Theory and practice. Knowledge Engineering Review **2** (1995)
3. Meltzoff, A.M.: Elements of a developmental theory of imitation. In:. Cambridge University Press, Cambridge (2002)
4. Butterworth, G.: Neonatal imitation: existence, mechanisms and motives. In Nadel, J., Butterworth, G., eds.: Imitation in Infancy. Cambridge University Press, Cambridge, UK (1999)
5. Vygostky, L.S.: Play and its role in the mental development of the child. Soviet Psychology **5** (1967) 6–18

6. de Boer, B.: Self organization in vowel systems. Journal of Phonetics **28** (2000) 441–465
7. Jansen, B., de Vylder, B., de Boer, B., Belpaeme, T.: Emerging shared action categories in robotic agents through imitation. In Dautenhahn, K., Nehaniv., C.L., eds.: Proceedings of the Second International Symposium on Imitation in Animals and Artifacts., University of Wales, Aberystwyth (2003) 145–152
8. Jansen, B.: An imitation game for emerging action categories. In Banzhaf, W.e.a., ed.: Proceedings of the 7th European Conference on Artificial Life, Lecture Notes in Artificial Intelligence., Berlin, Springer (2003) 800–809
9. Atkeson, C.G., Moore, A.W., Schaal, S.: Locally weighted learning. Artificial Intelligence Review **11** (1997) 11–73

Towards Imitation Learning from a Viewpoint of an Internal Observer

Yuichiro Yoshikawa[1], Minoru Asada[1,2], and Koh Hosoda[1,2]

[1] Dept. of Adaptive Machine Systems Frontier Research Center,
Graduate School of Engineering, Osaka University
[2] HANDAI Frontier Research Center,
Graduate School of Engineering, Osaka University
{yoshikawa, asada, hosoda}@er.ams.eng.osaka-u.ac.jp

Abstract. How an *internal observer*, that is not given any *a priori* knowledge or interpretation of what its sensors receives, learn to imitate seems a formidable issue from a viewpoint of a constructivist approach towards both establishing the design principle for an intelligent robot and understanding human intelligence. This paper argue two issues towards imitation by an internal observer: one concerns how to construct the self body representation of the robot with vision and proprioception and the other concerns how to construct a mapping of vocalization between agents with different articulation systems. Preliminary results with real robots are given.

1 Introduction

The ability of imitation has been focused in robotics – partially because learning by imitation is regarded as a promising way to accelerate the learning of a robot [1], and partially because it is also one of the most interesting cognitive issues to model human intelligence by a constructivist approach [2]. In the previous work, the designer usually provides specific knowledge to imitate a certain behavior (ex. [3]). However, to model how humans acquire the ability of imitation, we must also address the issue to design a robot that can imitate by itself. In this study, therefore, we assume that the robot is an *internal observer*. An internal observer is defined as an agent that is not given any *a priori* knowledge or interpretation of what the sensory signals it receives mean. By introducing the assumption that the agent can distinguish the different senor modalities, we can start to attack an issue how it can interpret its sensory signals by finding the relationships of its sensory data between different modalities. That is, association of the sensory data from different modalities.

For an internal observer to imitate, constructing a map between the observed demonstrator's body and its own one seems essential for a certain class of imitation where it can imitate through performing the mapped action of the other agent in the coordinate system of its own body based on this map. There are at least two issues to be addressed. First, it must possess the representation of its own body to associate it with other's body. This is not easy because the internal

F. Iida et al. (Eds.): Embodied Artificial Intelligence, LNAI 3139, pp. 278–283, 2004.

observer does not even know what its body is at the beginning. Another concerns how to construct a map between bodies of different agents without *a priori* knowledge about the relationship between them. To learn the map by itself, the robot needs to find references between them. We must consider the fact that the body of the robot is different from the other agent's one.

In the rest of this paper, we will introduce the preliminary results of our study. Concerning acquiring the representation of the body, we address the problem of finding its body in its uninterpreted sensory data [4]. A cross modal map is proposed as the learning structure based on the idea that the invariance in multisensory data represents the body. Concerning the construction of a mapping between different bodies, we address the problem of acquiring common vowels with the caregiver who has different articulation parameters from the robot [5]. We propose a model of interaction that guides a robot to acquire articulation to vocalize.

2 Acquisition of Body Representation [4]

One of the fundamental problems of acquiring a representation of the body is how to find the body in the receptive field without *a priori* knowledge from a viewpoint of external observer. Some previous studies proposed methods by which an agent distinguishes the body of the other agent and its own one based on the correlation between its motion and the motion-induced optical flow (e.g. [6,7]). However, the agent could not distinguish its body from the environment without *a priori* knowledge how its motor system affects its vision. Although another study proposed a method by which an agent finds the boundary of its tactile sensor in the vision based on experience of collision [8], the agent needs to be taught which object in the vision collides with its body.

Sensation of its body is considered to be invariant with its posture. For example, when it fixates one object in the environment, the view changes depending on the environmental changes. However, when it fixates its body, the view is independent of the environment. Therefore, it is suggested that such invariance in multimodal sensors can be used to define its body. The robot can find the invariance through the experiences of taking various postures.

As a structure to find the invariance in multiple modalities, we introduce a fully-connected network called *cross modal map* (see Fig. 1(a)). A cross modal map consists of various sensor nodes that are hardwired to real sensors and are activated when the hardwired sensors receive something in their receptive fields. After Hebbian learning, only the weights between the nodes that are simultaneously activated during a certain period of time increase. Since the same pair of nodes are considered to be simultaneously activated in the sensation of self-body, the connections which have large synaptic weights are regarded to represent the body.

2.1 Experiment

A preliminary experiment to learn to represent the body surfaces of the robot by the cross modal mapping between 15 nodes for binocular vision (disparity at the center region of the left camera) and 20×15 nodes for proprioception (joint angles) was performed. Fig. 1 (b) is a section of the acquired cross modal map in which the arms have a certain postural configuration after about six minutes learning. During the learning process, the robot keeps changing its posture at random. It shows which disparity node has the largest connection with which posture node as a function of the disparity with respect to pan and tilt angles of the camera head. The shape of the function resembles an egocentric view of the robot (see Fig. 1 (c)). The fixation areas of which disparity node have strong connections (large weights) to the posture nodes were parts of the robot body. Therefore, the robot succeed in learning the cross modal map that represents the body surface of the learner.

Since the sensors of the robot are embedded on its own rigid body, the sensation of self body is constrained to be invariant with its proprioception. However, by using the representation of the invariance, the robot can only judge whether the fixated point in the vision is its body or not. We should extend the proposed method for the concept of body part. Then, we should address many issues such as representing kinematics/dynamics, representing the reachable region by the robot movements, and the establishment of the correspondence between its own body and the other's.

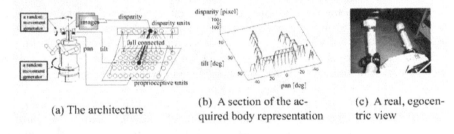

(a) The architecture

(b) A section of the acquired body representation

(c) A real, egocentric view

Fig. 1. The architecture of the robot with a cross-modal map (a), a section of the acquired cross modal map (b) and an example of the egocentric view of the robot (c)

3   Acquisition of Common Vowels [5]

Infants, who are internal observers, seem to acquire the phonemes of adults without *a priori* knowledge about the correspondence between its vocalization and the phonemes. Previous studies showed that computer simulated agents with a vocal tract and cochlea can acquire shared vowels in population by self-organization through interaction with other agents [9,10]. Although they didn't assume *a priori* knowledge about vowel, there was an assumption that the agents can reproduce the similar sounds of other agents' so that "imitation game [9]" or "magnet effect [10]" makes self-organized vowels shared in population. However, infants face with more difficult situations. First, they cannot reproduce the caregiver's utterances as they are because their vocalization system is not mature. Furthermore, even if they can imitate the adult phoneme, they perceive the reproduced sounds differently from the caregiver's original sounds because the sound wave of the former travels inside the body to the infant's auditory sensors. In this case, imitation cannot be equated with raw sensory similarity. To take infant's immaturity into account for modeling the infant's acquisition process of vowels, we use a robot that consists of an artificial articulatory system with a 5-DoFs mechanical system that can deform a silicon-made vocal tract connected to an artificial larynx (see Fig. 2). It vocalizes some sounds which can be interpreted as human vowels but are different from the human vowels from a viewpoint of low-level signal similarity.

It is reported that maternal imitation of a three-month-old infant's cooing (i.e., parrot-like utterances) increases the vocalization rates [11], and the infant's speech-like cooing tends to lead the mother's utterances [12]. Based on these observations, we conjecture that the caregiver's imitation of the robot's vocalizations plays an important role in the vowel acquisition process – in other words, a regular reaction (a parrot-like behavior), which can be regarded as action invariance, make it possible to acquire vowels instead of actions that produce similar sensory information. As a preliminary, constructive model of our conjecture, we design a random articulation mechanism and embed it in the robot so that an interaction can emerge between the robot and the caregiver who produces its own corresponding vowel when the robot's articulated sounds can be heard as the vowel.

The learning mechanism consists of auditory and articulation layers and connections between them. The auditory layer clusters formants (i.e., sound features) of the caregiver by self-organization while the articulation layer clusters its own articulation parameters. The connections between them are updated according to Hebbian learning. The robot learns through interaction to match its articulation with audition, that is, it acquires the vowel sounds of the caregiver. However, interactions may connect multiple articulation units with a corresponding vowel since the caregiver will interpret some vocalizations caused by different articulations as the same vowel. To match a listened vowel with a unique articulation, we introduce *subjective criteria*, that are evaluated only in terms of the robot's state, into the learning rule — that is, the articulation vectors with less torque and less intensity of deformation changes obtains stronger connection from auditory layer and vice versa.

3.1   Experiment

We examined whether the robot can acquire Japanese vowel sounds by interacting with a caregiver. After the robot vocalizes by a random articulation vector, the human caregiver determines whether the robot's vowel corresponds to the Japanese vowel and utters the corresponding vowel. The robot calculates formants of the caregiver's vocalization and updates connections between nodes to represent the caregiver's utterances and ones to represent the robot's articulation by Hebbian learning with and without subjective criteria. Each element of a code vectors in the articulation layer is quantized into five levels; these elements are

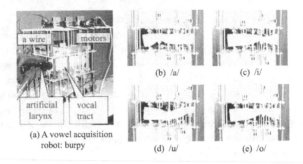

Fig. 2. The appearance of the test-bed robot (a) and the acquired vowels (b)–(e)

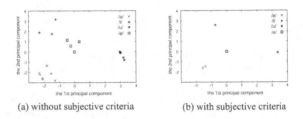

Fig. 3. The acquired clusters without subjective criteria (a) and with it (b)

the motor commands of the random articulation mechanism. Fig.2(b)–(e) shows the acquired articulations. The vocalized sound produced by these articulations can be interpreted as being Japanese vowels.

We observed which units in the articulation layer are activated by the propagation of the activation in the auditory layer when the caregiver utters one of vowels. The activated unit in the articulation layer can be regarded as the matched vowels with the caregiver's one. Fig. 3(a) shows the distribution of the matched articulation acquired by the normal learning rule without subjective criteria, while Fig. 3(b) shows one by the learning rule with subjective criteria. We can see that fewer articulations are selected in the learning with subjective criteria. Therefore, we confirmed that the subjective criteria decreased the number of units in the articulation layer that are activated by the auditory layer. The selected articulation were more facile to articulate.

4   Conclusion

As a preliminary work on understanding the mechanism of imitation by an internal observer, we studied the issues of acquiring the vowel sounds of a caregiver and acquiring a body representation based on constructing mappings between different modalities. Although the robots explored at random to construct the mappings in the both proposed model, they had better utilize their developing mappings to accelerate the learning process. Furthermore, they should learn to use the acquired mappings toward various cognitive functions. Therefore, how to motivate the robot to learn and use mapping is one of our future topic.

Acknowledgment. This study is partially supported by "the Advanced and Innovational Research program in Life Sciences" and "The 21st Century COE Program (Project: Center of Excellence for Advanced Structural and Functional Materials Design)" both of which are programs of the Ministry of Education, Culture, Sports, Science and Technology of Japan, and also by Research Fellowships for Young Scientists, a program of Japan Society for Promotion of Science.

References

1. Schaal, S.: Is imitation learning the route to humanoid robots? Trends in Cognitive Science **3** (1999) 233–242
2. Asada, M., MacDorman, K.F., Ishiguro, H., Kuniyoshi, Y.: Cognitive developmental robotics as a new paradigm for the design of humanoid robots. Robotics and Autonomous System **37** (2001) 185–193
3. Kuniyoshi, Y., Inaba, M., Inoue, H.: Learning by watching: Extracting reusable task knowledge from visual observation of human performance. IEEE Transaction on R&A **10** (1994) 799–821
4. Yoshikawa, Y., Hosoda, K., Asada, M.: Does the invariance in multi-modalities represent the body scheme? - a case study with vision and proprioception -. In: Proc. of the 2nd Intl. Symp. on Adaptive Motion of Animals and Machines. (2003)
5. Yoshikawa, Y., Asada, M., Hosoda, K., Koga, J.: A constructive approach to infant's vowel acquisition through mother-infant interaction. Connection Science **15** (2003) 245–258
6. Fitzpatrick, P., Metta, G.: Toward manipulation-driven vision. In: Proc. of the IROS'02. (2002) 43–48
7. Asada, M., Uchibe, E., Hosoda, K.: Cooperative behavior acquisition for mobile robots in dynamically changing real worlds via vision-based reinforcement learning and development. Artificial Intelligence **110** (1999) 275–292
8. MacDorman, K.F., Tatani, K., Miyazaki, Y., Koeda, M.: Proto-sysmbol emergence. In: Proc. of the Intl. Conf. on Intelligent Robot and Systems. (2000) 1619–1625
9. de Boer, B.: Self-organization in vowel systems. J. of Phonetics **28** (2000) 441–465
10. Oudeyer, P.Y.: Phonemic coding might result from sensory-motor coupling dynamics. In: Proc. of the 7th intl. conf. on simulation of adaptive behavior. (2002)
11. Peláez-Nogueras, M., Gewirtz, J.L., Markham, M.M.: Infant vocalizations are confitioned both by maternal imitation and motherese speech. Infant behavior and development **19** (1996) 670
12. Masataka, N., Bloom, K.: Accoustic properties that determine adult's preference for 3-month-old infant vocalization. Infant Behavior and Development **17** (1994) 461–464

On Evolutionary Design, Embodiment, and Artificial Regulatory Networks

Wolfgang Banzhaf

Department of Computer Science, Memorial University of Newfoundland
St. John's, NL, A1B 3X5, CANADA
banzhaf@cs.mun.ca

Abstract. In this contribution we consider the idea that successful evolutionary design is best achieved in a networked system. We exemplify this thought by a discussion of artificial regulatory networks, a recently devised method to model natural genome-protein interactions. It is argued that emergent phenomena in nature require the existence of networks in order to become permanent.

1 Introduction

Michael Conrad [1] is often cited with the following: „*In conventional design the vast majority of interactions that could possibly contribute to the problem are deliberately excluded*". As designers of system we often lean toward the easiest solution: Divide and conquer. I.e., we design a system using components proven to function as specified, with each of these components in turn being designed by the same process, but for a particular sub-task. Whereas there is nothing wrong with such a design methodology, the question is whether it will scale up. By scaling-up I mean whether it would be possible, using such a method, to design a system with, say, human-like complexity and sophistication. Essentially we are asking whether a complexity which rivals that of Life's creatures can be designed and constructed in this way. It might be conjectured, that this will not be possible [2].

Now that life has already entered the scene, we can put forward a different thesis: Life-like performance and complexity in the human (artificially designed) world will only be possible if we take inspiration from Biology. Alternatively, human-designed systems will unintentionally develop into life-like systems. The essence of this idea of bio-inspiration is emergence (of functionality) through (possibly unforeseen) interactions among components. Thus, instead of isolating the sub-parts of our systems in order to get "clean" functionality, we should rather count on the interactions for securing the functionality.

In order to stabilize emergent phenomena Nature uses networks. Networks are able to capture the interactions (links) of components (nodes), and through multiple connections from each component, become less prone to failures in components. This way networks allow the emergent phenomenon to embody itself (as the network). Examples from the natural world (including human activity) are

F. Iida et al. (Eds.): Embodied Artificial Intelligence, LNAI 3139, pp. 284–292, 2004.
© Springer-Verlag Berlin Heidelberg 2004

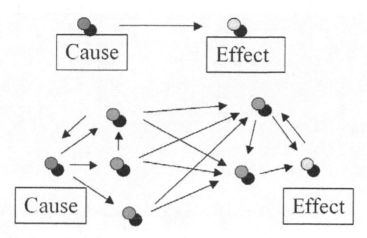

Fig. 1. Networks distribute the connection between cause and effect, and produce fault-tolerant mechanism.

- Elementary particles,
- Chemical reactions,
- Regulatory networks,
- Social interactions,

just to name a few. Elementary particles could be considered nodes in a network of elementary particle interactions, molecules could be considered nodes of a chemical reaction network, genes could be considered in the same way, interacting via their proteins and regulatory sites. Organisms could be considered nodes of a social network with communication links providing the edges of social interaction. What makes networks so fascinating and, at the same time so difficult to analyze, is their effect on simple cause-effect relations: They basically dissolve simple relations between causes and effects in favor of highly distributed networks of partial causes and partial effects. Figure 1 tries to sketch the situation: Assuming outgoing edges as causes and incoming edges as effects, one can see, that a simple relation of cause and effect could be substituted by a network. Nodes collect effects from incoming nodes and distribute there causes over outgoing edges. As a result, if analyzing form the point of view of the original nodes, it is difficult to understand how cause and effect nodes are actually connected. Natural evolution is not they only mechanism finding it useful to apply these systems. A single link (or more than one) can be broken without interrupting the cause-effect chain. The signals will simply flow via other edges. Thus networks provide a highly fault-tolerant environment for signal transfer.

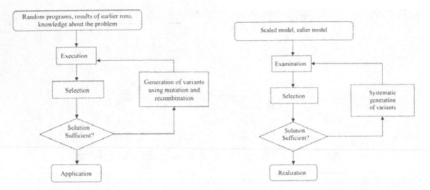

Fig. 2. Comparison of human and natural (evolutionary) design process

2 The Human Design Process Versus the Evolutionary Design Process

In this contribution, we are concerned with the design of systems, i.e. the intentional production of effects or function. Because networks are acting in a highly non-linear fashion and are difficult to analyze, evolutionary approaches to the design of those systems are considered. Here it might help to compare the human and the natural design process in more detail (see Fig. 2).

Traditional design consists of the application of complex principles and rules. It is usually a TOP-DOWN approach which begins with a high-level specification of the problem and moves down through a hierarchy of refinements until realization is reached. Only the best design is realized and closely examined for weaknesses. These weaknesses are then addressed in a separate step and weeded out.

Evolutionary design, on the other hand, consists of an often random combination of a large number of structural elements. If not random, the combination follows simple principles. It is a BOTTOM-UP approach which often passes through a more or less complex developmental process. In order to work properly, a multitude of designs must be examined by Nature. Examination takes the form of tests under "real" conditions.

The difference between the two design methods boils down mostly to cost considerations. If humans had the same cost structure as nature, we'd probably embark on an evolutionary design path towards new products and systems.

Natural design processes are said to be non-intentional, because no overarching plan to achieve certain functions can be identified. If something new is achieved, it is often by way of exaptation [3], a discovery process that works by exploiting side-effects to other functionally important (and selected for) features. In order for this process to be effective, functions should not be isolated from each other. In other words, a single element should have a potential for multiple functions and the partition of functions between components should be more "fluid" than we usually would tolerate. There should also be no boundaries between subsystems, at least no tight boundaries, because communication between subsystems is essential in order to benefit from discovered side-effects. All of this naturally connects to what has been said about networks.

It is also interesting to note, that living organisms are usually generated through a developmental process, i.e. a process of building while working. Again, networks are able to accommodate such a requirement, by providing functions while new nodes and connections are added in.

3 Approaches to Construction

Putting together an entity with of the order of
- 1000 parts   is possible, but difficult
- 1,000,000 parts   is possible in principle, but very difficult
- 1,000,000,000 parts   is impossible with conventional methods.

Humans construct machines by producing parts, putting them together and turning them on. Each of the parts has its independent existence and can be manufactured in isolation from other parts. An overall plan will make sure that the correct order of construction is obeyed, leading to an ultimately functional device.

This is fundamentally different from how Nature constructs systems. No fixed order of events in the construction of a system can be obeyed, due to the stochastic and distributed nature of interactions. Also, parts cannot be produced in isolation; neither can they be produced from a master plan. Finally, living organisms cannot be "turned on" once enough components are assembled. Rather, even a rudimentary system must live from the very beginning in order to be able to continue to live.

Besides problems of controllability of spatial and temporal flows, the sheer number of elements needed to build a living organism is substantial. This is one of the most daunting problems Nature has faced when designing and constructing organisms. Other methods of construction than those we apply in machine construction are needed if a working entity should result from the construction process. These new methods involve growing an entity from a single plan. But instead of having this single plan be always accessed from multiple sites, the plan itself becomes part of what is being built, by integrating it into the parts and subsystems being constructed.

The information dilemma of Nature's evolutionary design is substantial, and can be summarized with the following questions [4]:
- How to instruct bodies with so few genes?
- How to program brains for so many situations?

Nature's answer to the first question posed was to invent a developmental process of construction. Nature's answer to the second, more detailed question was to use an adaptive process which extends development from the ontogenetic level.

In order to appreciate the challenge Nature has been facing, I'll give a few numbers:

Take a single-cell organism like *E.Coli*. It has
- 300 x 10^6 molecules (excluding water)
- 3250 different varieties (proteins, mRNA, tRNA, DNA, lipids,)
- 4.6 x 10^6 base pair genome = 6 Mbits of information
- 4,300 protein coding genes (88 % of genome)
- 11 % of genome contains regulatory information.

If we want to go further and have a look at a multi-cellular organism like *H.Sapiens*:
- 50 x 10^12 cells
- Each cell of about the complexity of an *E.Coli*
- 3 x 10^9 base pairs = 4 Gbit of information
- 40,000 – 100,000 genes

The human brain consists of
- 10^9 neurons
- 1,000 – 10,000 synaptic connections per neuron
- 300 – 1,000 vesicles per synapse

As mentioned, the response of Nature was to invent development, a process, by which
- the required complexity is grown
- environmental complexity is channeled into the developing phenotype, i.e. the genotype only directs the assembly
- exploitation of side effects (through evolution) is possible
- open-ended evolution is brought about through adding layers of complexity
- the generation of modular structure comes for free due to its recursive nature
- coordination of cells is achieved via a chemical cell dialogue
- a switch into a mode of self-maintenance can be made after maturity has been reached
- fitness tests are punctual
- time and dynamics play an essential role
- and organism grows from a 1-cell stage which allows for sexuality.

Many of these features can be explored in computer models (see for example [5]), and a whole new area, computational development is presently forming itself. Before moving any further, we should check with biologists about what development is in their mind. I'd like to adopt the following definition of development for the purposes of this paper.
- Development is a differential transcription (and translation) process of genes in different cells and tissues at different times and with different rates.
- Each step in this sequence is ultimately initiated by the transcription and translation of the previous step.
- Diversity of body plans in all organisms is caused by
    - Interaction between gene products
    - Shifts of timing of gene expression: Heterochrony
    - Shifts in location of gene expression: Differentiation
- Implementation of development is realized using regulatory networks.

This last point is most notable, as it underlines the earlier reference to networks as the means for anchoring emergent phenomena. A closer look at the mechanisms of development and especially the role of genes in the process reveals:

- Genotypes are different from phenotypes
- Genes orchestrate the interaction of molecules (by regulation)
- Most molecules are available from the environment
- Some molecules are produced by genes
- Gene products are most often used as "amplifiers" injected at crucial branching points

The interface between genomes and bodies, and thus the implementation method for embodiment is provided by regulatory networks. They are the means of Nature to stabilize emergent phenomena of construction.   How does it work? In order to explore this question, we have suggested a model for artificial regulatory networks that builds on an earlier model proposed by T. Reil [6].

4 An Artificial Regulatory Network

The model (see [7, 8] for details) consists of a genome of bits generated by a random process. This process comes in two alternative implementations, one being a simple seeding of the bits by a process randomly determining the value in each bit position. The other process is more sophisticated and consists of 2 different phases that are iteratively applied, until the prescribed length of the genome is reached: Starting from a small genome kernel, again seeded by randomly choosing values, it loops through successive stages of duplication and divergent mutation until final length is reached.

In a second process, meaning is ascribed to bits of the genome. Figure 3 describes the situation. By scanning the genome, certain bit patterns are isolated which consist of sub-strings of small size. These sub-strings are called promoters, and they determine the open reading frame of genes. To make things simple, genes are of fixed length, and thus a fixed number of bits subsequent to the promoter are expressed by applying a genotype-phenotype mapping. After mapping, which results in another bit pattern of fixed length, a protein has been produced. The main feature of this protein is that it is mobile. Upon it wandering around it might encounter other proteins, or it might hit the genome at any place of chance. Depending on the pattern match between the protein and the genome at the position of encounter, an interaction occurs in the following way: The protein will attach at the genome, and will detach again after some time. The time of attachment will be all the longer, the better the (complementary) match between protein and genome is.

A third feature needs to be mentioned to understand what is happening in such a system: Upstream from the promoter site of a gene (we envisage only very few genes relative to the number of bits on a bit string genome), there are special sites called enhancer and inhibitor sites. Occupation of these sites with proteins will have a profound effect on the efficiency of expression of the corresponding gene. The effect will depend in a nonlinear way on the matching between genome and proteins trying

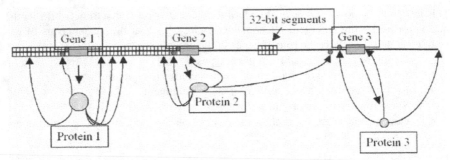

Fig. 3. The artificial regulatory network. A strand of bits, the genome contains short sequences signaling the beginning of a gene. Genes are expressed into proteins which subsequently can wander around and attach to the genome, specifically at regulatory sites upstream from genes, where they influence the rate of expression of other genes. Attachment is controlled in turn by the degree of complementarity between pattern on the genome and pattern on the protein.

to attach. The result of this process, in connection with a time-scale for production of new proteins via genotype-phenotype mapping is that particular inhibitor / enhancer sites are occupied more often and thus have more influence on the expression of the corresponding gene than others. Readers interested in details and quantitative considerations need to compare recent publications [7, 8].

It can be said, that the bit genome discussed and its gene products constitute an artificial regulatory network: The amount of each protein is determined by the matching between proteins and its regulatory site on the genome. We can imagine genes to be nodes in a network, with proteins building the links between those nodes, and the weight on links being the interaction strength determined by the degree of matching between a protein and the regulatory site of a gene.

Figure 4-6 show images of networks resulting from the second generation process (duplication and divergence) mentioned above at various stages of resolution. Figures have been drawn among network nodes that exceed a certain interaction threshold only. The lower the threshold, the more nodes (genes) come into the play, and the more intricate the connection pattern becomes. It is interesting – though not surprising - to see that above a certain threshold the network decays into unconnected components of smaller and smaller size.

At this point, my earlier statement regarding adaptation using side-effects (exaptation) probably becomes clearer: The strongest interactions basically show a network of unconnected components, mostly two genes interacting with each other. Suppose for the moment this would be all that is there. Thus, a modular structure is present in which each pair of genes can be put to independent, yet functionally similar use by evolution.

Lowering the threshold only a bit shows a different picture: New nodes come into play that did not have a role at the former threshold level. And there is a second effect: Independent modules become connected into larger units, thus there is crosstalk between modules. Both effects can be molded by evolution in an arbitrary way. If

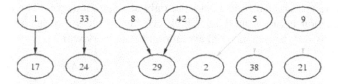

Fig. 4. ARN, generated by a process of duplication and divergence events. High threshold of matching between nodes required. Network decays into very simple modules of 2 to 3 nodes.

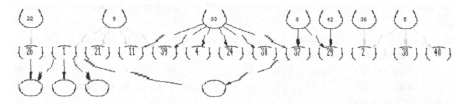

Fig. 5. ARN, generated by a process of duplication and divergence events. Lower threshold of matching between nodes required. Network contains both connected and non-connected components.

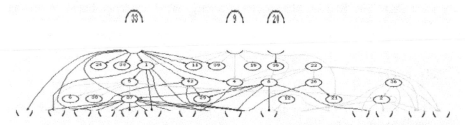

Fig. 6. ARN, generated by a process of duplication and divergence events. Low threshold of matching between nodes required. Network is fully connected and shows complex organization.

there is a need for new function one of the new genes in the game could be assigned to such use. If, on the other hand, it should turn out that crosstalk between "modules" has a beneficial effect on the overall system, this crosstalk could be elevated by increasing the interaction strength between the participating modules. This is the stuff evolution likes the most: Rich behavior yet smooth transitions between alternatives.

How would "molding" by evolution actually work? Simply by mutating the regulatory sites upstream from a gene, the interaction with particular proteins (and thus other genes) can be strengthened. A single bit flip would already elevate (or decrease) an interaction one step, allowing this interaction to become visible in a picture drawn again after the mutation. Possible side-effects of the mutation notwithstanding, very smooth transitions between network configurations are realized.

What has been numerously stated in regard to the evolution of modularity has probably become clearer with our example of an artificial regulatory network: Modules in nature are isolated from each other only to a certain degree. There is always a weak interaction between modules which effectively blurs the distinction between them and provides the rich material evolution likes to work with.

5 Embodiment

There are various notions of embodiment. In this contribution I have tried to argue that emergent phenomena can be "embodied" in networks which in turn are subject to evolutionary forces of variation and selection. If we step back a bit and look at bodies in the literal sense, we might adopt the same perspective: Bodies are so important for active entities, for adaptation, learning and intelligence, because bodies allow the environment to network with the system. I.e. bodies at least partially remove the isolation of an otherwise (machine-like) entity. Trying to achieve intelligent functions without this "crosstalk" between bodies and the environment is a typical human enterprise bound to fail.

Acknowledgments. I gratefully acknowledge discussions and joint work with Julian Miller on aspects of development and evolution. Discussions and joint work with my student P.Dwight Kuo at Memorial University have also contributed to posing questions raised here.

References

1. M. Conrad, Adaptability: The significance of variability from molecule to ecosystem. Plenum Press, New York, 1983.
2. R. Rosen, *Life Itself: A comprehensive inquiry into the nature, origin and fabrication of life.* Columbia University Press, New York,1991.
3. S.J. Gould, *The Structure of Evolutionary Thought.* Harvard-Belknap Press, Cambridge, 2003.
4. W. Banzhaf and J. Miller, *The challenge of complexity,* in A. Menon (Ed), Hilbert Challenges to Evolutionary Computing, Kluwer, Norwell, 2004.
5. J. Miller and W. Banzhaf , *Evolving the program for a cell: From French Flags to Boolean Circuits,* in S. Kumar and P. Bentley (Eds), On Growth, Form and Computers, Academic Press, New York, 2003.
6. T. Reil, *Dynamics of Gene Expression in an Artificial Genome,* in D. Floreano, J.-D. Nicoud, F. Mondada (Eds), Advances in Artificial Life, 5th European Conference ECAL-1999, Springer, Berlin, 1999.
7. W. Banzhaf, Artificial *Regulatory Networks and Genetic Programming,* in R. Riolo, B. Worzel (Eds), Genetic Programming in Theory and Application, Kluwer, Norwell, 2003.
8. W. Banzhaf, *On the dynamics of an artificial regulatory networks,* in W. Banzhaf, T. Christaller, P. Dittrich, J. Kim, J. Ziegler (Eds), Advances in Artificial Life, 7th European Conference ECAL-2003, Springer, Berlin, 2003.

Evolution of Embodied Intelligence

Dario Floreano, Francesco Mondada, Andres Perez-Uribe, and Daniel Roggen

Autonomous Systems Laboratory (ASL)
Institute of Systems Engineering (I2S)
Swiss Federal Institute of Technology (EPFL), Lausanne, Switzerland
http://asl.epfl.ch
dario.floreano@epfl.ch

Abstract. We provide an overview of the evolutionary approach to the emergence of artificial intelligence in embodied behavioral agents. This approach, also known as Evolutionary Robotics, builds and capitalizes upon the interactions between the embodied agent and its environment. Although we cover research carried out in several laboratories around the world, the choice of topics and approaches is based on work carried out at EPFL. We describe a large number of experiments including evolution of single robots in environments of increasing complexity, competitive and cooperative evolution, evolution of vision-based systems, evolution of learning, and evolution of electronics and morphologies for autonomous robots.

1 Introduction

For hundreds of years mankind has been fascinated with machines that display life-like appearance and behaviour. The early robots of the 19th century were anthropomorphic mechanical devices composed of gears and springs that would precisely repeat a pre-determined sequence of movements. Although a dramatic improvement in robotics took place during the 20th century with the development of electronics, computer technology, and artificial sensors, most of today robots used in factory floors are not significantly different from ancient automatic devices because they are still programmed to precisely execute a pre-defined series of actions. Are these machines intelligent? In our opinion they are not; they simply reflect the intelligence of the engineers that designed and programmed them. In the early 90's, we and other researchers started to address this issue by letting robots evolve, self-organise, and adapt to their environment in order to survive and reproduce, just like all life forms on Earth have done and keep doing. The name *Evolutionary Robotics* was coined to define the collective effort of engineers, biologists, and cognitive scientists to develop artificial robotic life forms that display the ability to evolve and adapt autonomously to their environment.

Within this perspective, artificial intelligence is a continuous and open-ended process that capitalises on physical interactions between the agent and its environment without human intervention. Embodiment does not only provide realism and semantic grounding to intelligent artefacts. It also provides opportunities that are unconceivable

F. Iida et al. (Eds.): Embodied Artificial Intelligence, LNAI 3139, pp. 293–311, 2004.

for bodiless systems. Embodied systems can tap upon a virtually infinite range of sensory cues and actions available in the physical world. Given the limits of their processing and behavioural abilities, they can be opportunistic and select only those sensory cues and actions that are necessary to carry on with the business of survival and reproduction.   For example, ants can build magnificent nests with differentiated space, climate control, and air filtering. They do so without resorting to a plan, but by executing an evolved set of simple, but highly specific, sensory-triggered actions. In embodied systems, computation, representation, and memory can be partially outsourced to the physical laws and material persistence of the world. Consider for example the task of goal-directed navigation. One option to achieve that task is to build, store, and use an internal model of the entire environment. Another option is to select and associate simple sensory cues with sequences of actions that will lead from one cue to the next until the goal is reached. Whereas man-made intelligent systems tend to use the first option, there is mounting evidence that animals (at least simple ones) exploit the second option.

In this chapter we will give an overview of some milestone experiments in evolution of physical robots and describe some examples of the intelligence that these robots develop.

2   Evolutionary Robotics

The possibility of evolving artificial creatures through an evolutionary process had already been evoked in 1984 by neurophysiologist Valentino Braitenberg in his truly inspiring booklet "Vehicles. Experiments in Synthetic Psychology". Braitenberg proposed a thought-experiment where one builds a number of simple wheeled robots with different sensors variously connected through electrical wires and other electronic paraphernalia to the motors driving the wheels. When these robots are put on the surface of a table, they will begin to display behaviours such as going straight, approaching light sources, pausing for some time and then rushing away, etc. Of course, some of these robots will fall off the table. All one needs to do is continuously pick a robot from the tabletop, build another robot just like one on the table, and add the new robot to the tabletop. If one wants to maintain a constant number of robots on the table, it is necessary to copy-build one robot for every robot that falls from the table. During the process of building a copy of robots, one will inevitably make some small mistakes, such as inverting the polarity of an electrical connection or using a different resistance. Those mistaken copies that are lucky enough to remain longer on the tabletop will have a high number of descendants, whereas those that fall off the table will disappear for ever from the population. Furthermore, some of the mistaken copies may display new behaviours and have higher chance of remaining for very long time on the tabletop. You will by now realise that the creation of new designs and improvements through a process of selective copy with random errors without the effort of a conscious designer was already proposed by Darwin to explain the evolution of biological life on Earth.

However, the dominant view by mainstream engineers that robots were mathematical machines designed and programmed for precise tasks, along with the technology available at that time, delayed the realisation of the first experiments in Evolutionary Robotics for almost ten years. In the spring of 1994 our team at EPFL (Swiss

Federal Institute of Technology in Lausanne) [8] and a team at the University of Sussex in Brighton [15] reported the first successful cases where robots evolved with minimal human intervention and developed neural circuits allowing them to autonomously move in real environments. The two teams were driven by similar motivations. On the one hand, we felt that a designer approach to robotics was inadequate to cope with the complexity of the interactions between the robot and its physical environment as well as with the control circuitry required for such interactions. Therefore, we decided to tackle the problem by letting these complex interactions guide the evolutionary development of robot brains subjected to certain selection criteria (technically known as fitness functions), instead of attempting to formalise the interactions and then designing the robot brains. On the other hand, we thought that by letting robots autonomously interact with the environment, evolution would exploit the complexities of the physical interactions to develop much simpler neural circuits than those typically conceived by engineers who use formal analysis methods. We had plenty of examples from nature where simple neural circuits were responsible for apparently very complex behaviours. Ultimately, we thought that Evolutionary Robotics would not only discover new forms of autonomous intelligence, but also generate solutions and circuits that could be used by biologists as guiding hypotheses to understand adaptive behaviours and neural circuits found in nature.

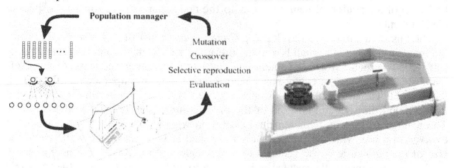

Fig. 1. Left: Artificial Evolution of neural circuits for a robot connected to a computer. Right: The miniature mobile robot Khepera in the looping maze used during an evolutionary experiment.

In order to carry out evolutionary experiments without human intervention, at EPFL we developed the miniature mobile robot *Khepera* [25] (6 cm of diameter for 70 grams) with eight simple light sensors distributed around its circular body (6 on one side and 2 on the other side) and two wheels (figure 1). Given its small size, the robot could be attached to a computer through a cable hanging from the ceiling and specially designed rotating contacts in order to continuously power the robot and let the computer keep a record of all its movements and neural circuit shapes during the evolutionary process, a sort of fossil record for later analysis. The computer generated an initial population of random artificial chromosomes composed of 0's and 1's that represented the properties of an artificial neural network. Each chromosome was then decoded, one at a time, into the corresponding neural network whose input neurons were attached to the sensors of the robots and the output unit activations were used to set the speeds of the wheels. The decoded neural circuit was tested on the robot for some minutes while the computer evaluated its performance (fitness). In these experiments, we wished to evolve the ability to move straight and avoid obstacles. Therefore, we instructed the computer to select for reproduction those individuals whose two wheels moved on the same direction (straight motion) and whose sensors had lower activation (far from obstacles). Once all the chromosomes of the population had been tested on the same physical robot, the chromosomes of selected individuals were organised in pairs and parts of their genes were exchanged (crossover) with small random errors (mutations) in order to generate a number of offspring. These offspring formed a new generation that was again tested and reproduced several times. After 50 generations (corresponding to approximately two days of continuous operation), we found a robot capable of performing complete laps around the maze without ever hitting obstacles. The evolved circuit was rather simple, but still more complex than hand-designed circuits for similar behaviours because it exploited nonlinear feedback connections among motor neurons in order to get away from some corners. Furthermore, the robot always moved in the direction corresponding to the higher number of sensors. Although the robot was perfectly circular and could move in both directions in the early generations, those individuals moving in the direction with fewer sensors tended to remain stuck in some corners because they could not perceive them properly, and thus disappeared from the population. This represented a first case of adaptation of neural circuits to the body shape of the robot in a specific environment.

The Sussex team instead developed a *Gantry robot* consisting of a suspended camera that could move in a small box along the x and y coordinates and also rotate on itself [15]. The image from the camera was fed into a computer and some of its pixels were used as input to an evolutionary neural circuit whose output was used to move the camera. The artificial chromosomes encoded both the architecture of the neural network and the size and position of the pixel groups used as input to the network. The team used a form of incremental evolution whereby the gantry robot was first evolved in a box with one painted wall and asked to go towards the wall. Then, the size of the painted area was reduced to a rectangle and the robot was incrementally evolved to go towards the rectangle. Finally, a triangle was put nearby the rectangle and the robot was asked to go towards the rectangle, but avoid the triangle. A remarkable result of these experiments was that evolved individuals used only two groups of pixels to recognise the shapes by moving the camera from right to left and using the time of pixel activation as an indicator of the shape being faced (for the triangle, both groups of pixels become active at the same time, whereas for the rectangle the top group of pixels becomes active before the lower group). This was compelling evidence that evolution could exploit the interaction between the robot and its environment to develop smart simple mechanisms that could solve apparently complex tasks.

The next question was whether more complex cognitive skills could be evolved by simply exposing robots to more challenging environments. To test this hypothesis, at EPFL we put the Khepera robot in an arena with a battery charger in one corner under a light source (figure 2) and let the robot move around as long as its batteries were discharged [9]. To accelerate the evolutionary process, the batteries were simulated and lasted only 20 seconds; the battery charger was a black painted area of the arena and when the robot happened to pass over it, the batteries were immediately recharged. The fitness criterion was the same used for the experiment on evolution of

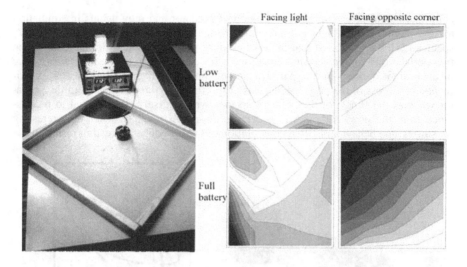

Fig. 2. Left: A Khepera robot is positioned in an arena with a simulated battery charger (the black-painted area on the floor). The light tower above the recharging station is the only source of illumination. Right: Activity levels of one neuron of the evolved individual. Each box shows the activity of the neuron (white = very active, black = inactive) while the robot moves in the arena (the recharging area is on the top left corner). The activity of the neuron reflects the orientation of the robot and its position in the environment, but is not affected by the level of battery charge.

straight navigation (figure 1), that is keep moving as much as possible while staying away from obstacles. Those robots that managed to find the battery charger (initially by chance) could live longer and thus accumulate more fitness points. After 240 generations, that is 1 week of continuous operation, we found a robot that was capable of moving around the arena, go towards the charging station only 2 seconds before the battery was fully discharged, and then immediately returning in the open arena. The robot did not simply sit on the charging area because it was too close to the walls and its fitness was very low (remember from the previous experiment that robots had higher fitness when its proximity sensors had lower activation). When we analysed the activity of the evolved neural circuit while the robot was freely moving in the arena, we discovered that the activation of one neuron depended on the position and orientation of the robot in the environment, but not on the level of battery charge (figure 2). In other words, this neuron encoded a spatial representation of the environment (sometime referred to as "cognitive map" by psychologists), computationally similar to some neurons that neurophysiologists discovered in the hippocampus of rats exploring an environment.

3 Competitive Co-evolution

Encouraged by these experiments, we decided to make the environment even more challenging by co-evolving two robots in competition with each other. The Sussex team had begun investigating co-evolution of predator and prey agents in simulation to see whether increasingly more complex forms of intelligence emerged in the two species [24]. They showed that the evolutionary process changed dramatically when two populations co-evolved in competition with each other because the performance of each robot depends on the performance of the other robot. In the Sussex experiments the fitness of the prey species was proportional to the distance from the predator whereas the fitness of the predator species was inversely proportional to the distance from the prey. Although in some evolutionary runs they observed interesting pursuit-escape behaviours, often co-evolution did not produce interesting result.

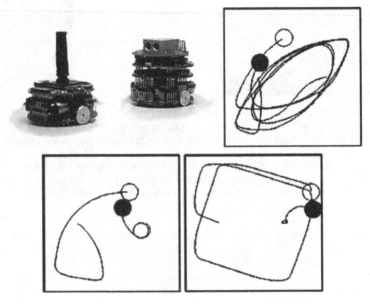

Fig. 3. Co-evolutionary prey (left) and predator (right) robots. Trajectories of the two robots (prey is white, predator is black) after 20, 45, and 70 generations.

At EPFL we wanted to use physical robots with different hardware for the two species and give them more freedom to evolve suitable strategies by using as fitness function the time of collision instead of the distance between the two competitors [10]. In other words we did not explicitly select predator robots for getting closer to the prey and prey robots for keeping at a distance from predators, but we let them choose the most suitable strategies to succeed the ultimate survival criterion: catch the prey and avoid the predator, respectively. We created a predator robot with a vision system spanning 36 degrees and a prey robot that had only simple sensors capable of detecting an object at 2 cm of distance, but that could move twice as fast as the predator (figure 3). These robots were co-evolved in a square arena and each pair of predator and prey robots were let free to move for 2 minutes (or less if the predator could catch the prey). The results were quite surprising. After 20 generations, the predators developed the ability to search for the prey and follow it while the prey escaped moving all around the arena. However, since the prey could go faster than the predator, this strategy did not always pay off for predators. 25 generations later we noticed that predators watched the prey from far and eventually attacked it anticipating its trajectory. As a consequence, the prey began to move so fast along the walls that often predators missed it and crashed into the wall. Again, 25 generations later we discovered that predators developed a "spider strategy". Instead of attempting to go after the prey, they quietly moved towards a wall and waited there for the prey to arrive. The prey moved so fast near the walls that it could not detect the predator early enough to avoid it!

However, when we let the two robot species co-evolve for more generations, we realised that the two species rediscovered older strategies that were effective against the current strategies used by the opponent. This was not surprising. Considering the simplicity of the environment, the number of possible strategies that can be effectively used by the two robot species is limited. Even in nature, there is evidence that co-evolutionary hosts and parasites (for example plants and insects) recycle old strategies over generations.

Stefano Nolfi, who worked with us on these experiments, noticed that by making the environment more complex (for example with the addition of objects in the arena) the variety of evolved strategies was much higher and it took much longer before the two species re-used earlier strategies [26]. We also noticed that the competing selection pressure on the two species generated much faster evolution and behavioural change than in robots evolved in isolation under an externally defined fitness function. These experiments never stopped surprising us and indeed turned out to be a source of inspiration for the best-selling novelist Michael Crichton in his latest science fiction book *Prey* [6]. We feel that this area of research has still much to deliver for the bootstrapping of machine intelligence.

4 Cooperative Co-evolution

Beside competition, living organisms display a sort of "collective intelligence", characterised by complex levels of cooperation that provide them with higher evolutionary advantage. For instance, it has been estimated that one-third of the animal biomass of the Amazon rain forest consists of social insects, like ants and termites [17]. The success of social insects might come from the fact that social interactions can compensate for limitations of the individual, both in terms of physical and cognitive capabilities.

A social insect colony is a complex system often characterised by division of labour and high genetic similarity among individuals [37]. Ants, bees, wasps, and termites provide some of the most remarkable examples of altruist behaviour with their worker caste, whose individuals forego their own reproduction to enhance reproduction of the queen. These and other examples of group harmony and cooperation show the colony as if it behaved as a "superorganism" where individual-level selection is muted, with the result that colony-level selection reigns.

Biologists agree that relatedness plays a major role in favouring the evolution of cooperation in social insects [19]. However, the concept of the colony as a superorganism has been challenged [19]. In collaboration with ant biologist Laurent Keller and robot designer Roland Siegwart, we are trying to determine whether the role of relatedness and the level of selection can be experimentally demonstrated using colonies of artificial ants implemented as small mobile robots with simple vision and communication abilities (figure 4). For this purpose, we have defined experimental settings where these robotic ants are supposed to look for food items randomly scattered in a foraging area. The robots are provided with artificial genomes that code for their behaviours in an indirect manner (i.e., the patterns of behaviour activation coded by the same genetic code vary according to the phenotype frequencies in the colony). There are two kinds of food items. Small food items can be transported by single robots to the nest. Large food items require two cooperating robot to be pushed away. By varying the energetic value of the food items, we can put more or less pressure on the advantage of cooperative behaviours.

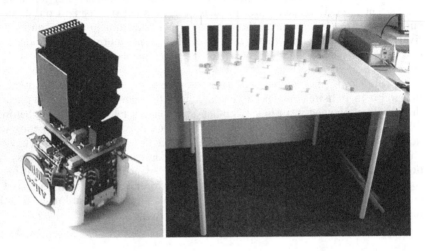

Fig. 4. Left: The sugarcube robot Alice equipped with vision system, distance sensors, communication sensors, and two frontal „mandibles" to better grasp objects. Right: The arena with small and large objects. The nest is under the textured wall where a small gap let objects –but not robots—fall on the floor.

In a first set of experiments carried out in simulation, we investigated how colony performance evolved under different levels of selection (individual and colony level) and under high versus low "genetic" relatedness between robots of the same colony. We ran experiments using a minimalist simulator of the collective robotics evolutionary setup [28], and found that "genetically" homogeneous colonies of foraging simulated robots performed better than heterogeneous ones. Moreover, our experiments showed that altruistic behaviours have low probability of emerging in heterogeneous colonies evolving under individual-level selection. Our current work is aimed at running these experiments in colonies of 20 sugar-cube robots in order to better study the role of physical interactions.

5 Physical Interactions

Collaboration among animals can also take place at a pure physical level. For instance, a mother can help her kids by pushing, pulling, or transporting them on the back. Human acrobats can build towers with their bodies, ants can build bridges, rafts, pulling chains or doors, and bees can build curtains or balls, for instance. In all these examples the group of individuals can achieve a task impossible for a single individual by dynamically aggregating into different and functional physical structures. To investigate this new research direction, in collaboration with other European partners [29], we are developing a new robotic concept, called *s-bot*, capable of physically interconnecting to other s-bots to form a *swarm-bot* (http://www.swarm-bots.org). Each s-bot is a fully autonomous mobile robot capable of performing basic tasks such as autonomous navigation, perception of the environment and grasping of objects (figure 5). Ants can lift each other and heavy objects with their mandibles and can establish flexible connections between each other with their legs. Similarly, each s-bot is equipped with a strong beak gripper that can lift heavy objects or another s-bot and with a flexible gripper that can grasp another s-bot on the belt to maintain physical contact. S-bots can organise in swarm-bot configuration by dynamically attaching to each other and form various shapes according to environmental constraints or task needs.

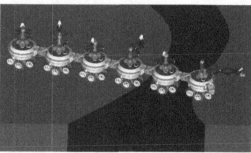

Fig. 5. Left: The prototype of the s-bot with the strong beak gripper and the flexible arm. Right: Several s-bots can self-connect to build a swarm-bot capable of passing obstacles one single s-bot cannot deal with.

In addition to these features, an s-bot is capable of communicating with other s-bots by emitting and receiving sounds. S-bots can also use body signals by changing the colour of their body belts to display their internal states. Other s-bots, with their vision system, can see this corporal expression and react, for instance helping the "red" robot, following the "blue" one, or connecting to the "green" one to form a swarm-bot configuration. Assembled in swarm-bot configuration, the robots are able to perform exploration, navigation and transport of heavy objects in very rough terrain, where a single s-bot could not possibly achieve the task.

The control of this hardware structure is very challenging and has implications on the whole design, from mechanics to software. In this project we resort to a combination of artificial evolution, behaviours inspired from the world of social insects, and standard engineering methodologies. Standard engineering methodologies are applied in all local sub-problems where classical approaches are well known, reliable and form a basic structure on top of which we can build the collective control. This is for instance the case of mechanical design, low-level motor control, sensor management (not processing), and low level communication procedures. Bio-inspired solutions are applied where natural mechanisms are well identified and can be translated into our robot design and control. Examples of bio-inspired design elements are the shape of the grippers and the interactive synchronisation of the robots when grasping an object. Another bio-inspired element is the general concept to solve complex tasks with the combination of many simple mechanisms. On the top of these two approaches we apply artificial evolution to exploit in the best way the specific properties of each part for a given behaviour.

Artificial evolution generated a set of simple rules capable of coordinating the movement of a group of connected s-bots [1]. In this particular case, evolution exploited the property of a force sensor within the body of each s-bot to integrate the behaviour of the whole group without need of external communication or additional coordination layers. These results indicate that physical interactions alone can provide useful information for coordination. Still, it is the responsibility of the engineer to provide sensors and actuators that can be handled efficiently by evolution. This illustrates a big difference with respect to natural evolution, where the behaviours and the body of organisms co-evolve.

6 Active Vision and Feature Selection

Brains are characterised by limited bandwidth and computational resources. At any point in time, we can focus our attention only to a limited set of features or objects. One of the most remarkable —and often neglected— differences between machine vision and biological vision is that computers are often asked to process an entire image in one shot and produce an immediate answer whereas animals are free to explore the image over time searching for features and dynamically integrating information over time.

We thought that the computational complexity of vision-based behaviour could be greatly simplified if the processes of active vision and of feature selection are co-evolved while the robot interacts with the environment. Each of these two processes has been investigated and adopted in machine vision. Active vision is the sequential and interactive process of selecting and analysing parts of a visual scene [2]. Feature selection instead is the development of sensitivity to relevant features in the visual scene to which the system selectively responds [14]. However, the combination of active vision and feature selection is still largely unexplored.

To investigate that hypothesis, we devised a very simple neural architecture composed of only one layer of synaptic connections (figure 6, left) that link visual neurons to two sets of motor outputs. One set of output units controls the behaviour of the system (for example, the movements of a robot or the categorisation of an image discrimination system). The other set controls the behaviour of the vision system (movement over the visual field, zooming factor, pre-filtering strategy). The synaptic weights, which are genetically encoded and evolved using a simple genetic algorithm, are responsible both for the visual features to which the system responds to and for the actions of the vision system.

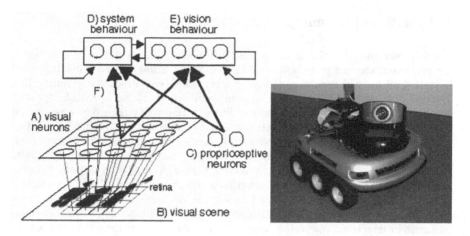

Fig. 6. Left: Architecture of the control system. The architecture is composed of A) a grid of visual neurons with non-overlapping receptive fields whose activation is given by B) the grey level of the corresponding pixels in the image; C) a set of proprioceptive neurons that provide information about the movement of the vision system; D) a set of output neurons that determine the behaviour of the system (pattern recognition, car driving, robot navigation); E) a set of output neurons that determine the behaviour of the vision system; F) a set of evolvable synaptic connections. The number of neurons in each sub-system can vary according to the experimental settings. Right: The Koala robot equipped with a mobile camera whose image is fed into the vision neurons of the neural architecture.

We carried out a series of experiments on co-evolution of active vision and feature selection for behavioural systems equipped with primitive retinal systems and deliberately simple neural architectures [7]. In a first set of experiments, we show that sensitivity to very simple features is co-evolved with, and exploited by, active vision to perform complex shape discrimination [18]. We also show that such discrimination problem is very difficult for a similar vision system without active behaviour because the architecture must solve non-linear transformations (position and size invariance) of the image in order to solve the task. Instead, the co-evolved active vision and feature selection system rely on linear transformations of parts of the image (oriented edges and corners), which are actively searched and sequentially scanned in order to provide the correct answer. In a second set of experiments, we applied the same co-evolutionary method and architecture for driving a simulated car over roads in the Swiss Alps and show that active vision is exploited to locate and fixate the edge of the road while driving the car. In a third set of experiments, we used once again the same co-evolutionary method and architecture for an autonomous robot equipped with a pan/tilt camera (figure 6, right) that is asked to navigate in an arena located in an office environment [22]. Evolved robots exploit active vision and simple features to direct their gaze at invariant parts of the environment (horizontal edge between the floor and furniture) and perform collision-free navigation. In a fourth set of experiments, we apply this methodology to an all-terrain robot with a static, but large, field of view that must navigate in a rugged terrain. Here again, the system becomes sensitive to a set of simple visual features that are maintained within the retina by the active vision mechanisms.

7 Evolution of Learning

Another interesting direction in Evolutionary Robotics is the evolution of learning. In a broad sense, learning is the ability to adapt during lifetime and we know that most living organisms with a nervous system display some type of adaptation during life. The ability to adapt quickly is crucial for autonomous robots that operate in dynamic and partially unpredictable environments, but the learning systems developed so far have many constraints that make them hardly applicable to robots interacting with an environment without human intervention. Of course, evolution is also a form of adaptation, but modifications occur only over several generations, and that may require too long time for a robotic system (for a comparative discussion of lifelong learning and evolution, see [27]). In order to compensate for the problems of both approaches, we decided to genetically encode and evolve the mechanisms of neural adaptation [11]. The idea was to exploit evolution to find good combinations of learning structures, rather than static controllers, and to evolve learning structures that without the constraints of off-the-shelf learning algorithms. The artificial chromosomes encoded a set of rules that were used to change the synaptic connections among the neurons while the robot moved in the environment. The results were very interesting.

A Khepera robot equipped with a vision system was put in an arena with a light bulb and a light switch (figure 7). The light switch is marked by a black stripe painted on the wall. The fitness is given by the amount of time spent by the robot under the light bulb when the light is on. Initially the light is off. Therefore, the robot must first go towards the black stripe to switch the light on (notice that the fitness function does not explicitly encourage this behaviour). The black and grey areas on the floor are used by the computer to detect through a sensor positioned under the robot when to switch the light on and when to accumulate fitness points, but this information is not given to the evolutionary controller. Evolved robot learned during their lifetime the sequence of behaviours necessary to increase their fitness. These included: wall avoidance, movement towards the stripe, movement towards the light, and resting under the light.

Not only the evolution of learning rules resulted in more complex skills, such as the ability to solve sequential tasks that simple insects cannot solve, but also the number of generations required was much smaller. However, the most important result was that evolved robots were capable of adapting during their lifetime to several types of environmental change that were never seen during the evolutionary process, such as different light conditions, environmental layouts, end even a different robotic body. Very recently, Akio Ishiguro and his team at the University of Nagoya used a similar approach for a simulated humanoid robot and showed that the evolved nervous system was capable of adapting the walking style to different terrain conditions that were never presented during evolution [13]. The learning abilities that these evolved robots display are still very simple, but current research is aimed at understanding under which conditions more complex learning skills could evolve in autonomous evolutionary robots.

Fig. 7. Left: A Khepera robot with a vision system is positioned in an arena with a light bulb and a light switch (black stripe on the wall). At the beginning of the robot life, the light bulb is off. The robot must develop from random synaptic connection using genetically determined learning rules how to switch the light on and stay under the light bulb. Right: Trajectory of an evolved robot with enabled synaptic adaptation.

8 Evolvable Hardware

In the experiment described so far, the evolutionary process operated on the features of the software that controlled the robot (in most cases, in the form of an artificial neural network). The distinction between software and hardware is quite arbitrary and in fact one could build a variety of electronic circuits that display interesting behaviours without any software. A few years ago, some researchers realised that the methods used by electronic engineers to build circuits represent only a minor part of all possible circuits that could be built out of a given number of components. Furthermore, electronic engineers tend to avoid circuits that display complex and highly non-linear dynamics, and more in general those which are hard-to-predict, which may be just the type of circuits that a behavioural machine requires. Adrian Thompson at the University of Sussex suggested the evolution of electronic circuits without imposing any design constraints [35]. Thompson used a type of electronic circuit, known as Field Programmable Gate Array (FPGA), whose internal wiring can be entirely modified in a few nanoseconds. Since the circuit configuration is a chain of 0's and 1's, he used this chain as the chromosome of the circuit and let it evolve for a variety of tasks, such as sound discrimination and even robot control. Some evolved circuits used 100 times less components than circuits conceived for similar tasks with conventional electronic design, and displayed novel types of wiring. Interestingly, evolved circuits were sensitive to environmental features, such as temperature, which is usually a drawback in electronic design practice, but is a common feature of all living organisms.

The field of Evolutionary Electronics was born and these days several researchers around the world use artificial evolution to discover new types of circuits or let circuits evolve to new operating conditions. For example, Adrian Stoica and his colleagues at NASA/JPL are designing evolvable circuits for robotics and space application [34], while Tetsuya Higuchi, another pioneer of this field, at the Electro-Technical Laboratory near Tokyo in Japan is already bringing to the market mobile phones and prosthetic implants with evolvable circuits [16].

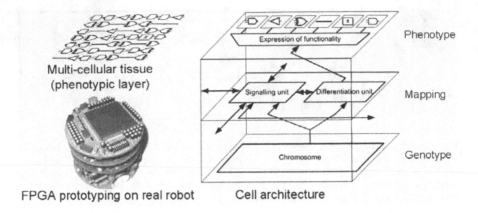

Fig. 8. A schematic representation of the electronic tissue. Each cell of the tissue is composed of three layers, a genotype layer to store the artificial genome of the entire tissue, a phenotype layer to express the functionality of the cell, and an intervening mapping layer to dynamically express the genes into functionalities according to gene expression and cell signalling processes. In addition, each cell of the circuit has input and output connections with the environment. Cells can be dynamically added or removed from the circuit at runtime. A prototype of the electronic tissue has been added on top of the Khepera robot and evolved to generate tissues of spiking neural controllers.

At EPFL, in collaboration with other European partners [36], we are pushing even further the analogy between silicon devices and biological cells in the attempt to create an electronic tissue capable of evolution, self-organisation, and self-repair (http://www.poetictissue.org). The electronic tissue is multi-cellular surface composed of several tiny re-configurable electronic circuits that can be attached or detached while the tissue is in operation. Similarly to a biological cell, each electronic cell is composed of three layers (figure 8). The genotype layer stores the artificial genome of the entire tissue. The phenotype layer expresses the functionality of the cell such as a neuron, a hair cell, a photoreceptor, a motor cell, etc. Finally, the mapping layer regulates the gene expression mechanisms depending on inter-cellular electronic signals. In addition, each electronic cell or group of cells can be attached to a sensor (a phototransistor, a whisker, a microphone, etc.) and/or to an actuator (a servomotor or an artificial muscle). An artificial genome is sent to a mother cell that sends it to all available cells, mimicking a process of cell duplication. As a cell receives a genome, a process of gene expression starts. The gene expression mechanism is affected by intercellular signals so that the functional property expressed by a cell partially depends on the type and intensity of received signals, on its position in the tissue, on the time of genome reception, and on environmental stimulation. For example, cells connected to photoreceptors may have a higher likelihood to process photons. Early prototypes of the system have been interfaced to a robot by connecting the sensors and actuators to cells. The tissue has been subjected to an evolutionary process where different genomes are sequentially tested, reproduced, crossed over and mutated until the robot displayed suitable navigation in a maze [32].

9   Evolutionary Morphologies

In the early experiments on evolution of navigation and obstacle avoidance (figure 1), the neural circuits adapted over generations to the distribution of sensors of the Khepera robot. However, in Nature also the body shape and sensory-motor configuration is subjected to an evolutionary process. Therefore, one may imagine a situation where the sensor distribution of the robot must adapt to a fixed and relatively simple neural circuit. The team of Rolf Pfeifer at the A.I. laboratory in Zurich developed Eyebot, a robot with an evolvable eye configuration, to study the interaction between morphology and computation for autonomous robots [20]. The vision system of Eyebot is similar to that of houseflies and is composed of several directional light receptors whose angle can be adjusted by motors. The authors evolved the relative position of the light sensors while using a simple and fixed neural circuit in a situation where the robot was asked to estimate distance from an obstacle while moving along a track. The experimental results confirmed the theoretical predictions: The evolved distribution of the light receptors displayed higher density of receptors toward in the frontal direction than on the sides of the robot. The messages of this experiment are quite important: on the one hand the body shape plays an important role in the behaviour of an autonomous system and should be co-evolved with other aspects of the robot; on the other hand, computational complexity can be traded with a morphology adapted to the environment.

Back in 1997, when quadruped robots where still an affair of research laboratories, we used a co-evolutionary approach to investigate the balance between morphology and control of a four leg robot [12] (figure 9). More specifically, we were interested in finding a good ratio between leg and body size as well as minimise the number of motorised degrees of freedom provided by a behaviour-based control system with a number of evolvable parameters. We carried out co-evolution of body and control in 3D simulations, but constrained the genetic representation of the robot morphology to a number of primitives that could be built using available technology. Evolved robots were capable of walking forward and turning very smoothly to avoid obstacles using an infrared sensor positioned in front of the robot. These robots used rotating joints only on the front legs. We then built a physical robot according to the dimensions found by the co-evolutionary process (figure 9, right) and downloaded the evolved control system for autonomous navigation. The physical robot displayed the same walking behaviour shown in simulation, although it had a noticeable trembling (which looked as if it was affected by the mad-cow disease) caused by the differences between simulations and physical reality. Since our purpose was to study the interactions between body and control co-evolution, we did not attempt to improve the walking behaviour of the physical robot. However, a possible strategy would be to evolve the learning rules (as described in a section above) and have the "newborn" physical robot adapt online to its own physical characteristics. Also adding some noise to the sensors and actuator while simulating the robot may help bridge the gap to reality [23] by avoiding that the controller over-specialises to the simulation.

Co-evolution of the body and controller has also been applied to biped robots [4]. The results showed better walking characteristics than when only the controller was evolved. The idea of co-evolving the body and the neural circuit of autonomous robots had already been investigated in simulations by Karl Sims [33], but only recently

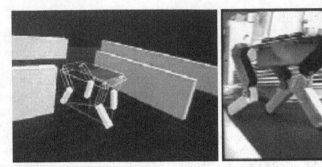

Fig. 9. An evolved 4 legged robot. The control system of the robot, its body size, and length of legs have been evolved in 3D simulations (left). The physical robot (right) has been built according to the evolved genetic specifications. The evolved control system is transferred from the simulated to the physical robot. Such evolved robot can walk and avoid obstacles. The robot is approximately 20 cm long and less than 1kg without batteries. Leg control performed by a set of HC11 microcontrollers.

this has been achieved in hardware. Jordan Pollack and his team at Brandeis University have co-evolved the body shape and the neurons controlling the motors of robots composed of variable-length sticks whose fitness criterion is to move forward as far as possible [21]. The chromosomes of these robots include specifications for a 3D printer that builds the bodies out of thermoplastic material. These bodies are then fitted with motors and let free to move while their fitness is measured. Artificial evolution generated quite innovative body shapes that resemble biological morphologies such as those of fishes.

10 A Look Ahead

Over the last 10 years, the role of embodiment and behavioural interaction has been increasingly recognised as a cornerstone of natural and artificial intelligence. New research initiatives in information technologies, neuroscience, and cognitive science sponsored by the European Commission, U.S. National Science Foundation, and a number of national programs explicitly emphasise these two aspects.

Many more examples of evolutionary robots exhibiting intelligent behaviours are available out there, too many to be covered in this short document. However, we are just scratching the surface of a radically new way of understanding how intelligent life emerged on this planet and could evolve in machines. There are a number of conceptual and technological challenges ahead. For example, evolution does not automatically lead to intelligent behaviours. A lot of prior knowledge and experience is still required to select appropriate parameters, such as the genetic encoding, the neural network architecture, the mapping of sensors and actuator to the network or even the fitness function. Developing better methodologies to select those parameters is an important aspect that needs to be tackled for evolving more complex systems. Also, we are facing what is called the "bootstrap problem". If the environment or the fitness function is too harsh for the evolving individual during the initial generations (so that all the individuals of the first generation have zero fitness), evolution cannot select good individuals and make any progress. A possible solution (and by far not the only one) is to start with environments and fitness functions that become increasingly more complex over time. However, this means that we must put more effort in developing methods for performing incremental evolution that, to some extent, preserve and capitalise upon previously discovered solutions. In turn, this implies that we should understand what are suitable primitives and genetic encoding upon which artificial evolution can generate more complex structures. A key aspect will most likely be the emergence of modular and hierarchical structures through mechanisms of genetic regulatory networks, cell differentiation, and inter-cellular signalling. Another challenge is hardware technology. Despite the encouraging results obtained in the area of evolvable hardware, many of us feel that we should drastically reconsider the hardware upon which artificial evolution operates. This means that maybe we should put more effort in self-assembling materials that give less constraints to the evolving system, facilitate the evolutionary process, and may eventually lead to truly self-reproducing machines.

Acknowledgements. This paper is an updated and extended version of *A.I. 101: Machine Self-evolution* by the same authors. The experiments carried out at EPFL were possible thanks to the collaboration with Stefano Nolfi, Joseba Urzelai, Jean-Daniel Nicoud, Andre Guignard, Laurent Keller, Roland Siegwart, Eduardo Sanchez, Marco Dorigo, Luca Gambardella, and Jean-Luis Deneubourg. The authors thank the Swiss National Science Foundation (grant nr. 620-58049.99) and the Future Emergent Technologies division of the European Commission for continuous support of this exciting research field (Swiss OFES grants 00.05291-1 and 01.0012-1).

References

1. Baldassarre, G., Nolfi, S. and Parisi, D. (2002) Evolving Mobile Robots Able to Display Collective Behaviours. In C. K. Hemelrijk and E. Bonabeau, (eds.) *Proceedings. of the International Workshop on Self-Organisation and Evolution of Social Behaviour*, 11-22.
2. Bajcsy, R. (1988) Active Perception. *Proceedings of the IEEE*, 76, 996-1005.
3. Ballard, D.H. (1991) Animate Vision. *Artificial Intelligence*, 48, 57-86.
4. Bongard, J.C. and Paul, C. (2001). Making Evolution an Offer It Can't Refuse: Morphology and the Extradimensional Bypass. In Keleman, J. and Sosik, P. (eds.) *Proc. of the 6th European Conf. on Artificial Life*, Prague, CZ, 401-412.
5. Braitenberg, V. (1984). *Vehicles. Experiments in Synthetic Psychology*. Cambridge, MA: MIT Press.
6. Crichton, M. (2002) *Prey*. New York: Harper Collins.
7. Floreano, D., Kato, T., Marocco, D. and Sauser, E. (2003) Co-evolution of Active Vision and Feature Selection. *Biological Cybernetics*. Submitted.
8. Floreano, D. and Mondada, F. (1994) Automatic Creation of an Autonomous Agent: Genetic Evolution of a Neural Network Driven Robot. In D. Cliff, P. Husbands, J.-A. Meyer, and S. Wilson (eds.), *From Animals to Animats III*, Cambridge, MA: MIT Press, 421-430.
9. Floreano, D. and Mondada, F. (1996) Evolution of Homing Navigation in a Real Mobile Robot. *IEEE Transactions on Systems, Man, and Cybernetics--Part B: Cybernetics*, 26(3), 396-407.

10. Floreano, D., Nolfi, S., and Mondada, F. (2001) Co-Evolution and Ontogenetic Change in Competing Robots. In M. Patel, V. Honavar, and K. Balakrishnan (eds.), *Advances in the Evolutionary Synthesis of Intelligent Agents*, Cambridge (MA): MIT Press.
11. Floreano, D. and Urzelai, J. (2000a) Evolutionary Robots with on-line self-organization and behavioral fitness. *Neural Networks*, 13, 431-443.
12. Floreano, D. and Urzelai, J. (2000b) Artificial Evolution of Adaptive Software: An Application to Autonomous Robots. *3D The Journal of Three dimensional images* (in japanese), 14(4), 64-69
13. Fujii, A., Ishiguro, A., Aoki, T. and Eggenberger, P. (2001) Evolving Bipedal Locomotion with a Dynamically-Rearranging Neural Network. In J. Kelemen and P. Sosik (eds.) *Advances in Artificial Life (ECAL 2001)*, Berlin: Springer Verlag, pp. 509-518.
14. Hancock, P.J., Baddeley, R.J. and Smith L.S. (1992) The principal components of natural images. *Network*, 3, 61-70.
15. Harvey, I., Husbands, Ph., and Cliff, D. (1994) Seeing the Light: Artificial Evolution, Real Vision. In D. Cliff, P. Husbands, J.-A. Meyer, and S. Wilson (eds.), *From Animals to Animats III*, Cambridge, MA: MIT Press, 392-401
16. Higuchi, T., Iwata, M., Keymeulen, D., Sakanashi, H., Murakawa, M., Kajitani, I., Takahashi, E., Toda, K., Salami, M., Kajihara, N. and Otsu, N. (1999) Real-World Applications of Analog and Digital Evolvable Hardware. *IEEE Transactions on Evolutionary Computation*, 3(3), 220-235.
17. Holldobler, B. and Wilson, E.O. (1990) *The Ants*. Berlin: Springer-Verlag.
18. Kato, T. and Floreano, D. (2001) An Evolutionary Active-Vision System. In *Proceedings of the Congress on Evolutionary Computation (CEC'01)*, Piscataway: IEEE Press.
19. Keller, L. and Reeve, H.K. (1999) Dynamics of conflicts within insect societies. In L. Keller (Ed.), *Levels of Selection in Evolution*, Princeton University Press.
20. Lichtensteiger, L. and Eggenberger, P. (1999). Evolving the morphology of a compound eye on a robot. In *Proceedings of the Third European Workshop on Advanced Mobile Robots (Eurobot '99)*. New York: IEEE Press, 127-134.
21. Lipson, H. and Pollack, J.B. (2000). Automatic design and manufacture of robotic life-forms. *Nature*, 406 (6799): 974-978.
22. Marocco, D. and Floreano, D. (2002) Active Vision and Feature Selection in Evolutionary Behavioral Systems. In Hallam, J. and Floreano, D. and Hayes, G. and Meyer, J. (eds.) *From Animals to Animats 7: Proceedings of the Seventh International Conference on Simulation of Adaptive Behavior*. Cambridge, MA: MIT Press, 247-255.
23. Miglino, O., Lund, H.H. and Nolfi, S. (1995). Evolving Mobile Robots in Simulated and Real Environments. *Artificial Life* 2(4), 417-434.
24. Miller, G.F. and Cliff, D. (1994). Protean Behavior in Dynamic Games: Arguments for the co-evolution of pursuit-evasion tactics. In Cliff, D. and Husbands, P. and Meyer, J. and Wilson, S. W. (eds.) *From Animals to Animats III*. Cambridge, MA: MIT Press, 411-420.
25. Mondada, F., Franzi, E. and Ienne, P. (1993). Mobile Robot Miniaturization: A Tool for Investigation in Control Algorithms. In Yoshikawa, T. and Miyazaki, F (eds.) *Proceedings of the Third International Symposium on Experimental Robotics*, Tokyo: Springer Verlag, 501-513.
26. Nolfi, S. and Floreano, D. (1998) Co-evolving predator and prey robots: Do 'arm races' arise in artificial evolution? *Artificial Life*, 4(4), 311-335.
27. Nolfi, S. and Floreano, D. (2000). *Evolutionary Robotics. The Biology, Intelligence, and Technology of Self-Organizing Machines*. Cambridge, MA: MIT Press.
28. Perez-Uribe, A., Floreano, D. and Keller, L. (2003) Effects of group composition and level of selection in the evolution of cooperation in artificial ants. In W. Banzhaf et al. (eds.) *Proceedings of the European Conference on Artificial Life*, Berlin: Springer Verlag, 128-137.

29. Pettinaro, G.C., Kwee, I., Gambardella, L.M., Mondada, F., Floreano, D., Nolfi, S., Deneubourg, J.-L. and Dorigo, M. (2002) SWARM Robotics: A Different Approach to Service Robotics. In *Proceedings of the 33rd International Symposium on Robotics*, Stockholm, Sweden, October 7-11, 2002. International Federation of Robotics.

30. Pfeifer R. (2000). On the relation among morphology and materials in adaptive behavior. In: J-A Meyer, A. Berthoz, D. Floreano, H.L. Roitblat, and S.W. Wilson (eds.) *From Animals to Animats 6. Proceedings of the VI International Conference on Simulation of Adaptive Behavior*. Cambridge, MA: MIT Press, 413-419.

31. Pfeifer, R. and Scheier, C. (1999) *Understanding Intelligence*. Cambridge, MA: MIT Press.

32. Roggen, D., Floreano, D. and Mattiussi, C. (2003) A Morphogenetic Evolutionary System: Phylogenesis of the POEtic Circuit . In Tyrrell, Haddow, Torresen (eds.) *Proceedings of the Fifth International Conference on Evolvable Systems*. Berlin: Springer Verlag, 153-164.

33. Sims, K. (1994) Evolving 3D Morphology and Behavior by Competition. In R. A. Brooks and P. Maes (eds.) *Artificial Life IV*, Cambridge, MA: MIT Press, 28-39.

34. Stoica, A., Zebulum, R., Keymeulen, D., Tawel, R., Daud, T. and Thakoor, A. (2001), Reconfigurable VLSI Architectures for Evolvable Hardware: from Experimental Field Programmable Transistor Arrays to Evolution-Oriented Chips. *IEEE Transactions on VLSI Systems, Special Issue on Reconfigurable and Adaptive VLSI Systems*, 9(1), 227-232.

35. Thompson, A. (1998) *Hardware Evolution: Automatic design of electronic circuits in reconfigurable hardware by artificial evolution*. Berlin: Springer Verlag.

36. Tyrrell, A.M., Sanchez, E., Floreano, D., Tempesti, G., Mange, D., Moreno, J.M., Rosenberg, J. and Villa, A. (2003) POEtic Tissue: An Integrated Architecture for Bio-inspired Hardware. In Tyrrell, Haddow, Torresen (eds.) *Proceedings of the International Conference on Evolvable Systems*. Berlin: Springer Verlag, 129-140.

37. Wilson, E.O. (1971) *The Insect Societies*. Cambridge, MA: Harvard University Press.

Self-Reconfigurable Robots: Platforms for Emerging Functionality

Satoshi Murata[1,2], Akiya Kamimura[2], Haruhisa Kurokawa[2], Eiichi Yoshida[2],
Kohji Tomita[2], and Shigeru Kokaji[2]

[1] Interdisciplinary Graduate School of Science and Engineering,
Tokyo Institute of Technology
4259 Nagatsuda, Midori, Yokohama, 226-8502 Japan
murata@dis.titech.ac.jp
http://www.mrt.dis.titech.ac.jp/
[2] National Institute of Advanced Industrial Science and Technology (AIST)
1-2 Namiki, Tsukuba, 305-8564 Japan
{kamimura.a, kurokawa-h, e.yoshida, k.tomita,
s.kokaji}@aist.go.jp
http://unit.aist.go.jp/is/dsysd/index.html

Abstract. We have studied modular self-reconfigurable robots that are capable of changing their overall shape and functionality by automatic recombination of homogenous robotic modules. Our latest model, called Modular Transformer (M-TRAN), is able to metamorphose into various 3-D configurations and generate robotic motions that are suitable to its configuration. This paper presents a review of hardware design of the module, some developed software for self-reconfiguration and motion generation, and some experimental results.

1 Introduction

Artificial systems have become increasingly complicated, engendering rapidly increasing design, production, and maintenance costs. In addition, the surrounding environment of such devices is becoming more complicated, varied and unpredictable. Therefore, the designers these devices cannot determine a complete set of desirable functions *a priori*. We believe that "emergent functionality" is an important concept to resolve this issue. Namely, devices that are able to produce necessary functions according to a situation are capable of coping with it. To produce, or "emerge", such functionality, a system must incorporate some mechanism that allows the independent change of its own functions. Physical reconfiguration of the device's hardware is necessary: so-called "self-reconfiguration."

We have studied modular self-reconfigurable robots that are capable of changing their overall shape and functionality through the recombination of homogenous robotic modules. The most distinguished property of self-reconfigurable robots is that both assembly and repair are performed autonomously without any external help. Self-reconfigurable robots have robotic modules that can change their local connectivity to form a specific shape (self-assembly) through cooperation of many modules using inter-module communication. Thereby, the whole robot can generate robotic

F. Iida et al. (Eds.): Embodied Artificial Intelligence, LNAI 3139, pp. 312–330, 2004.

motion through the cooperation of all modules. Module homogeneity allows self-repair through replacement of faulty modules with spares.

Numerous types of self-reconfigurable robots have been proposed. Among them, M-TRAN (modular transformer), developed by our group, is a superior realization of this kind of robot. This paper specifically addresses several research topics related to M-TRAN system hardware, software, and algorithms.

The following section reviews studies of various self-reconfigurable robots. Section 3 explains M-TRAN hardware. Section 4 describes issues of self-reconfiguration planning between different shapes. In Section 5, a motion generation method using a genetic algorithm (GA) is explained. We conclude this paper in the last section.

2 Related Work

Research activities addressing self-reconfigurable robots began in the late 1980s when Fukuda et al. proposed the "Cellular Robot" concept [1]. Subsequently, great efforts have been made to realize two-dimensionally self-reconfigurable robots [2–9]. Those studies proposed many two-dimensional hardware and various algorithms for self-assembly and self-repair. Thereafter, the self-reconfigurable robot concept was extended to three-dimensional (3D) robots [10–19]. Most hardware of 3D self-reconfigurable robots are classified into two types: "lattice type" [10–15,17] and "linear (or string or chain) type" [16,18,19]. The former corresponds to a system where each module has several fixed directions for connection similar to that of atoms; groups of them form various types of crystals. The resultant structure is static. Consequently, it is difficult to generate a group motion on the system. In contrast, the latter is fundamentally a robot with many joint modules that can easily generate various robotic motions similarly to a snake or caterpillar. However, self-reconfiguration is difficult for the latter type because it requires complicated control for reconnection.

3 Hardware of M-TRAN

We recently developed a novel self-reconfigurable system called a modular transformer (M-TRAN). It has both lattice and linear type features [20–26, 31]. The module has a simple bipartite composition. This module is suitable for forming various robotic shapes through self-reconfiguration. This section explains the M-TRAN module hardware and its control system.

The M-TRAN module comprises two semi-cubic blocks and a link (Fig. 1). It has two rotational actuation axes and six connection surfaces. Each rotational angle ranges ±90°. When all angles are restricted to 0° or ±90°, all cubic blocks are placed on a regular cubic grid system. By this property, precise positioning between two neighbor modules is unnecessary for self-reconfiguration as long as the angles are multiples of 90°. Once a specific structure is made by self-reconfiguration, each rotational angle can be controlled freely to form robotic motion as a whole.

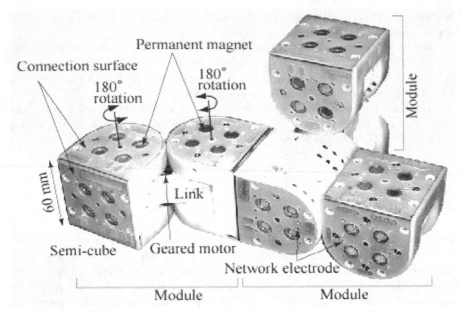

Fig. 1. M-TRAN II modules. This picture shows three inter-connecting modules. A module consists of two semi-cylindrical parts (60 × 60 × 60 mm) connected by a link. Each module weighs about 400 g. Two parallel rotation axes are driven by servomotors in the link. Each part has three magnetic connection faces equipped with electrical connection channels for inter-module communication. Compared to the first generation, it has smaller size, three more processors (one neuron chip and three PICs), and more powerful and faster operation.

Figures 1 and 2 show the second generation of the M-TRAN system called M-TRAN II. Its actuators and connections are sufficiently strong to lift one or two other modules under gravity. Both a reliable connection for self-reconfiguration and sufficient actuation power for robotic motion are realized. Wireless, stand-alone operation is possible using a battery in each module.

The module connection mechanism is designed based on the Internally Balanced Magnetic Unit (IBMU) principle proposed by Hirose et al. [27]. We adopt spring actuators made of shape memory alloy (SMA) to release the magnetic connection (Fig. 3).

The M-TRAN II control system contains two layers of multi-CPU system (Fig. 4). The upper layer is a computer network of modules; the lower layer is a microcontroller system inside each module.

The upper layer contains the main-CPUs (Neuron chip; Echelon Corp.) connected by a two-wired network bus (IEEE RS-485). The host PC is also identified as a node in this network when the tethered attachment is connected to a module. This network transfers programs and data to each module from the host PC. After transfer, the tether can be removed. Then the modules are activated by a transferred command that is received by one module via a radio communication channel. The communication

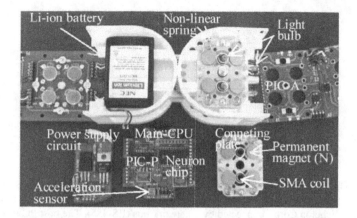

Fig. 2. M-TRAN II module interior view. The left part is a "passive" block and the right is "active." Each block has three connection surfaces. When the passive and the active blocks contact each other, they automatically connect by permanent magnets. The active one has a mechanism for releasing this connection (its principle described below). Connecting plates in the active block have several retractable electrodes for inter-module communication. A power supply circuit, a main CPU board, and a battery are inside the passive block. Power supply by a single battery is sufficient to drive two motors and for releasing one connection in a module. The link part contains two geared motors for rotation and potentiometers to measure rotation angles. The link part has a motor-driver circuit that realizes PID positioning control.

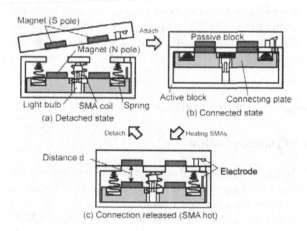

Fig. 3. Connection mechanism. When passive and active blocks are closely placed, the connecting plate is moved by magnetic force; thereby, the magnets of the two blocks make contact (a to b). Because the magnetic force and repulsive force by nonlinear springs are internally balanced inside the casing, a firm connection is achieved without reducing the force of magnets (b). When the SMA coils are heated, they generate sufficient force to push the connection plate down; eventually, the magnets are detached and connection force between two blocks becomes very small (c).

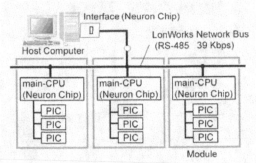

Fig. 4. M-TRAN II control system. The upper layer is a computer network of modules and the lower layer is a microcontroller system inside each module. The upper layer consists of the main CPUs (Echelon Corp.) connected by a network bus (RS-485). The host PC is also a node in this network with an interface board when the tethered attachment is connected to one module. This network allows transfer of programs and data to each module from the host PC.

protocol of the upper layer is the LON protocol, which is provided as firmware on the Neuron chip. All modules in this network, including the host PC, are equivalent in their priority. In these experiments, one module with the lowest ID number is selected as a master.

Inside each module, the main CPU (Neuron chip) and three microcontrollers (PIC16F873 and F877) compose the lower layer. The main CPU works as a master and sends commands to the others. The PIC in the link part controls two motors by PID control and the PIC in the active block controls the light bulbs (heater for SMAs) to release the connections. The PIC in the passive brick is for detection of a local connection and for local communication with connected modules. These four controllers communicate using asynchronous serial communication in the module.

4   Self-Reconfiguration Planning

4.1   Self-Reconfiguration Method

In the M-TRAN system, self-reconfiguration is achieved by repeating basic operations such as detaching a surface from the neighbor, rotating a semi-cylindrical box, and reconnecting the surface to another neighbor.

Figure 5 illustrates simplified self-reconfiguration schemes of M-TRAN system. A module on the floor tiled with passive and active connection surfaces, can rotate around the horizontal axis (a), or rotate around the vertical axis (b). The former is called "forward roll mode;" the latter is called "pivot translation mode." Although a single M-TRAN module does not have sufficient degrees of freedom (DOFs) to switch from one mode to another, it is possible using a partner module (c). Such switching is called mode conversion. The actual reconfiguration process is not as

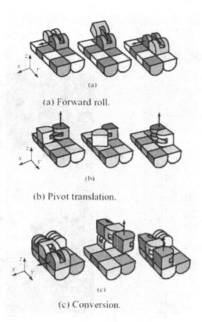

(a) Forward roll.

(b) Pivot translation.

(c) Conversion.

Fig. 5. Self-reconfiguration schemes. (a) Forward roll: The module travels on a line by rotating around the x-axis. Although it cannot change the direction of the line, the module can change the vertical level by climbing over another module. (b) Pivot translation: The module traverses the plane by rotation around the z-axis. The module can head to any direction but cannot change its vertical level in this mode. (c) Mode conversion: A module attached with an arrow is lifted up by a module behind (converter module) and placed back at the same position but in a different posture.

simple as shown in this illustration. We must combine these actions to design a self-reconfiguration path to a desirable configuration.

4.2 Difficulties in Self-Reconfiguration Planning

The central problem in self-reconfiguration planning is to find a general algorithm that can verify whether two arbitrary configurations A and B are reachable from one to the other. If they are it calculates an appropriate reconfiguration path between them which has the smallest transformation cost. However, such calculation is difficult even for two-dimensional systems [28,29]. Such difficulty has its origin in the nature of the many DOF searching problems of modular architecture.

The most important issue in designing a self-reconfiguration planner is defining the metric that indicates the difference between two configurations [30]. For most two-dimensional lattice systems and for isotropic three-dimensional modules, there is

Command sequence list
Motion command panel
Simulation panel

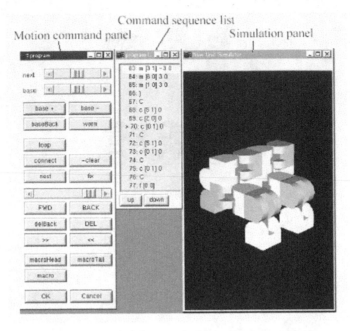

Fig. 6. Self-reconfiguration design interface. Left panel: motion command input, center panel: motion command sequence list, right panel: simulation result display.

a good correspondence between the lattice distance. The transformation cost is evaluated as the number of necessary motion steps. Therefore, the lattice distance can be used as a metric for those isotropic systems; it gives planning algorithms at reasonable cost. M-TRAN has less symmetry because of its non-isotropic shape and parallel axes. This fact implies that the lattice metric is not good for reconfiguration planning.

Along with this problem, some generic hardware constraints must be taken into account: (a) connectivity – all the modules must remain connected during self-reconfiguration; (b) collision avoidance – collisions between modules must be avoided; and (c) torque limit – one module can only lift a few modules depending on their posture. In the M-TRAN system, we must consider different modes (pivot or forward roll) and their mode conversion. This consideration further complicates the problem.

We have developed two types of software to cope with such self-reconfiguration planning complexity. The first is a motion design interface, which helps a human programmer to design a complex self-reconfiguration sequence and motion generation through a powerful graphical interface. The second is a locomotion planner for an M-TRAN cluster, which relaxes the above difficulties by assuming some periodic regularity of the structure.

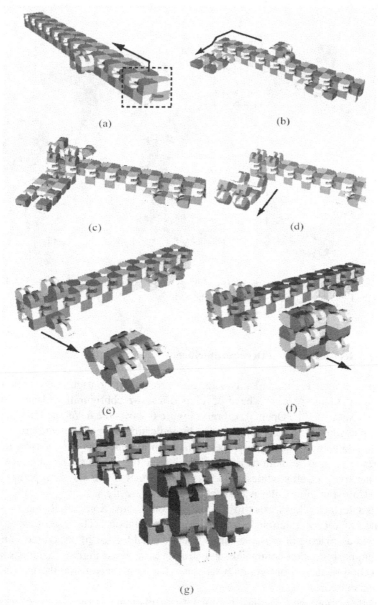

(a)

(b)

(c)

(d)

(e)

(f)

(g)

Fig. 7. Sequence planned using interface software with macro commands. The interface allows human programmers to design a complicated self-reconfiguration like this. (a) Four modules at one end of the cluster are lifted up on the cluster. (b) They walk along the cluster to the other end similarly to an inchworm. (c) They are placed on the floor and changed to a crawler configuration (d, e). Then it cuts the loop to become a four-legged walker (f, g).

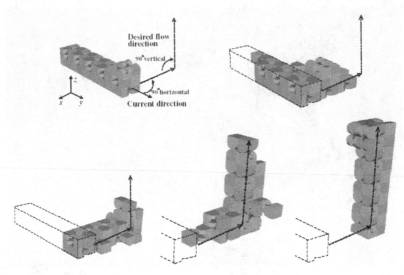

Fig. 8. An automatically generated cluster flow by two-layer control. The super-module cluster follows the flow that is placed by a human operator.

4.3 Self-Reconfiguration Design Interface

A graphical interface remarkably aids users to plan reconfiguration sequences in three-dimensional space. We have developed a self-reconfiguration design interface for M-TRAN using the OpenGL Library. Figure 6 shows the developed interface. The users can design any initial configuration by indicating the position and orientation of each module using simple mouse operation. Then the interface allows users to design a sequence of module motions in a similar interactive manner. The interface checks the connectivity of all modules and alerts the designer if some system part is disconnected. Collision of modules is also checked automatically.

This interface allows users to use macro commands. Sometimes, the same short sequence of reconfiguration is useful in many situations. Those sequences can be registered as macros in reusable form. Users can edit a list of structured commands including basic motion commands and structured macros. Figure 7 shows snapshots from a complicated reconfiguration sequence that is planned using the interface software and macros.

Another approach to reduce complexity of planning is to introduce regularities into module clusters. The cluster comprises four-module blocks (two pivot translation modules in two layers, in orthogonal directions). This block is regarded as a super module or meta-module. The advantage of introducing the super-modules is that any serial connection of the super-modules maintains connectivity of the whole cluster. The super modules simplify the reconfiguration problem considerably.

Fig. 9. An experiment assessing self-reconfiguration and motion generation of the M-TRAN II system. These photos are taken from video. The first frame is a walking configuration. It flattened in the following three frames to take an "H" configuration. Then the central unit, which serves as a manipulator, moves modules in the left side to the right (continues). Light bulbs are turned on when the connection is detached (second frame of second row).

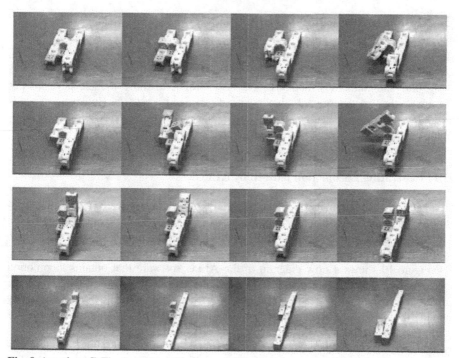

Fig. 9. (continued) The central manipulator module continues to carry modules from the left to the right until all modules are stacked in one line (from first to third row). Then the stack is untied to form a linear configuration (second frame of the fourth row). After the manipulator modules are disconnected (second from the last); it moves as a snake. Self-reconfiguration sequence from four-legged to snake configuration is designed manually. The walking gait of the first four-legged configuration and the snake motion are both generated by the method described in Section 5.

We have developed a centralized reconfiguration planning software utilizing the structure built by the super-modules. The planner comprises global and local planners. The global planner decides the global motion regarded as flow of the super-modules. The flow is a trajectory in 3-D space that the super-module must follow. For instance, a translation of the whole cluster can be realized by sending a super-block at the tail of the cluster toward the head. The role of local planner is to decompose the flow into precise motion commands (schemes) based on a rule database of admissible motion for local connectivity. Figure 8 shows one example that is designed automatically by this method.

We conducted a self-reconfiguration experiment using M-TRAN hardware (Fig. 9). It demonstrates M-TRAN's reliable self-reconfiguration capability and motion generation as a robotic system. This self-reconfiguration sequence is designed manually using the interface software. Walking motion is generated by ALPG software, which is explained in the following section.

5 Motion Generation

In this section, we consider motion generation for the self-reconfigurable system. Although motion generation and control have been subjects of robotics study for a long time and numerous methods have been proposed, those models have always assumed the predetermined dynamics of a fixed system. However, a self-reconfigurable system can change its shape. For this reason, we developed an automatic motion generator for arbitrary configuration.

5.1 Generating Locomotion Pattern by ALPG

We have built an automatic locomotion pattern generation software (ALPG) by combining a dynamics simulator (Vortex; CMLabs Simulations, Inc.), a dynamics model of a M-TRAN module, a neural oscillator which drives the modules, and a GA, which optimizes the neural oscillator network [31].

Figure 10 shows a flow diagram of the ALPG software that obtains locomotion patterns for various configurations. The initial configuration and system shape are given in this method. Note that "configuration" here means topological connectivity among the modules and that "shape" means joint angles of the modules. From this initial condition, each module's actuator is driven by a neural oscillator network, which controls the frequency, phase and amplitude of each joint angle. The locomotion pattern performance is evaluated using sequential dynamic simulations; the GA optimizes the parameters for the oscillators. The best result is transferred to the M-TRAN hardware for final evaluation.

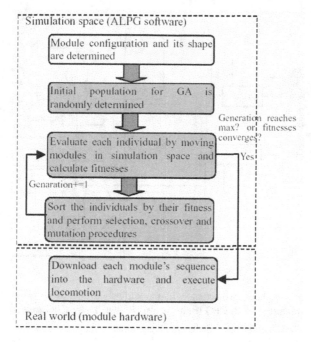

Fig. 10. A flow diagram of the ALPG software, which generates locomotion patterns for various configurations.

5.2 Central Pattern Generator

We applied a neural oscillator model known as the Central Pattern Generator (CPG) to control each module's motion. Each neuron in this model is represented by a set of differential equations, the basic structure of which is depicted in Fig. 11 [32–35].

Each module's motor has its own CPG and is controlled directly by the CPG output. The CPG oscillations are mutually entrained, which is caused by feedback signals from the rotation angles among connected neurons. Two kinds of entrainment exist. Each generates a cooperative motion with the modules. One is an entrainment among CPGs made by the CPG connection; the other is a more complicated entrainment among CPGs and mechanical system (called global entrainment caused by feedback from each joint angle to each CPG.).

5.3 Evolutionary Computation

We implemented a GA to search an optimal oscillator network for locomotion. The initial values of four state variables of each CPG and the connection weights among the CPGs are evolved together using the GA. The connection weights in the neural

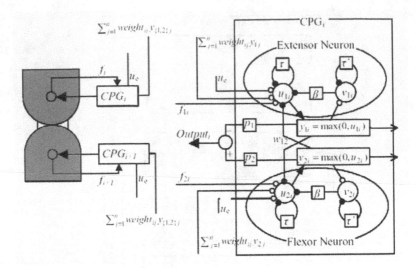

Fig. 11. CPG in a module. Two actuators in a module are driven by corresponding CPGs (left). The detailed structure of each CPG (right) is also shown. These CPGs are connected randomly to other CPGs by the weights at the initial time.

network determine the phase differences between the CPG oscillations. After several hundreds of generations, we succeeded in obtaining stable limit cycles. The following explains some important settings of the evolutionary computation. The GA process stops when the average fitness becomes constant, or the number of generations is greater than some maximum number of generations.

(1) Gene expression
Each individual has its gene, which is a set of parameters composed of initial state of neural oscillators and connectivity matrix (connection weights) of the CPG network.

(2) Fitness Evaluation
Locomotion by each individual is evaluated individually in 15 s in simulation space by the fitness function. That function evaluates velocity, straightness, and energy consumption of the locomotion.

(3) Selection, crossover and mutation procedures
Each individual is evaluated and sorted by its fitness; the low fitness valued group is eliminated first. Then, crossover operations are applied to produce new individuals to replace the eliminated ones. Because the initial state values are represented by real-numbers and the connection weights are represented by discrete numbers (1,0,–1), we applied two different algorithms for crossover, Unimodal Normal Distribution Cross-over (UNDX) for the former and the N-point crossover method for the latter.

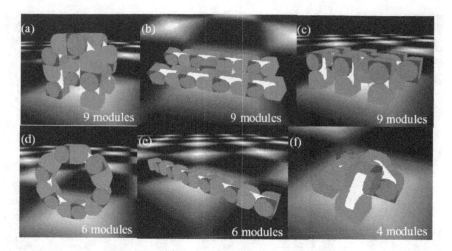

Fig. 12. Initial configurations given to ALPG software. All parameters in the software are identical for every trial throughout the simulation.

5.4 Simulation and Experiment

We applied the ALPG method to various module configurations shown in Fig. 12. Stable locomotion patterns were obtained for all the configurations using the ALPG. By those results, experiments of locomotion were carried out in which each module followed the joint angle trajectory by PID controller. Figure 13 shows the corresponding locomotion patterns. Configuration (d) of Fig. 12 has the highest fitness value: it is difficult to control the real hardware because such a rolling motion requires additional feedback control. Other gaits worked well in the hardware experiments. They also showed good relevance to simulation results. The measured fitness of locomotion was always 20–30% less than that of the simulated one.

The time required for ALPG computation to search a feasible locomotion pattern depends on the number of modules and the frequency of collisions in the dynamics simulation. For example, it took about six hours using a PC with a 2.53 GHz Pentium 4 processor to evolve a stable walking pattern for the nine-module configuration.

Figure 14 illustrates the locomotive motion in phase space (coordinates are the angle and angular velocity). All joint angles oscillate at a constant frequency (about 1.2 Hz), amplitude, and phase difference, indicating that the locomotion pattern is converged to a stable limit cycle.

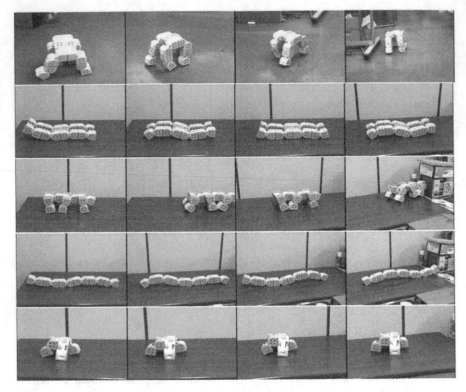

Fig. 13. Experimental results for some configurations. The solution by ALPG is downloaded to the hardware M-TRAN.

6 Conclusion

This paper summarized our research activities related to M-TRAN system development. A novel design of M-TRAN system realizes both reliable self-reconfiguration and robotic motion control. Using the improved M-TRAN II module hardware, we achieved high torque actuation, firm and reliable connection and detachment mechanisms, high speed computation by onboard processors, inter-module communication via network on the modules, and low energy consumption, which enables autonomous and tetherless operation, and so on.

Several important issues remain in the M-TRAN project. The first is hardware improvement intended for simpler and smaller design of module, faster computation, higher energy efficiency, and sensor implementation for system adaptation. The second issue is software for self-reconfiguration and motion generation. Various search and learning methods must be tried to solve those problems. The current hardware control system is centralized. It is not suitable to achieve self-repair. We must investi-

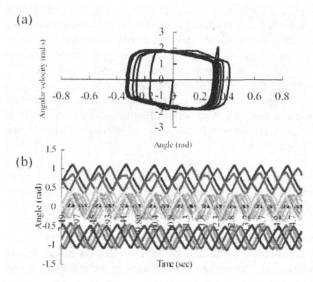

Fig. 14. Obtained walking gait of configuration (a) in Fig. 12. (a) Motion in the phase space (joint angle vs. joint angular velocity). (b) Angles of all joints.

gate a distributed control method for M-TRAN system. The last issue is the manner in which the target configuration can be defined automatically. This issue is more general and may apply to all self-reconfigurable systems. We must consider not only the optimal configuration to solve this problem of target configuration definition: we must also consider the evolution of system configuration and motion.

(Videos of M-TRAN simulations and experiments are available on WWW [37].)

Discussion

Here, we summarize discussion between the reviewers and us. We think it is helpful for the interdisciplinary audience of this volume.

Q. The salient contribution of this work is that a real-world reconfigurable hardware has actually been made. We also need to clarify the conceptual novelty of this work.

A. Heretofore, no devices have been capable of reconfiguring their own physical configurations in a self-contained form. M-TRAN realizes that concept on the basis of homogenous modular architecture.

Q. What do the authors mean by "emergent functionality"? Is the locomotion an emergent function because of the coordination of multiple modules? Alternatively, is the behavior an emergent function because the genetic algorithm revealed a solution with respect to fitness?

A. Generally speaking, when system optimization or system adaptation to the environment is achieved as a result of some self-organization process, we call it emergent functionality. Nevertheless, this is a narrow definition as long as it means optimization of system parameters. Although proposed methodologies for the M-TRAN system are in this category, those methodologies offer the potential of structural change through self-reconfiguration, which differs qualitatively from mere parameter optimization. We built a static structure by self-reconfiguration in the first series of research; now we have achieved locomotion by M-TRAN. We infer that more functions will be achieved through further research on self-reconfigurable artifacts.

Q. *Why does the system proposed in this paper have a distributed controller, distributed energy, and distributed actuation? To some extent, but not always, nature adopts this solution because of flexibility. Notwithstanding, this is achieved at the sacrifice of efficiency.*

A. In terms of efficiency, centralized architecture is usually better than distributed architecture. However, centralized architecture always results from top-down design. This research investigated a novel design methodology using a bottom-up approach. We believe that distributed architecture is the most feasible framework to pursuit this line because it offers advantages in various properties: 1) flexible functionality by module combination; 2) fault tolerance by redundancy; 3) scalability on the system level; 4) fast response by local control feedback, and so on. It is difficult to say what is the best architecture for emergent functionality. At least, we adopt a homogeneous distributed architecture for M-TRAN. To build a systematic methodology of bottom up design, we must first investigate homogeneous architecture completely. A study of heterogeneity should follow.

Q. *The paper addresses the duration of simulation to solve a problem. In relation to it, the authors could expand this issue a little toward problem solving in different time-scales with reference to evolution, development, and learning.*

A. Drawing an analogy between biological systems and M-TRAN, the CPG network optimization using GA can be regarded as individual adaptation. This could be accomplished by another approach such as reinforcement learning. Development of a biological system corresponds to the self-assembly of a desired robot shape for M-TRAN. Evolution of morphology requires numerous generations and is much slower than development and individual adaptation in biological systems. In the present M-TRAN system, a human designer gives the target shape. As mentioned in the Conclusion, we must consider the relation between morphology and its functionality. We would like to emphasize that a self-reconfigurable system like M-TRAN is a platform upon which we can investigate both individual adaptation and morphological evolution concurrently in a single framework. In this sense, the self-reconfigurable systems open the new possibility of artifacts beyond natural evolution.

References

1. Fukuda, T., and Nakagawa, S. Approach to the Dynamically Reconfigurable Robotic System. *J. Intell. Robot Sys.* 1 (1988) 55–72

2. Murata, S., Kurokawa, H., and Kokaji, S. Self-Assembling Machine. In *Proc. 1994 IEEE Int. Conf. on Robotics and Automation,* (1994) 441–448

3. Chirikjian, G., Pamecha, A., and Ebert-Uphoff, I. Evaluating Efficiency of Self-Reconfiguration in a Class of Modular Robots. *J. Robotic Systems* 12(5) (1996) 317–338

4. Tomita, K., Murata, S., Yoshida, E., Kurokawa, H., and Kokaji, S. Self-assembly and Self-Repair Method for Distributed Mechanical System. *IEEE Trans. Robotics Automation* 15(6) (1999) 1035–1045.

5. Hosokawa, K., Tsujimori, T., Fujii, T., Kaetsu, H., Asama, H., Kuroda, Y., and Endo, I. Self-organizing Collective Robots with Morphogenesis in a Vertical Plane. In *Proc. 1998 IEEE Int. Conf. on Robotics and Automation* (1998) 2858–2863

6. Walter, J., Welch, J., and Amato, N. Distributed Reconfiguration of Hexagonal Metamorphic Robots in Two Dimensions. In *Proc. SPIE, Sensor Fusion and Decentralized Control in Robotic Systems III* (2000) 441–453

7. Yoshida, E., Murata, S., Tomita, K., Kurokawa, H., and Kokaji, S. a. An Experimental Study on a Self-repairing Modular Machine. *Robotics and Autonomous Systems* 29 (1999) 79–89

8. Yoshida, E., Murata, S., Kurokawa, H., Tomita, K., and Kokaji, S. A Distributed Method for Reconfiguration of 3-D homogeneous structure. *Advanced Robotics* 13(4) (1999) 363–380

9. Yoshida, E., Kokaji, S., Murata, S., Tomita, K., and Kurokawa, H. Miniaturization of Self-Reconfigurable Robotic System using Shape Memory Alloy. *J. Robotics Mechatronics* 12(2) (2000) 1579–1585

10. Kotay, K., Rus, D., Vona, M., and McGray, C. The Self-Reconfiguring Robotic Molecule. In *Proc. 1998 IEEE Int. Conf. on Robotics and Automation* (1998) 424–431

11. Kotay, K., and Rus, D. Motion Synthesis for the Self-Reconfigurable Molecule. In *Proc. 1998 IEEE/RSJ Int. Conf. on Intelligent Robots and Systems* (1998) 843–851

12. Murata, S., Kurokawa, H., Yoshida, E., Tomita, K., and Kokaji, S. A 3-D Self-Reconfigurable Structure. In *Proc. 1998 IEEE Int. Conf. on Robotics and Automation* (1998) 432–439

13. Ünsal, C., K.l.ccote, H., and Khosla, P. A modular self-reconfigurable bipartite robotic system: Implementation and motion planning. *Autonomous Robots* 10(1) (2001) 23–40

14. Rus, D., and Vona, M. Crystalline Robots: Self-reconfiguration with Compressible Unit Modules. *Autonomous Robots* 10(1) (2001) 107–124

15. Kotay, K., and Rus, D. Scalable parallel algorithm for configuration planning for self-reconfiguring robots. In *Proc. SPIE, Sensor Fusion and Decentralized Control in Robotic Systems III* (2000)

16. Yim, M., Duff, D., and Roufas, K. PolyBot: a Modular Reconfigurable Robot. In *Proc. 2000 IEEE Int. Conf. on Robotics and Automation* (2000) 514–520

17. Yim, M., Zhang, Y., Lamping, J., and Mao, E. Distributed Control for 3D Metamorphosis. *Autonomous Robots* 10(1) (2001) 41–56

18. Nguyen, A., Guibas, L., and Yim, M. Controlled Module Density Helps Reconfiguration Planning. In *Workshop on Algorithmic Foundations of Robotics (WAFR)* (2000)23–35

19. Castano, A., Chokkalingam, R., and Will, P.. Autonomous and Self-Sufficient CONRO Modules for Reconfigurable Robots. In *Distributed Autonomous Robotics 4,* Springer, Berlin (2000) 155–164

20. Murata, S., Yoshida, E., Tomita, K., Kurokawa, H., Kamimura, A., and Kokaji, S. Hardware Design of Modular Robotic System. In *Proc. 2000 IEEE/RSJ Int. Conf. on Intelligent Robots and Systems*, F-AIII-3-5 (2000)

21. Murata, S., Yoshida, E., Kamimura, A., Tomita, K., Kurokawa, H., Kokaji, S.: Homogeneous Distributed Mechanical Systems, Proc. Morpho Functional Machine Workshop, Tokyo, (2001) 222–235

22. Yoshida, E., Murata, S., Kamimura, A., Tomita, K., Kurokawa, H., Kokaji, S.: Reconfiguration Planning for a Self-Assembling Modular Robot, In *Proc. Intl. Symp. on Assembly and Task Planning (ISATP, Fukuoka)* (2001) 276–281

23. Yoshida, E., Murata, S., Kamimura, A., Tomita, K., Kurokawa, H., Kokaji, S.: A Motion Planning Method for a Self-Reconfigurable Modular Robot, In *Proceedings of 2001 IEEE/RSJ International Conference on Intelligent Robots and Systems (IROS 2001, Hawaii)*, CD-ROM (2001) 590–597

24. Kamimura, A., Murata S., Yoshida, E., Kurokawa, H., Tomita, K., Kokaji, S.: Self-Reconfigurable Modular Robot, -Experiments on Reconfiguration and Locomotion -, *IEEE/RSJ International Conference on Intelligent Robots and Systems (IROS 2001, Hawaii)*, CD-ROM (2001)

25. Kamimura, A., Yoshida, E., Murata, S., Kurokawa, H., Tomita, K., Kokaji, S.: A Self-Reconfigurable Modular Robot (MTRAN) -Hardware and Motion Planning Software-, *Distributed Autonomous Robotic Systems* 5 (2002) 17–26

26. Kurokawa, H., Kamimura, A., Yoshida, E., Tomita, K., Murata, S., Kokaji, S.: Self-Reconfigurable Modular Robot (M-TRAN) and Its Motion Design, In *Proc. Intl. Conf. Control, Automation, Robotics and Vision (ICARCV 2002)* (2002) 51–56

27. Hirose, S., Imazato, M., Kudo, Y., and Umetani, Y. Internally-balanced Magnetic Unit. *Advanced Robotics*, 3(4) (1986) 225–242

28. Chirikjian, G., Pamecha, A., Ebert-Uphoff, I.: Evaluating efficiency of self-reconfiguration in a class of modular robots, *J. of Robotic Systems*, 13–5 (1996) 317–338

29. Chiang, C.-J., Chirikjian, G.: Modular robot motion planning using similarity metrics, *Autonomous Robots 10* (2001) 91–106

30. Miyashita, K. Kokaji, S.: Navigating modular robots in the face of heuristic depressions, *Proc. of Distributed Autonomous Robotic System 5 (DARS-2002)* (2002) 7–26

31. Kamimura, A., Kurokawa, H., Yoshida, E., Tomita, K., Murata, S., Kokaji, S.: Automatic Locomotion Pattern Generation for Modular Robots, In *Proc. IEEE Intl. Conf. on Robotics and Automation (ICRA)* (2003) 714–720

32. Matsuoka, K.: Mechanisms of frequency and pattern control in the neural rhythm generators, *Biolog. Cybern.*, 56 (1987) 345–353.

33. Taga, G.: A model of the neuro-musculo-skeletal system for human locomotion II -real-time adaptability under various constraints, *Biolog. Cybern.*, 73 (1995) 113–121

34. Kimura, K. et al.: Realization of dynamic walking and running of the quadruped using neural oscillator, *Autonomous Robots*, 7–3 (1999) 247–258

35. Hase, K. et al.: Development of three-dimensional whole-body musculoskeletal model for various motion analyses, *JSME Int. J.* C40 (1997) 25–32

36. Ono, I. and Kobayashi, S.: A Real-coded Genetic Algorithm for Function Optimization Using Unimodal Normal Distribution Crossover, In *Proc. 7th ICGA (1997)* 246–253

37. M-TRAN II Web site, URL http://unit.aist.go.jp/is/dsysd/mtran/English/index.html

Author Index

Lecture Notes in Artificial Intelligence (LNAI)

Vol. 2871: N. Zhong, Z.W. Raś, S. Tsumoto, E. Suzuki (Eds.), Foundations of Intelligent Systems. XV, 697 pages. 2003.

Vol. 2854: J. Hoffmann, Utilizing Problem Structure in Planing. XIII, 251 pages. 2003.

Vol. 2843: G. Grieser, Y. Tanaka, A. Yamamoto (Eds.), Discovery Science. XII, 504 pages. 2003.

Vol. 2842: R. Gavaldá, K.P. Jantke, E. Takimoto (Eds.), Algorithmic Learning Theory. XI, 313 pages. 2003.

Vol. 2838: N. Lavrač, D. Gamberger, L. Todorovski, H. Blockeel (Eds.), Knowledge Discovery in Databases: PKDD 2003. XVI, 508 pages. 2003.

Vol. 2837: N. Lavrač, D. Gamberger, L. Todorovski, H. Blockeel (Eds.), Machine Learning: ECML 2003. XVI, 504 pages. 2003.

Vol. 2835: T. Horváth, A. Yamamoto (Eds.), Inductive Logic Programming. X, 401 pages. 2003.

Vol. 2821: A. Günter, R. Kruse, B. Neumann (Eds.), KI 2003: Advances in Artificial Intelligence. XII, 662 pages. 2003.

Vol. 2807: V. Matoušek, P. Mautner (Eds.), Text, Speech and Dialogue. XIII, 426 pages. 2003.

Vol. 2801: W. Banzhaf, J. Ziegler, T. Christaller, P. Dittrich, J.T. Kim (Eds.), Advances in Artificial Life. XVI, 905 pages. 2003.

Vol. 2797: O.R. Zaïane, S.J. Simoff, C. Djeraba (Eds.), Mining Multimedia and Complex Data. XII, 281 pages. 2003.

Vol. 2792: T. Rist, R.S. Aylett, D. Ballin, J. Rickel (Eds.), Intelligent Virtual Agents. XV, 364 pages. 2003.

Vol. 2782: M. Klusch, A. Omicini, S. Ossowski, H. Laamanen (Eds.), Cooperative Information Agents VII. XI, 345 pages. 2003.

Vol. 2780: M. Dojat, E. Keravnou, P. Barahona (Eds.), Artificial Intelligence in Medicine. XIII, 388 pages. 2003.

Vol. 2777: B. Schölkopf, M.K. Warmuth (Eds.), Learning Theory and Kernel Machines. XIV, 746 pages. 2003.

Vol. 2752: G.A. Kaminka, P.U. Lima, R. Rojas (Eds.), RoboCup 2002: Robot Soccer World Cup VI. XVI, 498 pages. 2003.

Vol. 2741: F. Baader (Ed.), Automated Deduction – CADE-19. XII, 503 pages. 2003.

Vol. 2705: S. Renals, G. Grefenstette (Eds.), Text- and Speech-Triggered Information Access. VII, 197 pages. 2003.

Vol. 2703: O.R. Zaïane, J. Srivastava, M. Spiliopoulou, B. Masand (Eds.), WEBKDD 2002 - Mining Web Data for Discovering Usage Patterns and Profiles. IX, 181 pages. 2003.

Vol. 2700: M.T. Pazienza (Ed.), Extraction in the Web Era. XIII, 163 pages. 2003.

Vol. 2699: M.G. Hinchey, J.L. Rash, W.F. Truszkowski, C.A. Rouff, D.F. Gordon-Spears (Eds.), Formal Approaches to Agent-Based Systems. IX, 297 pages. 2002.

Vol. 2691: V. Mařík, J.P. Müller, M. Pechoucek (Eds.), Multi-Agent Systems and Applications III. XIV, 660 pages. 2003.

Vol. 2684: M.V. Butz, O. Sigaud, P. Gérard (Eds.), Anticipatory Behavior in Adaptive Learning Systems. X, 303 pages. 2003.

Vol. 2671: Y. Xiang, B. Chaib-draa (Eds.), Advances in Artificial Intelligence. XIV, 642 pages. 2003.

Vol. 2663: E. Menasalvas, J. Segovia, P.S. Szczepaniak (Eds.), Advances in Web Intelligence. XII, 350 pages. 2003.

Vol. 2661: P.L. Lanzi, W. Stolzmann, S.W. Wilson (Eds.), Learning Classifier Systems. VII, 231 pages. 2003.

Vol. 2654: U. Schmid, Inductive Synthesis of Functional Programs. XXII, 398 pages. 2003.

Vol. 2650: M.-P. Huget (Ed.), Communications in Multiagent Systems. VIII, 323 pages. 2003.

Vol. 2645: M.A. Wimmer (Ed.), Knowledge Management in Electronic Government. XI, 320 pages. 2003.

Vol. 2639: G. Wang, Q. Liu, Y. Yao, A. Skowron (Eds.), Rough Sets, Fuzzy Sets, Data Mining, and Granular Computing. XVII, 741 pages. 2003.

Vol. 2637: K.-Y. Whang, J. Jeon, K. Shim, J. Srivastava, Advances in Knowledge Discovery and Data Mining. XVIII, 610 pages. 2003.

Vol. 2636: E. Alonso, D. Kudenko, D. Kazakov (Eds.), Adaptive Agents and Multi-Agent Systems. XIV, 323 pages. 2003.

Vol. 2627: B. O'Sullivan (Ed.), Recent Advances in Constraints. X, 201 pages. 2003.

Vol. 2600: S. Mendelson, A.J. Smola (Eds.), Advanced Lectures on Machine Learning. IX, 259 pages. 2003.

Vol. 2592: R. Kowalczyk, J.P. Müller, H. Tianfield, R. Unland (Eds.), Agent Technologies, Infrastructures, Tools, and Applications for E-Services. XVII, 371 pages. 2003.

Vol. 2586: M. Klusch, S. Bergamaschi, P. Edwards, P. Petta (Eds.), Intelligent Information Agents. VI, 275 pages. 2003.

Vol. 2583: S. Matwin, C. Sammut (Eds.), Inductive Logic Programming. X, 351 pages. 2003.

Vol. 2581: J.S. Sichman, F. Bousquet, P. Davidsson (Eds.), Multi-Agent-Based Simulation. X, 195 pages. 2003.

Vol. 2577: P. Petta, R. Tolksdorf, F. Zambonelli (Eds.), Engineering Societies in the Agents World III. X, 285 pages. 2003.

Vol. 2569: D. Karagiannis, U. Reimer (Eds.), Practical Aspects of Knowledge Management. XIII, 648 pages. 2002.

Vol. 2560: S. Goronzy, Robust Adaptation to Non-Native Accents in Automatic Speech Recognition. XI, 144 pages. 2002.

Vol. 2557: B. McKay, J. Slaney (Eds.), AI 2002: Advances in Artificial Intelligence. XV, 730 pages. 2002.

Vol. 2554: M. Beetz, Plan-Based Control of Robotic Agents. XI, 191 pages. 2002.

Vol. 2543: O. Bartenstein, U. Geske, M. Hannebauer, O. Yoshie (Eds.), Web Knowledge Management and Decision Support. X, 307 pages. 2003.

Vol. 2541: T. Barkowsky, Mental Representation and Processing of Geographic Knowledge. X, 174 pages. 2002.

Vol. 2533: N. Cesa-Bianchi, M. Numao, R. Reischuk (Eds.), Algorithmic Learning Theory. XI, 415 pages. 2002.